Phylogeny and Evolution of Bacteria and Mitochondria

Editor

Mauro Degli Esposti

Center for Genomic Sciences
UNAM Campus de Cuernavaca
Cuernavaca, Morelos
Mexico

CRC Press
Taylor & Francis Group
Boca Raton London New York

CRC Press is an imprint of the
Taylor & Francis Group, an **informa** business

A SCIENCE PUBLISHERS BOOK

Cover illustration reproduced by kind courtesy of Dr. Mauro Degli Esposti (editor)

CRC Press
Taylor & Francis Group
6000 Broken Sound Parkway NW, Suite 300
Boca Raton, FL 33487-2742

First issued in paperback 2020

ISBN-13: 978-1-138-50168-3 (hbk)
ISBN-13: 978-0-367-78073-9 (pbk)

Library of Congress Cataloging-in-Publication Data

Names: Esposti, Mauro Degli, author, editor.
Title: Phylogeny and evolution of bacteria and mitochondria / [by] Mauro
  Degli Esposti, Center for Genomic Sciences, UNAM Campus de Cuernavaca,
  Cuernavaca, Morelos, Mexico.
Description: Boca raton, FL : CRC Press, [2018] | "A science publishers
  book." | Includes bibliographical references and index.
Identifiers: LCCN 2018029680 | ISBN 9781138501683 (hardback)
Subjects: LCSH: Bacteria--Evolution. | Bacteria--Phylogeny. |
  Mitochondria--Evolution.
Classification: LCC QR81.7 .E87 2018 | DDC 579.3--dc23
LC record available at https://lccn.loc.gov/2018029680

Visit the Taylor & Francis Web site at
http://www.taylorandfrancis.com

and the CRC Press Web site at
http://www.crcpress.com

Thanks to my daughters Emanuelle and Michelle for their inputs and suggestions on several aspects of this book, from the blurp to the cover, and my companion Rocio for her support

Preface

This book presents bacteria in a different way from classic texts of microbiology such as Bergey's manual, focusing on their functional properties and how they evolved along phylogeny. Paramount among the functional properties is the capacity of bacteria to extract bioenergy from their environment, from simple elements such as sulfur and iron to complex biopolymers such as chitin and cellulose. This capacity is usually referred to as central metabolism, which depends upon diverse electron transport systems that generate bioenergy by transporting protons across the cell membrane while transferring electrons from environmental donors to acceptors of various kinds. The core of bacterial life and survival thus resides on the genomic endowment of bioenergy-producing systems that cumulatively form the respiratory chain and define the bioenergetics of prokaryotes.

Respiration consists of reducing oxygen with electrons, a process that is chemically equivalent to the reduction of a variety of electron acceptors such as nitrate and sulfate, which are exploited by bacteria when oxygen is scarce or missing. Such redox processes are still defined as respiratory, since they are based upon the reduction of a terminal acceptor available from the environment like oxygen. However, bacteria evolved long before the current levels of ambient oxygen were established about two billion years ago; before that time, hydrogen, sulfur and reduced metals like Fe^{II} were most abundant. Ancestral bacteria exploited these elements in developing their bioenergy production machinery to survive and thrive with specialized, anaerobic respiratory chains. Many microbes that will be presented in this book have maintained such anaerobic respiratory chains even if they have become aerotolerant, or facultatively aerobic after oxygen reached today's levels in earth habitats. In practice, you cannot recognize bacteria that have different physiology, for example facultatively anaerobes and obligate aerobes, on the basis of their morphology, or even from their marker RNA gene—the molecular tool which has become the benchmark for bacterial classification. You can only distinguish them from the functional properties that you measure with microbiological and biochemical methods, or deduce from the genomic features that determine such properties. This book will present the core functional properties that distinguish various types of Gram negative bacteria, from the most ancestral with anaerobic respiratory chains to the progenitors of the aerobic mitochondrial organelles of our cells. Mitochondria derive from a particular brand of extinct bacteria whose identity, and relationship with extant bacteria, is beginning to emerge from the integration of genomic and functional analysis with molecular phylogeny, as will be discussed in the last two chapters of the book.

Writing the initial chapters of the book gradually unveiled the underlying thread of entwined function and phylogeny that links very diverse bacteria such as Bacteroidetes from our oral microbiota, predatory Myxobacteria in the soil and alphaproteobacteria forming symbiotic associations with plant roots. Fundamental bioenergy-producing redox systems have been inherited from phylogenetically ancient Bacteriodetes to plant symbionts producing functionally equivalent respiratory chains, despite enormous differences in microbiological characters and taxonomic classification. Ultimately, function has been conserved along evolution much more than the individual genes of bacteria and their characters that we can empirically evaluate, also in phylogenetic terms. This book then offers a new fascinating perspective to look at bacteria through the angle of their functional inheritance that evolved along billions of years, which we can now discern by genomic analysis. In this perspective, we shall also discover the living bacterial relatives of our mitochondria. Indeed, paraphrasing Woese (1987), bacterial evolution holds the key to the origin of the eukaryotic cell, which is deeply interconnected with the acquisition of mitochondria.

Although the initial idea for this book was to have several scientists contributing a chapter on their area of expertise, at the end I preferred to write multiple chapters myself. I am very grateful to Bill Martin, who has written the first, breathtaking chapter on the origin of bacterial life and physiology, to Esperanza Martínez-Romero, who has been so kind to help me writing three chapters, and also Paola Bonfante, who has contributed an incisive and informative part to chapter six. I also thank other co-authors and colleagues who have variably contributed to chapters of this book, which hopefully will constitute a reference source of information, and inspiration, for scientists interested on how bacteria evolved into the mitochondria of our cells.

Mauro Degli Esposti

Contents

1

Early Microbial Evolution from a Physiological Perspective

William F. Martin

Introduction to Early Microbial Evolution

Most thinking about microbial evolution is currently couched in terms of molecular evolutionary trees, the subject matter of a discipline called phylogeny. As such, few scientists would have difficulties imagining how one might study microbial evolution and more generally the evolution of life: make trees. Life, however, is a chemical reaction. All cells rely upon a main exergonic reaction that fuels ATP synthesis to run all other life processes. The nature of that main exergonic reaction, which has driven the evolutionary process forward in uninterrupted continuity since the origin of the first cells, is a central topic of investigation in the field of physiology or, more specifically, bioenergetics. It is not immediately obvious how one might study the evolution of a chemical reaction. It is, however, obvious that in order to investigate the evolution of life's common thread, the harnessing of exergonic chemical reactions for the purpose of conserving energy as ATP, one has to think outside the extremely narrow confines of phylogenetic trees, sequence alignments, and models that compare different trees by statistical criteria such as log likelihood values. When we consider the overall course of microbial evolution, physiology and phylogeny do not correspond well, because lateral gene transfer is very real and very prevalent among prokaryotes, and lateral gene transfer decouples physiology from phylogeny. This chapter presents a view of early microbial evolution from the author's non-mainstream perspective. The text is based upon recently published papers about physiology, phylogeny, and evolution (Martin 2016), early CO_2

Institute for Molecular Evolution, University of Düsseldorf, Germany.
Email: bill@hhu.de

reduction (Sousa et al. 2018) and aspects concerning the advent of photosynthesis (Martin et al. 2018).

Physiology and Phylogeny

Before the days of molecular phylogenies, the standard way of viewing microbial evolution was as a process of physiological evolution (Eck and Dayhoff 1968, Decker et al. 1970, Dickerson 1980). The object of investigating physiological evolution was to order the sequence of events in which the different pathways arose that modern microbes use to harness carbon and energy. The physiological view of microbial evolution was, of course, replaced in the 1980s by a gene centered view of microbial evolution that was constructed around the ribosomal RNA tree of life, also called the universal tree or the three domain tree (Woese et al. 1990). The rRNA tree installed the order into microbial systematics that microbiologists had sought for decades (Stanier and van Niel 1962), but did not explain much about physiological evolution because physiology never mapped properly onto the rRNA tree. That was not because the branching pattern in rRNA tree missed the true branching order for microbial lineages but because microbial physiology never mapped cleanly onto any phylogenetic tree for prokaryotes, regardless of its topology, because of lateral gene transfer (LGT). Dickerson (1980) suggested that LGT could possibly even transform non-respiring lineages into respiring lineages via the transfer of many genes, something that we now know actually occurs during microbial evolution (Nelson-Sathi et al. 2012).

Is the evolutionary history of microbes a history of ribosome phylogenies, or is it a history of physiological processes? Knoll has said that Earth records its own history (Gaidos and Knoll 2012). So do genomes. Putting geological and genomic evidence into a consistent picture of microbial evolution is a challenging task, especially if energetics is to enter into the picture as well (Shock and Boyd 2015). The only connection between rocks and genes is physiology. Physiology is arguably what life is all about, because if the core ATP generating reaction of an organism comes to a halt, so does the life process and all else in the evolutionary process including population genetics, bottlenecks, drift and the like. The only microbial processes that have left an interpretable trace in the geochemical record are physiological.

Rocks preserve evidence of microbial activity in the form of carbon (Ueno et al. 2006), sulfur (Poulton et al. 2004) and nitrogen (Stüecken et al. 2015) isotopes, in addition to evidence for molecular oxygen (Fischer et al. 2016). The oldest sedimentary rocks, which are ca. 3.8 billion years of age, harbor traces for life in the form of light carbon isotopes, which is generally interpreted as evidence for biological CO_2 fixation at that time (Mojzsis et al. 1996, Ueno et al. 2002), although new findings have it that biological CO_2 fixation goes back as far as 3.95 Ga (Tashiro et al. 2017). Geologists also tell us that nitrogen fixation has been around for at least 3.2 billion years (Stüecken et al. 2015) and that molecular oxygen has been around for about 2.4 billion years (Fischer et al. 2016, Lyons et al. 2014). Rocks date the existence of microbial physiological processes, the distribution of which is rarely restricted to specific phylogenetic groups.

In the geochemical record that extends beyond roughly 1.5 billion years ago, there are only two kinds of geochemical traces that correspond to evidence for the existence of any modern prokaryotic group. One is the presence of molecular oxygen, which indicates the existence of cyanobacteria 2.4 billion years ago (Fischer et al. 2016). The other is biogenic methane in rocks 3.5 billion years of age, which provides evidence for the antiquity of archaea (Ueno et al. 2006) because methanogenesis is restricted to the archaea. The antiquity of biological methane does not, however, indicate which groups of methanogens are ancient because new phylogenetic depictions of the tree of life have methanogens basal among the archaea, with loss of methanogenesis having occurred in many independent groups (Evans et al. 2015, Williams et al. 2017), those losses corresponding to gene acquisitions from bacteria and changes in physiology in some cases (Nelson-Sathi et al. 2012, 2015). LGT thus decouples phylogeny from physiology in the methanogens, too, which now appear to be the most ancient archaea, but no longer appear as a monophyletic group (Fig. 1).

Most physiological traits that are preserved as isotopic evidence in ancient rocks, whether CO_2 reduction (Fuchs 2011, Berg et al. 2010), N_2 reduction (Zehr et al. 2001), or sulfur reduction (Rabus et al. 2015), are present in many different prokaryotic groups, both among the archaea and among the bacteria. There can also be little doubt that those physiological traits, regardless of whether they were present in the last universal common ancestor (LUCA) or not, have been distributed by LGT among prokaryotic groups. A clear example is anoxygenic photosynthesis, which operates in conjunction with at least three different CO_2 fixation pathways (Fischer et al. 2016, Sousa et al. 2013b). Anoxygenic photosynthesis is a major physiological trait, the coding capacity for which entails dozens of genes for the type II photosystem, chlorophyll, and carotenoids. This complex trait can, however, be condensed into a large but mobile plasmid that is colinear with the *Rhodobacter* photosynthesis operon and that is mobile among marine proteobacteria (Petersen et al. 2013). In microbial evolution, LGT decouples physiology from phylogeny, especially when it comes to photosynthesis (Martin et al. 2018).

What Kind of Tree?

The rRNA tree (Fox et al. 1980, Woese et al. 1990) installed direly needed order into microbial systematics. But in doing so, it severed the connections to physiological evolution because physiology never mapped properly onto any incarnation of the rRNA tree. A recent paper (Jelen et al. 2016) shows the distribution of some microbial metabolisms mapped out on the rRNA tree of life. Though not stated in that paper, the reason that the physiological reactions do not map cleanly on the rRNA tree is because of LGT. It is hard to imagine a kind of prokaryotic systematics now or in the future that is not based on ribosomal phylogeny. There are also good reasons to think that the ribosome is, in the main, inherited vertically

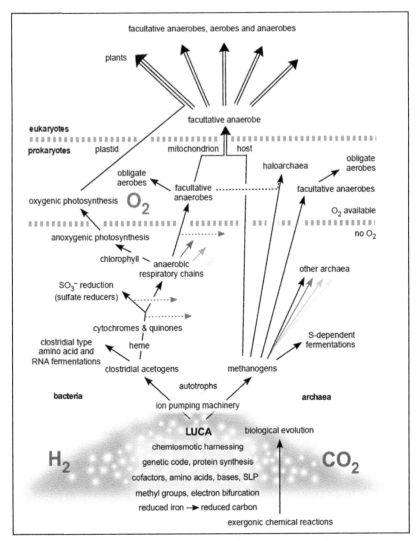

Fig. 1. Some aspects of early physiological evolution. The figure is reprinted with permission from Martin (2016). The figure combines aspects of several papers: Decker et al. (1970), Dickerson (1980), Martin and Russell (2003, 2007), Martin (2012), Müller et al. (2012), Lane and Martin (2012), Nelson-Sathi et al. (2012), Sousa et al. (2013b), Martin et al. (2015), Martin and Sousa (2016), Schönheit et al. (2016), Fischer et al. (2016), Sousa et al. (2016), Weiss et al. (2016). Line lengths are not proportional to time. Dotted lines symbolize lateral gene transfers. Parallel lines symbolize endosymbiotic events.

in prokaryotes (Jain et al. 1999). After all, an *E. coli* cell is about 16% ribosomal RNA by dry weight (Neidhardt et al. 1990) and a ribosome is about 40% protein by weight, such that if we take *E. coli* as a typical prokaryote, ribosomes make up about one third of the dry weight of a cell. If we get the phylogeny of a cell's ribosomes right, we would have a picture of how that major component of the cell evolved over time. Yet the synthesis of ribosomes requires the carbon and energy metabolism (physiology). Ribosomes consist of a large and a small subunit and about 50 proteins (Wang et al. 2009).

Newer versions of the rRNA tree are based on large concatenated alignments of ribosomal proteins, rather than ribosomal RNA (Hug et al. 2016), an approach to microbial phylogeny that has become quite popular lately (Rinke et al. 2013, Spang et al. 2015, Zaremba-Niedzwiedzka et al. 2017). Though unimportant, on a personal note, although the practice of concatenating large data sets for phylogeny started in my laboratory, it was the PhD work of Vadim Goremykin (Goremykin et al. 1997, Martin et al. 1998). We used such data sets for some time before it caught on in other groups. There are some problems with large concatenated datasets though, mainly that they give very high bootstrap or similar support values for individual branches, but in a manner that is more dependent on the model used to infer the tree than on the data itself (Lockhart et al. 1999). Another problem is that in long alignments with many sites, one has the luxury of removing sites and we devised methods to do that systematically (Stoebe et al. 1999, Hansmann and Martin 2000), an approach that is also commonplace now. The most severe problem with concatenated data is that the individual proteins, analyzed individually, never give the same tree at least when 20 species or more are being surveyed and the individual trees can be very different, but when lumped together in a concatenated alignment they generate a signal of what appears to be "strong support", whereby it is not at all clear what it is that is being supported (Thiergart et al. 2014).

It has become very easy and affordable to sequence genomes and even genomes of environments. The computer power to handle such data has kept pace with the leaps in sequencing technology accordingly. But the pursuit of genes and genomes has led us into a situation where, thanks to modern environmental sequencing, we now have genome-based trees for hundreds and thousands of previously undiscovered prokaryotes, but we do not know how they grow (their physiology). Environmental genomes reflect fascinating organisms in fascinating environments, but in order to understand their environmental significance, we need to know what they are doing for a living (Evans et al. 2015, Hug et al. 2016). It is hard to piece together what the organisms that belong to environmental sequences are doing in nature based on the information in their genome. One example is the methanogen-like metabolisms that people are finding in marine sediments. Evans et al. (2015) reported new archaeal lineages (Bathyarchaeota) from marine sediment that have the archaeal version of the acetyl-CoA pathway, but have no clear evidence for known forms of archaeal energy conserving metabolism. He et al. (2016) reported Bathyarchaeota lineages that appear to be performing a very simple and suspected primitive form of energy metabolism that is otherwise only known from bacteria so far—or from archaea under very specific kinds of culture conditions (Rother and Metcalf 2004)—acetogenesis. What kind of carbon and energy metabolism existed at the root of life's chemical reaction?

Fermentation at the Root?

Scientists interested in the first kinds of metabolism have long thought, since Haldane (1929) at least, that fermentations are the logical starting point for physiological evolution (Haldane 1929, Sagan 1967). Of course in Haldane's day, nobody knew how fermentations worked. Fermentations of the kind that anyone ever had in mind in an early metabolism context are disproportionation reactions of carbon in an organic compound like glucose. Starting from an intermediate oxidation state, the carbon is converted into a more oxidized form, like CO_2 (or the carboxyl group in lactate) and a more reduced form like ethanol (or the methyl group in lactate). Because glucose is an energy rich compound, there is energy to be gleaned from ethanol or lactate fermentations, enough to generate one ATP per CO_2 (or carboxylate) produced, the energy being conserved at an oxidative step in which an aldehyde group is converted into a carboxyl group via an enzyme bound hemithioacetal, its oxidation to an energy rich thioester, phosphorolysis of which releases an acyl phosphate with a free energy of hydrolysis high enough to readily phosphorylate any number of compounds, including the beta phosphate on ADP (Martin and Thauer 2017). Many other kinds of fermentations exist (Barker 1961, Wood 1961, Decker et al. 1970), but can such a reaction sequence reside at life's root?

A very surprising development in fermentations concerns the mechanisms of energy conservation. In strictly anaerobic prokaryotes (the ones that are likely to be ancient, also in Haldane's view), fermentations involve chemiosmotic coupling (Mai and Adams 1996, Adams et al. 2001, Buckel and Thauer 2013, 2018, Schuchmann and Müller 2014, Schönheit et al. 2016). Fermentations (disproportionations) are an altogether problematic starting point for physiological evolution. The main problem with the concept that fermentations came first concerns substrate. Fermentations involve channeling a very small number of substrates (often just one) present in the environment in large amounts through a very specific series of conversions leading to one highly tuned reaction sequence catalyzed by a handful of 3–4 enzymes that conserve energy as a compound that can phosphorylate ATP (Decker et al. 1970, Martin and Thauer 2017). At the onset of evolution, there were no environmental reserves of reduced carbon compound present in large amounts as specific isomers that could have fueled energy conservation via fermentations or promoted their origin from the elements (Schönheit et al. 2016).

On the early, uninhabited Earth, the main form of carbon was not glucose, glucose phosphate, ribose, glycogen, or starch. It was CO_2 (Sleep et al. 2011). That is because the early Earth went through a phase where the planet was molten rock and metal, which is typically hotter than 1000°C and carbon in contact with the elements on the early Earth at such temperatures will exist as CO_2, not as glucose or anything similar. Critics will interject, but what about carbon from space? Yes, there was a lot of carbon brought to Earth from meteorites, maybe 10% of Earth's total carbon (Chyba et al. 1990), but the carbon in meteorites has a different problem: it is too reduced. Carbon in meteorites is typically on the order of 98% polyaromatic hydrocarbons (Sephton 2002), which is basically graphite, a nonfermentable substrate. The remaining 2% is distributed across hundreds (or thousands) of different compounds, mostly various isomers of aliphatic acids, each present in parts per million or parts per billion concentrations. Straight or branched chain organic acids are also nonfermentable substrates; they are too reduced for disproportionation. Remaining components of stardust, present in trace concentrations each, might be fermentable in terms of their redox state—amino acids and bases are fermentable for example (Barker 1961)—but would require dozens or hundreds of preexisting isomerases and mutases in order to channel substrates into reactions suitable for harnessing via substrate level phosphorylation (SLP). Stardust delivers reduced carbon compounds, but not fermentable substrates (Schönheit et al. 2016).

Whence Energy?

If substrate level phosphorylation via fermentations at the onset of energy metabolism won't work, what will? Since mid 1980s, Georg Fuchs had repeatedly made the case that the acetyl-CoA pathway was the most ancient pathway of CO_2 fixation (Fuchs and Stupperich 1985, 1986, Fuchs 1994, 2011). But the acetyl-CoA pathway is also a pathway of energy metabolism. Rolf Thauer, in an enlightening conversation, said it this way: "Herr Martin, in the reaction of H_2 and CO_2, the equilibrium lies on the side of acetate." I had the great fortune to have had occasional discussions with Thauer and Fuchs during the 1990s, those discussions had strong influence on me and were the reason that when Mike Russell and I started publishing papers together about hydrothermal vents and life's origin in 2002, I was thinking about anaerobic autotrophs (acetogens and methanogens). Everett Shock et al. (1998) stated the situation this way: organisms that use acetyl-CoA pathway "get a free lunch that they are paid to eat".

Hydrothermal vents are attractive as the site of physiological origin. H_2 is continuously generated in the crust via the process of serpentinization (Sleep et al. 2004, Bach et al. 2006, Russell et al. 2010); it exits in hydrothermal vents at concentrations exceeding 10 mM, interfaces with large amounts of CO_2 in the ancient oceans, which contained more CO_2 than today, perhaps 1000 times as much (Zahnle et al. 2007). Some modern cells harness energy from the reaction of H_2 with CO_2, satisfying their carbon and energy needs from H_2 and CO_2 alone: acetogens (bacteria) and methanogens (archaea). They are both chemiosmotic in that they reduce CO_2 with electrons from H_2, pumping protons or sodium ions in the process. Their ATPases are related, their mechanisms of pumping are not (Thauer et al. 2008, Buckel and Thauer 2013, Schuchmann and Müller 2014, Basen and Müller 2016). Serpentinization also generates alkaline water (Schrenk et al. 2013) such that at the vent-ocean interface, a proton gradient exists. A case can be made that early replicating chemical systems possessing genes and the genetic code were not free living, but could harness geochemical ion gradients at hydrothermal vents (Lane and Martin 2012). But that still cannot be the beginning of physiology because if a replicating system had genes and the code, then it had to have possessed a form of physiology that supported its heritable replication before it came to possess ATPases that could harness a geochemical ion gradient.

How did physiology get going? Neither SLP via fermentations can be the first energy metabolism for lack of fermentable substrates (Schönheit et al. 2016), nor can chemiosmotic harnessing to conserve energy as ATP because that requires proteins as energy converters in all forms of life known (the only ones we have to explain). What possibilities are then left to choose from? If we look around at what is known among real microbes, the only alternative that I can see that exists in modern metabolism is that physiology got started through a kind of substrate level phosphorylation (SLP) that does not depend upon fermentation: the kind of SLP that is manifest in some autotrophs that employ the acetyl-CoA pathway, also called the Wood-Ljungdahl pathway (Martin and Russell 2007). Under standard physiological conditions, the acetyl-CoA pathway generates a thioester in the exergonic reaction of H_2, CO_2 and a thiol

$$2CO_2 + 4H_2 + CoASH \rightarrow CH_3COSCoA + 3H_2O$$

with $\Delta G_0' = -59.2$ kJ/mol (Fuchs 1994). In acetogens (Fuchs 2011) and in the methanogen *Methanosarcina mazei,* under specific growth conditions, namely on CO (Rother and Metcalf 2004), the thioester undergoes phosphorolysis to acetyl phosphate

$$CH_3CO\sim SCoA + HOPO_3^{2-} \rightarrow CH_3CO\sim OPO_3^{2-} + CoASH$$

with $\Delta G_0' = +10.5$ kJ/mol (Decker et al. 1970). The mixed anhydride is an excellent phosphoryl donor with a free energy of hydrolysis

$$CH_3CO\sim OPO_3^{2-} + H_2O \rightarrow CH_3COO^- + HOPO_3^{2-}$$

free energy of hydrolysis of $\Delta G_0' = -43$ kJ/mol (Thauer et al. 1977), higher than that of ATP ($\Delta G_0' = -31$ kJ/mol (Thauer et al. 1977), such that the ATP is readily synthesized from acetyl-phosphate

$$CH_3CO\sim OPO_3^{2-} + ADP \rightarrow CH_3COO^- + ATP$$

with $\Delta G_0' = -10.5$ kJ/mol via SLP (Decker et al. 1970). The two enzyme system that catalyzes this reaction in acetogens and many bacteria, phosphotransacetylase and acetate kinase, also occurs in *Methanosarcina mazei* (Rother and Metcalf 2004), yet is missing in most methanogens and in most archaea studied so far (Schönheit et al. 2016). Archaea tends to use a different and unrelated enzyme for ATP synthesis via SLP from acetyl-CoA. The archaeal enzyme, acetyl-CoA synthase (ADP forming), follows the same simple chemistry as the bacterial route, but the acyl phosphate remains enzyme bound throughout the reaction (Weiss et al. 2016).

Several papers have made the case for a role for this kind of reaction sequence in early biochemical evolution (Schock et al. 1998, Sleep et al. 2004, Martin and Russell 2003, 2007, Ferry and House 2006, Berg et al. 2010, Fuchs 2011), whereby acyl phosphates may serve as the universal phosphate-based energy currency before the advent of ATP as the universal energy currency (Martin and Russell 2007,

Whicher et al. 2018), such that the rise of ATP as the universal energy currency occurred only after the origin of genes and proteins. It should be kept in mind that there are several biochemical energy currencies other than ATP, including thioesters, reduced ferredoxin, and NADH (Buckel and Thauer 2018).

Acyl phosphates (and the thioesters from which they are derived in metabolism) could have served as the universal energy currency that energetically underpinned, hence permitted, the origin genes and proteins (and the genetic code). The rotor stator ATP synthase (or ATPase) is clearly an invention of the world of genes and proteins and, furthermore, was present as a set of genes and protein subunits in the last universal common ancestor because it is as universal as the ribosome and the genetic code itself (Martin and Russell 2007, Lane et al. 2010, Sousa et al. 2013b). In that view, the chemical sequence from thioesters to acyl phosphates to ATP as it occurs in autotrophically growing acetogens would recapitulate a temporal sequence of phases in early bioenergetic evolution. However, acetogens growing on H_2 and CO_2 cannot obtain net ATP synthesis via SLP; they require chemiosmotic coupling (ion pumping and gradient harnessing via the ATPase) at the plasma membrane, which they achieve via coupling the exergonic reduction of CO_2 with electrons from H_2 via low potential reduced ferredoxins to energy conservation in the form of ion gradients (Buckel and Thauer 2013, Schuchmann and Müller 2014, Basen and Müller 2016).

The Problem with H_2: Electron Bifurcation

As recently summarized elsewhere (Sousa et al. 2018), H_2 is generated today in abundance in hydrothermal systems through the process of serpentinization (Sleep et al. 2004, 2011, Bach et al. 2006, Martin and Russell 2007, Russell et al. 2010, Schrenk et al. 2013). Because of its abundance in hydrothermal systems, both today and on the early Earth (Holm and Charlou 2001, Sleep et al. 2004), H_2 generally looks interesting as a source of early reduction potential at the origin of CO_2 fixation. But H_2 has two rather severe problems as a source of electrons for CO_2 reduction at origins.

First, its midpoint potential is unfavorable for CO_2 reduction beyond the near-equilibrium reaction with formate (Schuchmann and Müller 2013). That is why microbes that reduce CO_2 with H_2 employ a newly discovered mechanism of energy conservation called flavin based electron bifurcation (Herrmann et al. 2008). Electron bifurcation generates reduced ferredoxins (Fd^-) with a midpoint potential on the order of -500 mV from H_2, which has a more positive midpoint potential of only -414 mV (Basen and Müller 2016, Buckel and Thauer 2018). Electrons from H_2 have to flow energetically uphill to reduce low potential Fd. How do electrons flow energetically uphill? One electron from H_2 is transferred energetically downhill to a higher potential acceptor like NAD^+ or heterodisulfide CoB-S-S-CoM (Kaster et al. 2011), while the other is transferred uphill to Fd so that the overall energetics of the reaction are favorable (Buckel and Thauer 2013). By transferring one electron from H_2 downhill to NAD^+ with $E_0' = -320$ mV or a heterodisulfide acceptor with $E_0{}' = -140$ mV, the overall energetics of the reaction become favorable (Kaster et al. 2011). Electrons can also confurcate, which is how *Thermotoga* generates H_2 from NADH (which thermodynamically should not work), the trick being that the hydrogenase obtains one electron from NADH and the other from a low potential ferredoxin, which makes the overall reaction energetically favorable (Schut and Adams 2009).

Second, H_2 never interacts directly with any substrate in metabolism, CO_2 or otherwise (Thauer 2011, Lubitz et al. 2014). It interacts with Fe or Ni atoms at the active site of hydrogenases, of which there are several kinds known, to yield protons and electrons. The electrons reduce FeS clusters. In the case of electron bifurcation in anaerobic autotrophs, the electrons ultimately end up reducing CO_2. The lack of direct interactions between H_2 molecules and CO_2 molecules in metabolism suggests to me that such interactions never existed at the origin of metabolism because enzymes, when they came along in evolution, did not (in my view) invent biochemical reactions, they just accelerated natural, spontaneous reactions that tend to occur anyway. One might counter that if early CO_2 fixation took place at hydrothermal vents, then electron bifurcation was not needed because there was plenty of natural FeS around that could do the job of Fd⁻ when it comes to fixing CO_2. But laboratory experiments using reduced FeS have generally not been successful when it comes to reducing CO_2, except when external voltage as an electron source is applied (Roldan et al. 2015), the most likely reason being that FeS clusters in proteins and Fe^{2+} ions in FeS minerals are one electron donors, with iron undergoing Fe^{2+} to Fe^{3+} valence changes. The steps of biological CO_2 reduction in autotrophs are always two electron reactions (Fuchs 2011). In biology, the electrons from Fd⁻ are donated to C in CO_2 via metals that readily undergo two electron reactions, such as Ni, Mo, or W atoms coordinated in proteins or cofactors, or electron pairs are donated via hydride transfer from organic cofactors like NAD(P)H (Fuchs 2011).

The discovery of electron bifurcation (Buckel and Thauer 2018) was a significant step in understanding the physiology of anaerobes because it solved many nagging but crucial problems in their physiology, from odd stoichiometries among end products (Buckel 2001) to pressing questions of how methanogens and acetogens generate low potential ferredoxins having a midpoint potential around −500 mV, from H_2, the midpoint potential of which is not sufficiently negative, as mentioned above, to allow the reaction to go forward (Kaster et al. 2011). Flavin based electron bifurcation is a mechanism of energy conservation because it generates low potential ferredoxin, which is a currency of biochemical energy (Herrmann et al. 2008, Martin and Thauer 2017, Buckel and Thauer 2018), like acyl phosphates (Lipmann 1941) or thioesters (De Duve 1991).

Whence Electrons for Early CO_2 Reduction?

Why does electron bifurcation pose problems in an early evolution context? The problem is that flavin based electron bifurcation is an elaborate physiological process that requires sophisticated proteins (Lubner et al. 2017, Wagner et al. 2017) working in concert with other proteins (Basen and Müller 2016) as an energy metabolic pathway to reduce Fd (for reducing CO_2). This presents a chicken-and-egg paradox: How was CO_2 reduced with H_2 before there were flavins and proteins to catalyze electron bifurcation? Was something other than H_2 doing the job of CO_2 reduction instead, and if so, then what?

As outlined elsewhere (Sousa et al. 2018), native transition metals now seem extremely interesting in an early evolution context. A very notable and very commonplace component of hydrothermally altered rocks is the mineral awaruite, an intermetallic compound with the formula Ni_3Fe (also with stoichiometry Ni_2Fe) (Schwarzenbach et al. 2014, McCollom 2016). It is comprised solely of native transition metals with oxidation state zero. Though sometimes designated as an alloy, it is technically not an alloy because its Fe and Ni atoms are not arranged in a crystalline or regularly repeating arrangement, for which reason it is an intermetallic compound. It is a natural, normal and common constituent of serpentinizing hydrothermal systems. Awaruite (named after the city Awarua in New Zealand) is formed during serpentinization, probably by reduction of the divalent metal ions the under conditions of high H_2 activities on the order of $\sim$ 200 mmol/kg (Klein and Bach 2009).

Awaruite was reported to catalyze CO_2 reduction to methane using H_2 as the reductant (Horita and Berndt 1999), but more recent findings indicate that native metals serve not as the catalyst but as the reductant in organic synthesis from CO_2. Guan et al. (2003) reported that Fe^0 in the presence of salts will reduce CO_2 to CH_4, C_3H_8, CH_3OH and C_2H_5OH in the 10–70 μM range at room temperature. He et al. (2010) reported reduction of CO_2 to formate and acetate in the 1–10 mM range using nanoparticular Fe^0 at 80–200°C. Muchowska et al. (2017) reported that native iron will accelerate and promote six out of the eleven reactions of the reverse citric acid cycle (rTCA, see Chapter three). More dramatic is the finding that native transition metals will also reduce CO_2 to formate, acetate, and pyruvate, the intermediates and end products of the acetyl-CoA pathway (Varma et al. 2018). The latter result, in particular, indicates two things. First, it indicates that native metals are very interesting as the primordial source of electrons for CO_2 reduction at the onset of organic synthesis for life. Second, it indicates that the acetyl-CoA pathway is probably as ancient as some have been saying (Fuchs 1994, 2011, Martin and Russell 2007, Sousa and Martin 2014).

The recognition that native metals efficiently reduce CO_2 to biologically relevant compounds under mild hydrothermal conditions significantly narrows the gap between geochemistry and biochemistry in my view (many will disagree). Native metals appear to solve the otherwise severe problems concerning the source of sufficient reducing power for organic synthesis from CO_2 at the onset of physiological evolution that were discussed as the 'early formyl pterin problem' (Martin and Russell 2007) or just summarized (in despair) as 'reduced iron' in Fig. 2 of an earlier paper (Martin 2012). Suggestions that ion gradients might have been chemically involved in primordial CO_2 reduction (Lane 2017) now appear unnecessarily complicated in light of the ability of native metals to reduce CO_2 to intermediates and end products of the acetyl-CoA pathway (Varma et al. 2018); yet there is every reason to think that the harnessing of geochemical ion gradients preceded the origin of free living cells (Martin and Russell 2007, Lane and Martin 2012). That is to say, the ability to harness preexisting ion gradients was very likely a property of stages in the evolution of life that had not reached the level of a free living cell. Evidence for that view comes from the observation that the rotor stator ATPase is as universal as the genetic code itself, yet the ability to generate ion gradients with a chemistry that is specified by genes was a prerequisite for the transition to the state of free-living cells, and was invented independently in the non free living ancestors of the archaeal and bacterial lineages (Martin and Russell 2007, Sousa et al. 2013b), because the mechanisms and proteins used by acetogens and methanogens lacking cytochromes and quinones to generate ion gradients are unrelated, hence they are best understood as independent inventions in the ancestors of the non free living ancestors of archaea and bacteria, respectively.

Do Genomes Harbor any Evidence of Very Early Evolution?

Many investigators have looked at genomes to explore early microbial evolution, particularly the nature of the last universal common ancestor, LUCA (Woese et al. 1990, Pace 1991, Woese 1998, Forterre and Philippe 1999, Kyprides et al. 1999, Penny and Poole 1999, Koonin 2003, Fox 2010, Di Giulio 2011). The standard approach to such undertakings has been to look for genes that are universally present in genomes, an approach that typically uncovers the same 30 or so genes for ribosomal proteins and subunits of the ATPase (Dagan and Martin 2006). If one relaxes the criteria for "universally" present, allowing an occasional gene loss, one can find about 100 proteins that trace to LUCA by the criterion of universal distribution (Puigbò et al. 2009). That tells us that LUCA had ribosomes and the genetic code, which we already knew, because the code is universal. We recently used a different approach to investigate the nature of LUCA. Rather than to look for the genes that are universal among genomes, we decided to look for the genes that were ancient. That is, we looked for genes that are present in archaea and bacteria but that have been vertically inherited from LUCA rather than having been laterally transferred from one domain to the other (Weiss et al. 2016). That delivered a list of 355 genes that appear to have been vertically inherited from LUCA.

Those genes held clues as to how and where LUCA lived. We found that LUCA was an anaerobe and its metabolism was replete with O_2-sensitive enzymes. These include proteins rich in O_2-sensitive FeS clusters and enzymes that entail the generation of radicals (unpaired electrons) via *S*-adenosyl methionine (SAM) in their reaction mechanism. That fits well with the 50-year-old (Eck and Dayhoff 1966), but still modern, view that FeS clusters represent very ancient cofactors in metabolism (Russell and Hall 1997, Camprubi et al. 2017). It also fits with newer insights about the ancient and spontaneous (non-enzymatic) chemistry underlying SAM synthesis (Laurino and Tawfik 2017). LUCA lived from gases (H_2, CO_2, CO,

N_2). For carbon assimilation, LUCA used the acetyl-CoA pathway. Central to its metabolism were thioesters, which harbor chemically reactive bonds (Semenov et al. 2016) that play a crucial role in metabolism, both modern and ancient (Goldford et al. 2017, Martin and Thauer 2017, Goldorf and Segrè 2018). LUCA's environment was furthermore rich in sulfur (Weiss et al. 2016). Taken together, LUCA's requirement for gases (CO_2, H_2, CO, N_2), its affinity to high temperature and metals, plus the ability to use ion gradients but not generate them, all point to the same environment: alkaline hydrothermal vents.

The presence of nitrogenase subunits among LUCA's gene set (Weiss et al. 2016) is of interest because it argues in favor of the antiquity of N_2 fixation, tracing it back to the very first cells. This contributes to recent debates about the age of N_2 fixation. Stüecken et al. (2015) found isotope evidence for the antiquity of N_2 fixation, while Boyd and Peters (2013) suggested that nitrogenase might have arisen late in evolution, subsequent to chlorophyll biosynthesis, based on phylogenetic tree branching pattern interpretations. A caveat to the inference of Boyd and Peters (2013) is that the tree of the nitrogenase subunit (NifD) and the related chlorophyll biosynthesis subunit (BchN) is unrooted; their interpretation that nitrogenase is younger than its chlorophyll counterpart is based on the (tenuous) premise that BchN is older solely for the reason that it is "simpler", lacking the molybdenum cofactor present in nitrogenase. It is also possible that molybdenum utilizing nitrogenase is ancestral and that chlorophyll biosynthesis stems from a substantially later stage in evolution, chlorophyll biosynthesis being derived from heme biosynthesis and cobalamin biosynthesis (Martin et al. 2018), as microbiologists have traditionally thought (Granick 1965, Decker et al. 1970). I would argue that the cofactors and the metals clusters of nitrogenase are older than the protein itself and that the metal free version of NifD-related proteins are the derived, not the ancestral state.

If Life Started at Vents, What Happened Next?

Up to here, the text has dealt mainly with the advent of the very first cells, how they fixed carbon and how they conserved energy. If the first cells arose at hydrothermal vents, then vents would have been the site of the first primary production because they were the source of H_2. The first bacteria would have been acetogens that lack cytochromes and quinones and the first archaea would have been methanogens that lack cytochromes and quinones (Martin and Russell 2007, Weiss et al. 2016). What came next? As outlined elsewhere, the next step was probably amino acid fermentations (Schönheit et al. 2016) because the first cells that accumulated via hydrothermal vent primary production were a rich substrate for the first heterotrophs, since cells are 50% protein. Protein is an excellent source of carbon and energy, and amino acid fermentations would require an absolute minimum in terms of biochemical innovation in order to glean ATP from their breakdown (Schönheit et al. 2016). Amino acid fermentations often entail simple SLP (Barker 1961, Schönheit et al. 2016) for energy conservation and would not require the advent of cytochromes or quinones. They, furthermore, would simply require pre-existing amino acid synthesis pathways to run in the reverse direction to satisfy carbon and nitrogen needs (Schönheit et al. 2016). Cells are about 25% RNA, which is also a rich source of carbon, nitrogen and energy for the first strictly anaerobic heterotrophs (Schönheit et al. 2016).

At some point in early evolution, some bacterial lineage invented cytochromes and quinones (Martin and Sousa 2016), but it is not clear which, possibly clostridial heterotrophs that found ways first to utilize environmental sulfur as a terminal electron acceptor for heterotrophic energy metabolism (Sousa et al. 2013b). The synthesis of heme stems from corrin (cobalamine) biosynthesis (Decker et al. 1970), and corrins are required both by acetogens and methanogens, corrin biosynthesis also tracing to LUCA's gene set (Weiss et al. 2016). The typical quinone of anaerobes, menaquinone, derives from chorismate, an intermediate of aromatic amino acid biosynthesis (see Chapter two).

With the advent of hemes and quinones, there could have been some modest diversification of early anaerobic respiratory chains, although the diversity of environmentally available terminal acceptors beyond S and sulfite (dissolved volcanic SO_2) might have been quite limited. However, primary production would have been restricted to sites of geochemical H_2 exhalation (vents). That is because primordial cells required reduced ferredoxin to fix CO_2, and before the advent of photosynthesis, the only way to generate reduced ferredoxins was through electron bifurcation with electrons stemming from H_2 (Martin et al. 2018). An exception would be the oxidation of native metals whose genesis would also have been restricted to vents as sites of sufficient H_2 via serpentinization to generate the extremely reducing conditions required to generate native transition metals such as awaruite (Sousa et al. 2018).

Escape from Vents

Following this line of reasoning, with the first free living cells being H_2 dependent acetogens and methanogens, the first habitat to be colonized outside the vent(s) where life arose would not be the open ocean or any land based habitat (if there was any land, the primordial oceans were twice as deep as today's) or even the open sea floor because what the cells needed for growth was not there: H_2. The recognition that about half of the Earth's biomass lives below the surface of the Earth (Whitman et al. 1998) should have a substantial impact on the way we view early evolution. The first new habitat to be colonized on the early Earth would be the crust, inoculated with the same convective currents of water through the crust that drive the serpentinization process. That would mean that the Earth's crust would become inhabited and populated with life before the oceans or the land. Sleep et al. (2004) pointed out that the crust would be a rich habitat for early life because of H_2 generated through serpentinization, a suggestion that fits well with the idea that life arose at vents in the first place (Baross and Hoffman 1985, Russell and Hall 1997, Martin and Russell 2003) and that H_2 dependent autotrophs are ancestral (Russell and Martin 2004).

Independence from Vents: Photosynthesis

The thermodynamic constraints imposed by electron bifurcation kept primary production tied to the Earth's crust until the advent of chlorophyll based photosynthesis (chlorophototrophy). The issue of where and how chlorophyll based photosynthesis arose was the topic of a recent paper (Martin et al. 2018) to which the interested reader is referred. The essentials of the inferences set forth in that paper are as follows. The first step in the origin of photosynthesis is chlorophyll (Chl) biosynthesis. Chl biosynthesis arose from the heme pathway, and the critical intermediate is protoporphyrinogen IX (proto IX), a tetrapyrrole and the last common intermediate in heme and chlorophyll synthesis (Gomez Maqueo Chew and Bryant 2007). It is a noteworthy observation that *Rhodobacter* mutants lacking Mg chelatase spontaneously accumulate Zn-protoporphyrinogen IX (Jaschke et al. 2011), Zn-PPIX. Another noteworthy observation is that Zn-PPIX is photochemically active, absorbing visible light which generates an excited triplet state (^{3}Zn-PPIX). The redox potential of photo-excited Zn-PPIX triplet is about −1.6 V (Dixit et al. 1984), which is sufficient to reduce low potential ferredoxin. The suggestion is that a heme containing bacterium that possessed a quinone based electron transport chain, if it accumulated Zn-PPIX, could reduce small amounts of ferredoxin in a manner that is independent of electron bifurcation (Martin et al. 2018). The oxidized tetrapyrrole, Zn-PPIX$^+$, could obtain its missing electron from reductants of physiological electron transport and a small, cytosolic (soluble) pathway of ferredoxin reduction would have been the result. Zn-PPIX would probably not exist unbound, but incorporated into abundant heme binding proteins, such as cytochromes. In support of that suggestion, synthetic Zn-tetrapyrroles and engineered, cytochrome bound Zn-tetrapyrroles do indeed perform light-dependent redox reactions and have been used in the study of artificial photosynthesis (Razeghifard and Wydrzynski 2003, Hay et al. 2004). Synthetic protein bound Zn-tetrapyrroles continue to be of interest in that context (Cohen-Ofri et al. 2011) and Zn-tetrapyrroles furthermore function in nature. Zn-BChl a is used in the RC2 of *Acidiphilium rubrum*, an alphaproteobacterium (Wakao et al. 1996), and Zn-BChl a′ appears to form a special pair of the type-1 RC in *Chloracidobacterium thermophilum* (Tsukatani et al. 2012).

The excited triplet state, Zn-PPIX, has roughly a one million fold longer lifespan than Mg-Chl excited states, which are in the range of a few nanoseconds for the photochemically active singlet species (Björn et al. 2009). Seven to 15 ms brings possible activities into the time domain of diffusion rate-limited chemical reactions with other molecules in the cell. This is ample time for an exited state 3Zn-PPIX 'cytochrome' to find by diffusion a cytosolic electron acceptor, such as soluble Fd, which not only is a ubiquitous protein in anaerobes (Buckel and Thauer 2013), but is also extremely abundant—in a typical anaerobe, Fd has cytosolic concentrations on the order of 80–400 μM (Thamer et al. 2003). Photoexcited ^{3}Zn-PPIX(cyt) is a strong reductant (Shen and Kostic 1996) that readily reduces both plastocyanin and ferricytochrome b5 (Qin and Kostic 1994). The redox potentials from ^{3}Zn(cyt) to Zn(cyt)$^+$ and back to Zn(cyt) would fit with Fd reduction, and with reduction of ZnCyt$^+$ by a respiratory chain (Shen and Kostic 1996). Hence, a protein-bound, photoexcited ^{3}Zn-porphyrin, ^{3}Zn-PPIX, or a ^{3}Zn-PPIX-cytochrome should have more than sufficient driving potential to reduce soluble Fd. These features could constitute the core of a primordial phototrophic Fd reduction pathway (Martin et al. 2018).

A primitive photochemical pathway with very low redox potentials of ^{3}Zn-PPIX would provide a unit of selectable function that would make natural variation enhancing its accumulation beneficial among individuals possessing that variation. The evolutionary steps to Chl biosynthesis in a corrin- and heme-possessing diazotrophic cell are few in number and simple in nature. They involve no enzymatic innovations, just the recruitment of preexisting enzyme activities from nitrogenase (BchNBL), hydrogenase (BciB), a radical SAM-dependent reaction (BchE), a Co chelatase from corrin biosynthesis (BchHDI) and a methyl transferase (BchM) (Martin et al. 2018). Chlorophyll a would have been the first chlorophyll (Martin et al. 2018), in line with the Granick hypothesis (Granick 1965) where the growth of biochemical pathways followed the evolution of lineages.

Incorporation of Chl into protein followed by insertion into the plasma membrane would deliver a basic reaction center. The physiological significance of such a reaction center would have been access to electron donors (reductants) other than H$_2$ that could reduce ferredoxin. Freedom from H$_2$ would allow cells to leave the vent and give them the opportunity to use sunlight, such that primary production could occur at the ocean surface. But if primary production was limited to vents before the onset of Chl-based photosynthesis, where was the light coming from? When we think of the origin of photosynthesis, we typically think of harnessing sunlight, but sunlight is very intense and has a very harmful *uv* component that was much stronger before the origin of the ozone layer (a latecomer in evolution). At the Earth's surface, there was nothing to support a stable ecosystem within which photosynthesis could evolve anyway. Light emitted from hydrothermal vents is a better source of light for the origin of photosynthesis because it has no *uv* component and because cells were alive in small stable ecosystems at vents and hence could evolve.

It is little known, or little appreciated, or both, that hydrothermal vents emit light in the visible spectrum. There are even two different kinds of light at vents: ambient and thermal (infrared black body). Ambient light at hydrothermal vents has a component with wavelengths in the visible range (van Dover et al. 1996, White et al. 2002). The first hints for visible light at vents came from studies of a shrimp that inhabits the dark abyss near vents and possesses unusual photoreceptive organs (van Dover et al. 1989). The mechanism(s) generating light in the visible spectrum at vents that exceeds the contribution from thermal light are still unknown; possible sources include sonoluminescence (the collapse of small bubbles) and triboluminescence (light emission from small ZnS crystals), as well as bioluminescence (White et al. 2000b) and O_2-dependent chemiluminescence (Tapley et al. 1999).

Thermal light is black body radiation emitted from vents with very hot (> 400°C) effluent (van Dover 1996, White et al. 2000a,b, 2002). At such temperatures, the emitted light is mainly of wavelengths > 900 nm, but lesser amounts of light extend down to ~ 750 nm, which can be absorbed by chlorophyll. Black body light from a > 400°C source typically has a vanishingly small component of what is commonly thought of as photosynthetically active radiation (400 to 700 nm). Because of its low flux in the Chl a absorption range, thermal light is not widely discussed as a source of photosynthetically relevant radiation. However, the idea has been alive and well in the literature for over 20 years, starting with the suggestion that the pathway to anoxygenic photosynthesis emerged as a heat- and light-sensing mechanism to guide motile prokaryotes to sources of chemical energy (Nisbet et al. 1995).

Beatty et al. (2005) isolated and cultivated an obligatory photoautotrophic, H_2S-dependent green sulfur bacterium (GSB) from a hydrothermal vent sample, showing that light emission at hydrothermal vents could indeed be sufficient to support photosynthesis. The hydrothermal vent GSB uses chlorobactene and BChl c as pigments, which are well suited to harnessing black body radiation because the *in vivo* BChl c absorbance peak is maximal at 750 nm, tailing off around 850 nm (Beatty et al. 2005). The photon flux measured at the vent where the GSB was isolated would permit a doubling time of about 3 years (Martin et al. 2018), which seems very slow, but in low energy environments (Whitmann et al. 1998), prokaryotes can have generation times on the order of 1000 years (Hoehler and Joergensen 2012). The photon fluxes at vents are low, but they can support growth rates that exceed those in some low energy environments by orders of magnitude.

Photosynthesis could have gotten started that way, in which case the first reaction centers would have been functioning like the type in reaction center in *Chlorobium* (Bryant and Liu 2013) that generates reduced ferredoxin for CO_2 fixation, and supported by chlorosomes (in essence, rods of cylindrically stacked paracrystalline bacteriochlorophyll molecules, surrounded by a lipid monolayer) as antenna complexes exquisitely well suited to low light fluxes (Bryant and Liu 2013). The origin of two photosystems as found in cyanobacteria likely occurred via gene duplication in the ancestor of cyanobacteria (Allen 2005, Sousa et al. 2013a), the bacterial lineages that have only one photosystem probably acquired them via lateral gene transfer, which would explain why photosystems are combined with different pathways of CO_2 fixation and sometimes found in lineages with no CO_2 fixation (Martin et al. 2018).

Oxygenic photosynthesis was in operation since 2.7–2.4 billion years ago (Fischer et al. 2016). That puts a lower bound on the date for the origin of mitochondria because the ancestral mitochondrion, in addition to being able to perform anaerobic fermentations, was able to respire (Müller et al. 2012—see Chapter seven).

What did Archaea Invent?

From my (admittedly narrow) perspective, it seems that archaea did not invent much in the way of novel physiological tools or pathways beyond methanogenesis. If methanogenesis without cytochromes and quinones truly is the ancestral state of physiology in the archaea, then the existence of cytochromes and quinones in archaeal lineages such as methanosarcinales (Thauer et al. 2008) and haloarchaea (Nelson-Sathi et al. 2012) could have come about by one of two routes. Either archaeal lineages independently invented cytochromes and quinones (or functional analogues of quinones such as methanophenazine), or the genes for those traits were acquired from bacteria. The latter seems to be the case (Nelson Sathi et al. 2012, 2015). The host for the origin of mitochondria was also an archaeaon according to newer findings (Zaremba-Niedzwiedzka et al. 2017), and that archaeal host acquired so many genes via the symbiotic origin of mitochondria at eukaryote origin that bacterial genes outnumber archaeal genes in the eukaryotic lineage (Ku et al. 2015). That however brings us to the origin of eukaryotes, which is a topic for a different contribution (Martin et al. 2015, Gould et al. 2016—see Chapter eight) and a good place to end this chapter.

Acknowledgement

I thank Verena Zimorski for help in preparing the manuscript, the ERC (AdvGr 666053) for financial support, many friends and colleagues for discussions, and Mauro Degli Esposti for the invitation to contribute to this book.

References

Adams, M.W.W., J.F. Holden, A. Lal Menon, G.J. Schut, A.M. Grunden, C. Hou et al. 2001. Key role for sulfur in peptide metabolism and in regulation of three hydrogenases in the hyperthermophilic archaeon *Pyrococcus furiosus*. J. Bacteriol. 183: 716–724. doi:10.1128/JB.183.2.716-724.2001.

Allen, J.F. 2005. A redox switch hypothesis for the origin of two light reactions in photosynthesis. FEBS Lett. 579: 963–968. doi:10.1016/j.febslet.2005.01.015.

Bach, W., H. Paulick, C.J. Garrido, B. Ildefonse, W.P. Meurer and S.E. Humphris. 2006. Unraveling the sequence of serpentinization reactions: Petrography, mineral chemistry, and petrophysics of serpentinites from MAR 15°N (ODP Leg 209, Site 1274). Geophys. Res. Lett. 33: L13306. doi:10.1029/2006GL025681.

Barker, H.A. 1961. Fermentation of nitrogenous compounds. pp. 151–207. In: Gunsalus, I.C. and Stanier, R.Y. (eds.). The Bacteria. A Treatise on Structure and Function. Academic Press, New York, USA.

Baross, J.A. and S.E. Hoffman. 1985. Submarine hydrothermal vents and associated gradient environments as sites for the origin and evolution of life. Orig. Life Evol. Biosph. 15: 327–345. doi:10.1007/BF01808177.

Basen, M. and V. Müller. 2016. "Hot" acetogenesis. Extremophiles. 1–12. doi:10.1007/s00792-016-0873-3.

Beatty, J.T., J. Overmann, M.T. Lince, A.K. Manske, A.S. Lang, R.E. Blankenship et al. 2005. An obligately photosynthetic bacterial anaerobe from a deep-sea hydrothermal vent. Proc. Natl. Acad. Sci. USA 102: 9306–9310. doi:10.1073/pnas.0503674102.

Berg, I.A., D. Kockelkorn, W.H. Ramos-Vera, R.F. Say, J. Zarzycki, M. Hügler et al. 2010. Autotrophic carbon fixation in archaea. Nat. Rev. Microbiol. 8: 447–460. doi:10.1038/nrmicro2365.

Björn, L.O., G.C. Papageorgiou, R.E. Blankenship and Govindjee. 2009. A viewpoint: Why chlorophyll a? Photosynth. Res. 99: 85–98. doi:10.1007/s11120-008-9395-x.

Boyd, E.S. and J.W. Peters. 2013. New insights into the evolutionary history of biological nitrogen fixation. Front. Microbiol. 4: 201. doi:10.3389/fmicb.2013.00201.

Bryant, D.A. and Z. Liu. 2013. Green bacteria: insights into green bacterial evolution through genomic analyses. Adv. Bot. Res. 66: 99–150. doi:10.1016/B978-0-12-397923-0.00004-7.

Buckel, W. 2001. Unusual enzymes involved in five pathways of glutamate fermentation. Appl. Microbiol. Biotechnol. 57: 263–273. doi:10.1007/s002530100773.

Buckel, W. and R.K. Thauer. 2013. Energy conservation via electron bifurcating ferredoxin reduction and proton/Na$^+$ translocating ferredoxin oxidation. Biochim. Biophys. Acta. 1827: 94–113. doi:10.1016/j.bbabio.2012.07.002.

Buckel, W. and R.K. Thauer. 2018. Flavin-based electron bifurcation, ferredoxin, flavodoxin, and anaerobic respiration with Protons (Ech) or NAD$^+$ (Rnf) as electron acceptors: A historical review. Front. Microbiol. 9: 401. doi:10.3389/fmicb.2018.00401.

Camprubi, E., S.F. Jordan, R. Vasiliadou and N. Lane. 2017. Iron catalysis at the origin of life. IUBMB Life. 69: 373–381. doi:10.1002/iub.1632.

Chyba, C.F., P.J. Thomas, L. Brookshaw and C. Sagan. 1990. Cometary delivery of organic molecules to the early Earth. Science 249: 366–373. doi:10.1126/science.11538074.

Cohen-Ofri, I., M. van Gastel, J. Grzyb, A. Brandis, I. Pinkas, W. Lubitz et al. 2011. Zinc-bacteriochlorophyllide dimers in de novo designed four-helix bundle proteins. A model system for natural light energy harvesting and dissipation. J. Am. Chem. Soc. 133: 9526–9535. doi:10.1021/ja202054m.

Dagan, T. and W. Martin. 2006. The tree of one percent. Genome Biol. 7: 118. doi:10.1186/gb-2006-7-10-118.

De Duve, C. 1991. Blueprint for a Cell: The Nature and Origin of Life. Neil Patterson Publishers, Burlington, North Carolina, USA.

Decker, K., K. Jungermann and R.K. Thauer. 1970. Energy production in anaerobic organisms. Angew. Chem. Int. Ed. Engl. 9: 138–158. doi:10.1002/anie.197001381.

Di Giulio, M. 2011. The last universal common ancestor (LUCA) and the ancestors of archaea and bacteria were progenotes. J. Mol. Evol. 72: 119–126. doi:10.1007/s00239-010-9407-2.

Dickerson, R.E. 1980. Cytochrome c and the evolution of energy metabolism. Sci. Am. 242: 98–110. doi:10.1038/scientificamerican0380-136.

Dixit, B.P.S.N., V.T. Moy and J.M. Vanderkooi. 1984. Reactions of excited-state cytochrome c derivatives. Delayed fluorescence and phosphorescence of zinc, tin, and metal-free cytochrome c at room temperature. Biochemistry 23: 2103–2107. doi:10.1021/bi00304a035.

Eck, R.V. and M.O. Dayhoff. 1966. Evolution of structure of ferredoxin based on living relics of primitive amino acid sequences. Science 152: 363–366. doi:10.1126/science.152.3720.363.

Evans, P.N., D.H. Parks, G.L. Chadwick, S.J. Robbins, V.J. Orphan, S.D. Golding et al. 2015. Methane metabolism in the archaeal phylum *Bathyarchaeota* revealed by genome-centric metagenomics. Science 350: 434–438. doi:10.1126/science.aac7745.

Ferry, J.G. and C.H. House. 2006. The step-wise evolution of early life driven by energy conservation. Mol. Biol. Evol. 23: 1286–1292. doi:10.1093/molbev/msk014.

Fischer, W.W., J. Hemp and J.E. Johnson. 2016. Evolution of oxygenic photosynthesis. Ann. Rev. Earth Planet. Sci. 44: 647–683. doi:10.1146/annurev-earth-060313-054810.

Forterre, P. and H. Philippe. 1999. Where is the root of the universal tree of life? Bioessays 21: 871–879. doi:10.1002/(SICI)1521-1878(199910)21:10<871::AID-BIES10>3.0.CO;2-Q.

Fox, G., E. Stackebrandt, R. Hespell, J. Gibson, J. Maniloff, T. Dyer et al. 1980. The phylogeny of prokaryotes. Science 209: 457–463. doi:10.1126/science.6771870.

Fox, G.E. 2010. Origin and evolution of the ribosome. Cold Spring Harb. Perspect. Biol. 2: a003483. doi:10.1101/cshperspect.a003483.

Fuchs, G. and E. Stupperich. 1985. Evolution of autotrophic CO$_2$ fixation. pp. 235–251. *In*: Schleifer, K.H. and E. Stackebrandt (eds.). Evolution of Prokaryotes. FEMS Symposium No. 29, Academic Press, London, UK.

Fuchs, G. and E. Stupperich. 1986. Carbon assimilation pathways in archaebacteria. System. Appl. Microbiol. 7: 364–369. doi:10.1016/S0723-2020(86)80035-2.

Fuchs, G. 1994. Variations of the acetyl-CoA pathway in diversely related microorganisms that are not acetogens. pp. 506–538. *In*: Drake, G. (ed.). Acetogenesis. Chapman and Hall, New York, NY.

Fuchs, G. 2011. Alternative pathways of carbon dioxide fixation: Insights into the early evolution of life? Annu. Rev. Microbiol. 65: 631–658. doi:10.1146/annurev-micro-090110-102801.

Gaidos, E. and A.H. Knoll. 2012. Our evolving planet: From dark ages to evolutionary renaissance. pp. 132–153. *In*: Impey, C., J. Lunine and J. Funes (eds.). Frontiers of Astrobiology. Cambridge University Press, Cambridge, UK. doi:10.1017/CBO9780511902574.011.

Goldford, J.E., H. Hartman, T.F. Smith and D. Segrè. 2017. Remnants of an ancient metabolism without phosphate. Cell. 168: 1126–1134. doi:10.1016/j.cell.2017.02.001.

Goldford, J.E. and D. Segrè. 2018. Modern views of ancient metabolic networks. Curr. Opin. Syst. Biol. in press. doi:10.1016/j.coisb.2018.01.004.

Gomez Maqueo Chew, A. and D.A. Bryant. 2007. Chlorophyll biosynthesis in bacteria: The origins of structural and functional diversity. Annu. Rev. Microbiol. 61: 113–129. doi:10.1146/annurev. micro.61.080706.093242.

Goremykin, V.V., S. Hansmann and W. Martin. 1997. Evolutionary analysis of 58 proteins encoded in six completely sequenced chloroplast genomes: Revised molecular estimates of two seed plant divergence times. Pl. Syst. Evol. 206: 337–351. doi:10.1007/BF00987956.

Gould, S.B., S.G. Garg and W.F. Martin. 2016. Bacterial vesicle secretion and the evolutionary origin of the eukaryotic endomembrane system. Trends Microbiol. 24: 525–534.

Granick, S. 1965. Evolution of heme and chlorophyll. pp. 67–88. *In*: Bryson, G. and H.J. Vogel (eds.). Evolving Genes and Proteins. Academic Press, New York, USA.

Guan, G., T. Kida, T. Ma, K. Kimura, E. Abe and A. Yoshida. 2003. Reduction of aqueous CO_2 at ambient temperature using zero-valent iron-based composites. Green Chemistry 5: 630–634. doi:10.1039/B304395A.

Haldane, J.B.S. 1929. The origin of life. Rationalist Annu. 3–10.

Hansmann, S. and W. Martin. 2000. Phylogeny of 33 ribosomal and six other proteins encoded in an ancient gene cluster that is conserved across prokaryotic genomes: Influence of excluding poorly alignable sites from analysis. Int. J. Syst. Evol. Microbiol. 50: 1655–1663. doi:10.1099/00207713-50-4-1655.

Hay, S., B.B Wallace, T.A. Smith, K.P. Ghiggino and T. Wydrzynski. 2004. Protein engineering of cytochrome b562 for quinone binding and light-induced electron transfer. Proc. Natl. Acad. Sci. USA 101: 17675–17680. doi:10.1073/pnas.0406192101.

He, C., G. Tian, Z. Liu and S. Feng. 2010. A mild hydrothermal route to fix carbon dioxide to simple carboxylic acids. Org. Lett. 12: 649–651. doi:10.1021/ol9025414.

He, Y., M. Li, V. Perumal, X. Feng, J. Fang, J. Xie et al. 2016. Genomic and enzymatic evidence for acetogenesis among multiple lineages of the archaeal phylum *Bathyarchaeota* widespread in marine sediments. Nat. Microbiol. 1: 16035. doi:10.1038/nmicrobiol.2016.35.

Herrmann, G., E. Jayamani, G. Mai and W. Buckel. 2008. Energy conservation via electron-transferring flavoprotein in anaerobic bacteria. J. Bacteriol. 190: 784–791. doi:10.1016/j.bbabio.2012.07.002.

Hoehler, T.M. and B.B. Jörgensen. 2013. Microbial life under extreme energy limitation. Nat. Rev. Microbiol. 11: 83–94. doi:10.1038/nrmicro2939.

Holm, N.G. and J.L. Charlou. 2001. Initial indications of abiotic formation of hydrocarbons in the Rainbow ultramafic hydrothermal system, mid-atlantic ridge. Earth Planet. Sci. Lett. 191: 1–8. doi:10.1016/S0012-821X(01)00397-1.

Horita, J. and M.E. Berndt. 1999. Abiogenic methane formation and isotopic fractionation under hydrothermal conditions. Science 285: 1055–1057. doi:10.1126/science.285.5430.1055.

Hug, L.A., B.J. Baker, K. Anantharaman, C.T. Brown, A.J. Probst, C.J. Castelle et al. 2016. A new view of the tree of life. Nat. Microbiol. 1: 16048. doi:10.1038/nmicrobiol.2016.48.

Jain, R., M.C. Rivera and J.A. Lake. 1999. Horizontal gene transfer among genomes: The complexity hypothesis. Proc. Natl. Acad. Sci. USA 96: 3801–3806. doi:10.1073/pnas.96.7.3801.

Jaschke, P.R., A. Hardjasa, E.L. Digby, C.N. Hunter and J.T. Beatty. 2011. A bchD (magnesium chelatase) mutant of *Rhodobacter sphaeroides* synthesizes zinc bacteriochlorophyll through novel zinc-containing intermediates. J. Biol. Chem. 286: 20313–20322. doi:10.1074/jbc.M110.212605.

Jelen, B.I., D. Giovannelli and P.G. Falkowski. 2016. The role of microbial electron transfer in the coevolution of the biosphere and geosphere. Annu. Rev. Microbiol. 70: 45–62. doi:10.1146/annurev-micro-102215-095521.

Kaster, A.-K., J. Moll, K. Parey and R.K. Thauer. 2011. Coupling of ferredoxin and heterodisulfide reduction via electron bifurcation in hydrogenotrophic methanogenic archaea. Proc. Natl. Acad. Sci. USA 108: 2981–2986. doi:10.1073/pnas.1016761108.

Klein, F. and W. Bach. 2009. Fe-Ni-Co-O-S phase relations in peridotite-seawater interactions. J. Petrol. 50: 37–59. doi:10.1093/petrology/egn071.

Koonin, E.V. 2003. Comparative genomics, minimal gene-sets and the last universal common ancestor. Nat. Rev. Microbiol. 1: 127–136. doi:10.1038/nrmicro751.

Ku, C., S. Nelson-Sathi, M. Roettger, F.L. Sousa, P.J. Lockhart, D. Bryant et al. 2015. Endosymbiotic origin and differential loss of eukaryotic genes. Nature 524: 427–432. doi:10.1038/nature14963.

Kyprides, N., R. Overbeek and C. Ouzounis. 1999. Universal protein families and the functional content of the last universal common ancestor. J. Mol. Evol. 49: 413–423. doi:10.1007/PL00006564.

Lane, N. 2017. Proton gradients at the origin of life. Bioessays 39: in press. doi:10.1002/bies.201600217.

Lane, N. and W.F. Martin. 2012. The origin of membrane bioenergetics. Cell 151: 1406–1416. doi:10.1016/j.cell.2012.11.050.

Lane, N., J.F. Allen and W. Martin. 2010. How did LUCA make a living? Chemiosmosis in the origin of life. BioEssays 32: 271–280. doi:10.1002/bies.200900131.

Laurino, P. and D.S. Tawfik. 2017. Spontaneous emergence of S-adenosylmethionine and the evolution of methylation. Angew. Chem. Int. Ed. Engl. 56: 343–345. doi:10.1002/anie.201609615.

Lipmann, F. 1941. Metabolic generation and utilization of phosphate bond energy. Adv. Enzymol. 1: 99–162. doi:10.1002/9780470122464.ch4.

Lockhart, P.J., C.J. Howe, A.C. Barbrook, A.W.D. Larkum and D. Penny. 1999. Spectral analysis, systematic bias, and the evolution of chloroplasts. Mol. Biol. Evol. 16: 573–576. doi:10.1093/oxfordjournals.molbev.a026139.

Lubitz, W., H. Ogata, O. Rüdiger and E. Reijerse. 2014. Hydrogenases. Chem. Rev. 114: 4081–4148.

Lubner, C.E., D.P. Jennings, D.W. Mulder, G.J. Schut, O.A. Zadvornyy, J.P. Hoben et al. 2017. Mechanistic insights into energy conservation by flavin-based electron bifurcation. Nat. Chem. Biol. 13: 655–659. doi:10.1038/nchembio.2348.

Lyons, T.W., C.T. Reinhard and N.J. Planavsky. 2014. The rise of oxygen in Earth's early ocean and atmosphere. Nature 506: 307–315. doi:10.1038/nature13068.

Mai, X. and M.W. Adams. 1996. Purification and characterization of two reversible and ADP-dependent acetyl coenzyme A synthetases from the hyperthermophilic archaeon *Pyrococcus furiosus*. J. Bacteriol. 178: 5897–5903. doi:10.1128/jb.178.20.5897-5903.1996.

Martin, W., B. Stoebe, V. Goremykin, S. Hansmann, M. Hasegawa and K.V. Kowallik. 1998. Gene transfer to the nucleus and the evolution of chloroplasts. Nature 393: 162–165. doi:10.1038/30234.

Martin, W. and M.J. Russell. 2003. On the origins of cells: A hypothesis for the evolutionary transitions from abiotic geochemistry to chemoautotrophic prokaryotes, and from prokaryotes to nucleated cells. Philos. Trans. R. Soc. Lond. B 358: 59–83. doi:10.1098/rstb.2002.1183.

Martin, W. and M.J. Russell. 2007. On the origin of biochemistry at an alkaline hydrothermal vent. Phil. Trans. R. Soc. Lond. B 362: 1887–1926. doi:10.1098/rstb.2006.1881.

Martin, W., S. Garg and V. Zimorski. 2015. Endosymbiotic theories for eukaryote origin. Philos. Trans. R. Soc. Lond. B. 370: 20140330. doi:10.1098/rstb.2014.0330.

Martin, W.F. 2012. Hydrogen, metals, bifurcating electrons, and proton gradients: The early evolution of biological energy conservation. FEBS Lett. 586: 485–493. doi:10.1016/j.febslet.2011.09.031.

Martin, W.F. 2016. Physiology, phylogeny, and the energetic roots of life. Period. Biol. 118: 343–352. doi:10.18054/pb.v118i4.4737.

Martin, W.F. and F.L. Sousa. 2016. Early microbial evolution: The age of anaerobes. Cold Spring Harb. Perspect. Biol. 8: a018127. doi:10.1101/cshperspect.a018127.

Martin, W.F. and R.K. Thauer. 2017. Energy in ancient metabolism. Cell. 168: 953–955. doi:10.1016/j. cell.2017.02.032.

Martin, W.F., D.A. Bryant and J.T. Beatty. 2018. A physiological perspective on the origin and evolution of photosynthesis. FEMS Microbiol. Rev. 42: 205–231. doi:10.1093/femsre/fux056.

McCollom, T.M. 2016. Abiotic methane formation during experimental serpentinization of olivine. Proc. Natl. Acad. Sci. USA 113: 13965–13970. doi:10.1073/pnas.1611843113.

Mojzsis, S.J., G. Arrhenius, K.D. McKeegan, T.M. Harrison, A.P. Nutman and C.R. Friend. 1996. Evidence for life on Earth before 3,800 million years ago. Nature 384: 55–59. doi:10.1038/384055a0.

Muchowska, K.B., S.J. Varma, E. Chevallot-Beroux, L. Lethuillier-Karl, G. Li and J. Moran. 2017. Metals promote sequences of the reverse Krebs cycle. Nat. Ecol. Evol. 1: 1716–1721. doi:10.1038/s41559-017-0311-7.

Müller, M., M. Mentel, J.J. van Hellemond, K. Henze, C. Woehle, S.B. Gould et al. 2012. Biochemistry and evolution of anaerobic energy metabolism in eukaryotes. Microbiol. Mol. Biol. Rev. 76: 444–495. doi:10.1128/MMBR.05024-11.

Neidhardt, F.C., J.L. Ingraham and M. Schaechter (eds.). 1990. Physiology of the Bacterial Cell—A Molecular Approach. Sinauer Associates, Sunderland, MA, USA.

Nelson-Sathi, S., T. Dagan, G. Landan, A. Janssen, M. Steel, J.O. McInerney et al. 2012. Acquisition of 1,000 eubacterial genes physiologically transformed a methanogen at the origin of Haloarchaea. Proc. Natl. Acad. Sci. USA 109: 20537–20542. doi:10.1073/pnas.1209119109.

Nelson-Sathi, S., F.L. Sousa, M. Roettger, N. Lozada-Chávez, T. Thiergart, A. Janssen et al. 2015. Origins of major archaeal clades correspond to gene acquisitions from bacteria. Nature 517: 77–80. doi:10.1038/nature13805.

Nisbet, E.G., J.R. Cann and C.L. van Dover. 1995. Origins of photosynthesis. Nature 373: 479–480. doi:10.1038/373479a0.

Pace, N.R. 1991. Origin of life-facing up to the physical setting. Cell. 65: 531–533. doi:10.1016/0092-8674(91)90082-A.

Penny, D. and A. Poole. 1999. The nature of the last universal common ancestor. Curr. Opin. Genet. Dev. 9: 672–677. doi:10.1016/S0959-437X(99)00020-9.

Petersen, J., O. Frank, M. Göker and S. Pradella. 2013. Extrachromosomal, extraordinary and essential—The plasmids of the Roseobacter clade. Appl. Microbiol. Biotechnol. 97: 2805–2815. doi:10.1007/s00253-013-4746-8.

Poulton, S.W., P.W. Fralick and D.E. Canfield. 2004. The transition to a sulphidic ocean similar to 1.84 billion years ago. Nature 431: 173–177. doi:10.1038/nature02912.c.

Puigbò, P., Y.I. Wolf and E.V. Koonin. 2009. Search for a 'Tree of Life' in the thicket of the phylogenetic forest. J. Biol. 8: 59. doi:10.1186/jbiol159.

Qin, L. and N.M. Kostic. 1994. Photoinduced electron transfer from the triplet state of zinc cytochrome c to ferricytochrome b5 is gated by configurational fluctuations of the diprotein complex. Biochemistry 33: 12592–12599. doi:10.1021/bi00208a009.

Rabus, R., S.S. Venceslau, L. Wöhlbrand, G. Voordouw, J.D. Wall and I.A.C. Pereira. 2015. A post-genomic view of the ecophysiology, catabolism and biotechnological relevance of sulphate-reducing prokaryotes. Adv. Microb. Physiol. 66: 55–321. doi:10.1016/bs.ampbs.2015.05.002.

Razeghifard, A.R. and T. Wydrzynski. 2003. Binding of Zn-chlorin to a synthetic four-helix bundle peptide through histidine ligation. Biochemistry 42: 1024–1030. doi:10.1021/bi026787i.

Rinke, C., P. Schwientek, A. Sczyrba, N.N. Ivanova, I.J. Anderson, J.–F. Cheng et al. 2013. Insights into the phylogeny and coding potential of microbial dark matter. Nature 499: 431–437. doi:10.1038/nature12352.

Roldan, A., N. Hollingsworth, A. Roffey, H.-U. Islam, J.B.M. Goodall, C.R.A. Catlow et al. 2015. Bio-inspired CO_2 conversion by iron sulfide catalysts under sustainable conditions. Chem. Commun. (Camb.) 51: 7501–7504. doi:10.1039/c5cc02078f.

Rother, M. and W.W. Metcalf. 2004. Anaerobic growth of *Methanosarcina acetivorans* C2A on carbon monoxide: An unusual way of life for a methanogenic archaeon. Proc. Natl. Acad. Sci. USA 101: 16929–16934. doi:10.1073/pnas.0407486101.

Russell, M.J. and A.J. Hall. 1997. The emergence of life from iron monosulphide bubbles at a submarine hydrothermal redox and pH front. J. Geol. Soc. Lond. 154: 377–402. doi:10.1144/gsjgs.154.3.0377.

Russell, M.J. and W. Martin. 2004. The rocky roots of the acetyl-CoA pathway. Trends Biochem. Sci. 29: 358–363. doi:10.1016/j.tibs.2004.05.007.

Russell, M.J., A.J. Hall and W. Martin. 2010. Serpentinization as a source of energy at the origin of life. Geobiol. 8: 355–371. doi:10.1111/j.1472-4669.2010.00249.x.

Sagan, L. 1967. On the origin of mitosing cells. J. Theoret. Biol. 14: 225–274.

Schönheit, P., W. Buckel and W.F. Martin. 2016. On the origin of heterotrophy. Trends Microbiol. 24: 12–25. doi:10.1016/j.tim.2015.10.003.

Schrenk, M.O., W.J. Brazelton and S.Q. Lang. 2013. Serpentinization, carbon, and deep life. Rev. Mineral. Geochem. 75: 575–606. doi:10.2138/rmg.2013.75.18.

Schuchmann, K. and V. Müller. 2013. Direct and reversible hydrogenation of CO_2 to formate by a bacterial carbon dioxide reductase. Science 342: 1382–1385. doi:10.1126/science.1244758.

Schuchmann, K. and V. Müller. 2014. Autotrophy at the thermodynamic limit of life: A model for energy conservation in acetogenic bacteria. Nat. Rev. Microbiol. 12: 809–821. doi:10.1038/nrmicro3365.

Schut, G.J. and M.W.W. Adams. 2009. The iron-hydrogenase of *Thermotoga maritima* utilizes ferredoxin and NADH synergistically: A new perspective on anaerobic hydrogen production. J. Bacteriol. 191: 4451–44657. doi:10.1128/JB.01582-08.

Schwarzenbach, E.M., E. Gazel and M.J. Caddick. 2014. Hydrothermal processes in partially serpentinized peridotites from Costa Rica: Evidence from native copper and complex sulfide assemblages. Contrib. Mineral. Petrol. 168: 1079. doi:10.1007/s00410-014-1079-2.

Semenov, S.N., L.J. Kraft, A. Ainla, M. Zhao, M. Baghbanzadeh, V.E. Campbell et al. 2016. Autocatalytic, bistable, oscillatory networks of biologically relevant organic reactions. Nature 537: 656–660. doi:10.1038/nature19776.

Sephton, M.A. 2002. Organic compounds in carbonaceous meteorites. Nat. Prod. Rep. 19: 292–311. doi:10.1039/B103775G.

Shen, C. and N.M. Kostic. 1996. Reductive quenching of the triplet state of zinc cytochrome c by the hexacyanoferrate(II) anion and by conjugate bases of ethylenediaminetetraacetic acid. Inorg. Chem. 35: 2780–2784. doi:10.1021/ic9510270.

Shock, E.L., T. McCollom and M.D. Schulte. 1998. The emergence of metabolism from within hydrothermal systems. pp. 59–76. *In*: Wiegel, J. and M.W.W. Adams (eds.). Thermophiles: The Keys to Molecular Evolution and the Origin of Life. Taylor and Francis, London, UK.

Shock, E.L. and E.S. Boyd. 2015. Principles of geobiochemistry. Elements 11: 395–401. doi:10.2113/gselements.11.6.395.

Sleep, N.H., A. Meibom, T. Fridriksson, R.G. Coleman and D.K. Bird. 2004. H_2-rich fluids from serpentinization: Geochemical and biotic implications. Proc. Natl. Acad. Sci. USA 101: 12818–12823. doi:10.1073/pnas.0405289101.

Sleep, N.H., D.K. Bird and E.C. Pope. 2011. Serpentinite and the dawn of life. Philos. Trans. R. Soc. Lond. B Biol. Sci. 366: 2857–2869. doi:10.1098/rstb.2011.0129.

Sousa, F.L., L. Shavit-Grievink, J.F. Allen and W.F. Martin. 2013a. Chlorophyll biosynthesis gene evolution indicates photosystem gene duplication, not photosystem merger, at the origin of oxygenic photosynthesis. Genome Biol. Evol. 5: 200–216. doi:10.1093/gbe/evs127.

Sousa, F.L., T. Thiergart, G. Landan, S. Nelson-Sathi, I.A.C. Pereira, J.F. Allen et al. 2013b. Early bioenergetic evolution. Philos. Trans. R. Soc. Lond. B. 368: 20130088. doi:10.1098/rstb.2013.0088.

Sousa, F.L. and W.F. Martin. 2014. Biochemical fossils of the ancient transition from geoenergetics to bioenergetics in prokaryotic one carbon compound metabolism. Biochim. Biophys. Acta. 1837: 964–981. doi:10.1016/j.bbabio.2014.02.001.

Sousa, F.L., M. Preiner and W.F. Martin. 2018. Native metals, electron bifurcation, and CO_2 reduction in early biochemical evolution. Curr. Opin. Microbiol. 43: 77–83. doi:10.1016/j.mib.2017.12.010.

Spang, A., J.H. Saw, S.L. Jørgensen, K. Zaremba-Niedzwiedzka, J. Martijn, A.E. Lind et al. 2015. Complex archaea that bridge the gap between prokaryotes and eukaryotes. Nature 521: 173–179. doi:10.1038/nature14447.

Stanier, R.Y. and C.B. van Niel. 1962. The concept of a bacterium. Archiv. Mikrobiol. 42: 17–35. doi:10.1007/BF00425185.

Stoebe, B., S. Hansmann, V. Goremykin, K.V. Kowallik and W. Martin. 1999. Proteins encoded in sequenced chloroplast genomes: An overview of gene content, phylogenetic information, and endosymbiotic gene transfer to the nucleus. pp. 327–352. *In*: Hollingsworth, C., R. Batemann and M. Gornall (eds.). Advances in Plant Molecular Systematics. Taylor and Francis, Andover, UK.

Stüecken, E.E., R. Buick, B.M. Guy and M.C. Koehler. 2015. Isotopic evidence for biological nitrogen fixation by molybdenum-nitrogenase from 3.2 Gyr. Nature 520: 666–669. doi:10.1038/nature14180.

Tapley, D.W., G.R. Buettner and J.M. Shick. 1999. Free radicals and chemiluminescence as products of the spontaneous oxidation of sulfide in seawater, and their biological implications. Biol. Bull. 196: 52–56. doi:10.2307/1543166.

Tashiro, T., A. Ishida, M. Hori, M. Igisu, M. Koike, P. Méjean et al. 2017. Early trace of life from 3.95 Ga sedimentary rocks in Labrador, Canada. Nature 549: 516–518. doi:10.1038/nature24019.

Thamer, W., I. Cirpus, M. Hans, A.J. Pierik, T. Selmer, E. Bill et al. 2003. A two [4Fe-4S]-cluster-containing ferredoxin as an alternative electron donor for 2-hydroxyglutaryl-CoA dehydratase from *Acidaminococcus fermentans*. Arch. Microbiol. 179: 197–204. doi:10.1007/s00203-003-0517-8.

Thauer, R.K., K. Jungermann and K. Decker. 1977. Energy-conservation in chemotropic anaerobic bacteria. Bacteriol. Rev. 41: 100–180.

Thauer, R.K., A.K. Kaster, H. Seedorf, W. Buckel and R. Hedderich. 2008. Methanogenic archaea: Ecologically relevant differences in energy conservation. Nat. Rev. Microbiol. 6: 579–591. doi:10.1038/nrmicro1931.

Thauer, R.K. 2011. Hydrogenases and the global H_2 cycle. Eur. J. Inorg. Chem 2011: 919–921.

Thiergart, T., G. Landan and W.F. Martin. 2014. Concatenated alignments and the case of the disappearing tree. BMC Evol. Biol. 14: 266. doi:10.1186/s12862-014-0266-0.

Tsukatani, Y, S.P. Romberger, J.H. Golbeck and D.A. Bryant. 2012. Isolation and characterization of homodimeric type-I reaction center complex from Candidatus *Chloracidobacterium thermophilum*, an aerobic chlorophototroph. J. Biol. Chem. 287: 5720–5732. doi:10.1074/jbc.M111.323329.

Ueno, Y., H. Yurimoto, H. Yoshioka, T. Komiya and S. Maruyama. 2002. Ion microprobe analysis of graphite from ca. 3.8 Ga metasediments, Isua crustal belt, West Greenland: Relationship between metamorphism and carbon isotopic composition. Geochim. Cosmochim. Acta. 66: 1257–1268. doi:10.1016/S0016-7037(01)00840-7.

Ueno, Y., K. Yamada, N. Yoshida, S. Maruyama and Y. Isozaki. 2006. Evidence from fluid inclusions for microbial methanogenesis in the early archaean era. Nature 440: 516–519. doi:10.1038/nature04584.

Van Dover, C.L., E.Z. Szuts, S.C. Chamberlain and J.R. Cann. 1989. A novel eye in eyeless shrimp from hydrothermal vents of the mid-atlantic ridge. Nature 337: 458–460. doi:10.1038/337458a0.

Van Dover, C.L., G.T. Reynolds, A.D. Chave and J.A. Tyson. 1996. Light at deep-sea hydrothermal vents. Geophys. Res. Lett. 23: 2049–2052. doi:10.1029/96GL02151.

Varma, S.J., K.B. Muchowska, P. Chatelain and J. Moran. 2018. Native iron reduces CO_2 to intermediates and end-products of the acetyl CoA pathway. Nat. Ecol. Evol. 2: in press. doi:10.1101/235523.

Wagner, T., J. Koch, U. Ermler and S. Shima. 2017. Methanogenic heterodisulfide reductase (HdrABC-MvhAGD) uses two noncubane [4Fe-4S] clusters for reduction. Science 357: 699–703. doi:10.1126/science.aan0425.

Wakao, N., N. Yokoi, N. Isoyama, A. Hiraishi, K. Shimada, M. Kobayashi et al. 1996. Discovery of natural photosynthesis using Zn-containing bacteriochlorophyll in an aerobic bacterium *Acidiphilium rubrum*. Plant Cell Physiol. 37: 889–893. doi:10.1093/oxfordjournals.pcp.a029029.

Wang, J., I. Dasgupta and G.E. Fox. 2009. Many nonuniversal archaeal ribosomal proteins are found in conserved gene clusters. Archaea 2: 241–251. PMC2686390.

Weiss, M.C., F.L. Sousa, N. Mrnjavac, S. Neukirchen, M. Roettger, S. Nelson-Sathi et al. 2016. The physiology and habitat of the last universal common ancestor. Nat. Microbiol. 1: 16116. doi:10.1038/NMICROBIOL.2016.116.

Whicher, A., E. Camprubi, S. Pinna, B. Herschy and N. Lane. 2018. Acetyl phosphate as a primordial energy currency at the origin of life. Orig. Life Evol. Biosph. In press. doi:10.1007/s11084-018-9555-8.

White, S.N., A.D. Chave and G.T. Reynolds. 2000a. Investigations of ambient light emission at dep sea hydrothermal vents. J. Geophys. Res. 107: EPM1–EPM13. doi:10.1029/2000JB000015.

White, S.N., A.D. Chave, G.T. Reynolds, E.J. Gaidos, J.A. Tyson and C.L. van Dover. 2000b. Variations in ambient light emissions from black smokers and flange pools on the Juan de Fuca Ridge. Geophys. Res. Lett. 27: 1151–1154. doi:10.1029/1999GL011074.

White, S.N., A.D. Chave, G.T. Reynolds and C.L. van Dover. 2002. Ambient light emission from hydrothermal vents on the Mid-Atlantic Ridge. Geophys. Res. Lett. 29: 1–4. doi:10.1029/2002GL014977.

Whitman, W.B., D.C. Coleman and W.J. Wiebe. 1998. Prokaryotes: The unseen majority. Proc. Natl. Acad. Sci. USA 95: 6578–6583.

Williams, T.A., G.J. Szöllősi, A. Spang, P.G. Foster, S.E. Heaps, B. Boussau et al. 2017. Integrative modeling of gene and genome evolution roots the archaeal tree of life. Proc. Natl. Acad. Sci. USA 114: E4602–E4611. doi:10.1073/pnas.1618463114.

Woese, C. 1998. The universal ancestor. Proc. Natl. Acad. Sci. USA 95: 6854–6859.

Woese, C.R., Kandler and M.L. Wheelis. 1990. Towards a natural system of organisms: Proposal for the domains Archaea, Bacteria, and Eucarya. Proc. Natl. Acad. Sci. USA 87: 4576–4579. doi:10.1073/pnas.87.12.4576.

Wood, W.A. 1961. Fermentation of carbohydrates and related compounds. pp. 59–149. *In*: Gunsalus, I.C. and R.Y. Stanier (eds.). The Bacteria. A Treatise on Structure and Function. Academic Press, New York, USA.

Zahnle, K., N. Arndt, C. Cockbell, A. Halliday, E. Nisbet, F. Selsis et al. 2007. Emergence of a habitable planet. doi:10.1007/s11214-007-9225-z.

Zaremba-Niedzwiedzka, K., E.F. Caceres, J.H. Saw, D. Bäckström, L. Juzokaite, E. Vancaester et al. 2017. Asgard archaea illuminate the origin of eukaryotic cellular complexity. Nature 541: 353–358. doi:10.1038/nature21031.

Zehr, J.P., J.B. Waterbury, P.J. Turner, J.P. Montoya, E. Omoregie, G.F. Steward et al. 2001. Unicellular cyanobacteria fix N_2 in the subtropical North Pacific Ocean. Nature 412: 635–638. doi:10.1038/35088063.

2

Bioenergetic Function of Gram Negative Bacteria— From Anaerobes to Aerobes

Mauro Degli Esposti

Introduction

This chapter will present the major aspects of bacterial bioenergetics, without the use of chemical formulae and thermodynamic balance equations. The aim of the chapter is to introduce, in simple terms, key functional traits that are present in the different types of Gram negative bacteria introduced in the following chapters. It will provide a sort of crash course in microbial energy metabolism and physiology to complement the enormous advances in bacterial genomics (Anton et al. 2014). Functional traits of energy metabolism constitute fundamental pieces of information to compare and distinguish how bacteria obtain the bioenergy they need for living, and therefore integrate functional information with genomic and phylogenetic data. The purpose is to provide a broad but meaningful view of the functional evolution of bacteria into mitochondria. The functional traits considered are often confined to groups of phylogenetically distant bacteria, reflecting evolutionary trajectories of gene loss and dispersal from the large genetic expansion that appears to have occurred around two to three billion years ago (Battistuzzi et al. 2004, David and Alm 2011), just before and during the great oxygenation event (GOE) that changed earth and its living organisms forever (Holland 2006, Anbar 2008, Lyons et al. 2014, Ducluzeau et al. 2014, Moore et al. 2017).

Center for Genomic Sciences. UNAM Campus de Cuernavaca, Cuernavaca, 62130 Morelos, Mexico.
Email: maurolitalia@gmail.com

As mentioned in the premise and Chapter one of the book, the most important functional aspect of bacteria is their capacity to produce bioenergy from environmental sources, from simple elements such as sulfur and iron to complex biopolymers such as cellulose. This capacity depends upon diverse electron transport (redox) systems that generate bioenergy by transporting protons across the cell membrane while transferring electrons from environmental donors to acceptors of various kinds. Collectively, these systems form respiratory chains that drive the energy metabolism of prokaryotes. Many of the bacterial traits of energy metabolism have been lost after the bacterial ancestors of mitochondria, the proto-mitochondria, became symbionts of other prokaryotes, thus producing the eukaryotic cell, as will be discussed in Chapter seven and eight. However, the Tricarboxylic Acid cycle (TCA) remains a central hub of metabolic pathways in mitochondria and many bacteria, especially proteobacteria (Chapter four). The reverse TCA cycle (rTCA), which runs in the opposite direction by reducing acetyl-CoA and fixing CO_2 (Buchanan 1990, Campbell et al. 2004), is among the most ancient pathways of bioenergy production and CO_2 fixation and was initially discovered in *Chlorobium*, a green sulfur bacterium growing autotrophically using photosynthesis and oxidizing sulfur compounds (Evans et al. 1964, Buchanan 1990). We shall encounter *Chlorobium* and its relatives in Chapter three.

Chlorobi are photosynthetic organisms and photosynthesis has been presented already in Chapter one; it will be mentioned multiple times in this and other chapters of the book, but only cursorily because it is involved in the evolution of chloroplasts of algae and plants, an aspect of eukaryogenesis which is beyond the scope of the book. Following the philosophy of the book, I present here fundamental aspects of the critical transition from anaerobic to aerobic metabolism (Fig. 1) that occurred in primordial prokaryotes after oxygen became abundant on earth due to

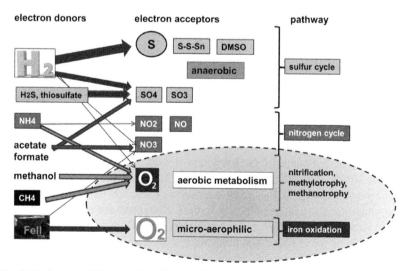

Fig. 1. Environmental bioenergetics of bacteria. See text for details of the various redox pathways.

oxygenic photosynthesis carried out by cyanobacteria (Lyons et al. 2014, Soo et al. 2017, Martin et al. 2018). The transition towards aerobic metabolism, namely the utilization of oxygen as terminal acceptor of bioenergetic electron transport chains, was probably a gradual process, enabled by ancestral membrane proteins belonging to two superfamilies of terminal oxidases (also called oxygen reductases). These proteins may have evolved before GOE, possibly to deal with local niches enriched in geochemically produced oxygen (Catresana and Saraste 1995, Anbar et al. 2007, Lyons et al. 2014, Ducluzeau et al. 2014). Here, I shall present a comprehensive overview of these terminal oxidases, considering the impact of metagenomic information on the possible trajectories of their evolution (Gupta 2016, Soo et al. 2017, Becraft et al. 2017). The overview provides a reference framework for the respiratory chains of various bacterial groups that will be discussed and illustrated in the following chapters of the book.

Before introducing the principal aspects of anaerobic and aerobic metabolism, I summarize in Table 1 the terminology frequently used to define the environmentally modulated metabolism of bacteria, which initially were predominantly autotrophic (capable of fixing CO_2 for synthesis of organic compounds) and subsequently became also heterotrophic (using organic material produced by other organisms as bioenergy fuels), as well as parasitic (directly exploiting the biological resources of host organisms). This offers a minimal glossary to the reader for following the literature on bacterial physiology and bioenergetics; the terms of reference in the text of next section are in **bold** when listed in Table 1.

Bacteria Exploit Environmental Sources of Bioenergy

Life on earth started with **chemosynthesis,** as described in Chapter one. While **photosynthesis** is the well-known process of harnessing the power of light to feed electrons into **protonmotive respiratory chains** (Chapter one), **chemosynthesis** is the equivalent process in which environmentally available electron donors are used to feed electrons into **protonmotive respiratory chains**. These are formed by membrane-bound electron transfer enzymes that couple redox reactions with charge separation and proton pumping across the lipid membrane. As a more general definition, **chemosynthesis** is the biological conversion of carbon-containing molecules (usually carbon dioxide, CO_2) and nutrients into organic matter using the oxidation of inorganic compounds (in particular hydrogen gas, Chapter one) as a source of energy. Sergei Winogradsky is considered the father of **chemosynthesis** and one of the founders of **environmental microbiology** (Waksman 1953, Dworkin 2012). Winogradsky discovered the first known form of **lithotrophy** during research on *Beggiatoa*—a sulfur-oxidizing bacterium belonging to the class of gammaproteobacteria (see Chapter five). **Lithotrophs** are a diverse group of organisms using inorganic substrates (usually of mineral origin) to obtain reducing equivalents to drive CO_2 fixation, biosynthesis of organic compounds and energy conservation (i.e., ATP production) via aerobic or anaerobic respiration (Lwoff et al. 1946). Chemosynthetic mats involved in cycling sulfur compounds are often

Table 1. Minimal glossary for the microbiological terms used to define nutrition and physiology of bacteria.

Term	Functional significance
Autotrophy	Capacity of producing organic material from only CO_2
Chemosynthesis	Process of exploiting environmental electron donors for bioenergy production
Photosynthesis	Process of extracting electrons from light to support bioenergy production
Heterotrophy	Living on organic material produced by other organisms
Fermentation	Metabolic process not associated with membranes and protomotive reactions that uses substrate level phosphorylation to produce bioenergy
Lithotrophs	Organisms that exploit mineral or gaseous sources of electrons in the environment
Chemolithotrophs	Organisms that use inorganic reduced compounds as a source of bioenergy
Chemolithoautotrophs	A type of (chemo)lithotrophs using autotrophic metabolism
Chemolithoheterotrophs	A type of (chemo)lithotrophs using heterotrophic metabolism
Oligotrophs	Organisms living in environments with limited levels of nutrients
Ectoparasitic	Organisms that parasitize others from the outside
Endocellular symbionts	Organisms that establish mutualistic symbiosis inside host cells
Motility	Capacity of movement driven by flagellar or gliding mechanisms
Outer membrane	External lipid membrane present, generally, in Gram negative bacteria
Nitration	Oxidation of nitrite (N cycle)
Nitrogen fixation	Reduction of N_2 to ammonia (N cycle)
Denitrification	Reduction of nitrate to N_2 (N cycle)
Ammonification	Oxidation of ammonia (N cycle)
Anammox	Anaerobic oxidation of ammonia coupled to nitrite reduction (N cycle)
Methylotrophy	Oxidation of C1 compounds such as methanol
Methanotrophy	Oxidation of methane, subtype of methylotrophy
Thiotrophy	Autotrophy using pathways of sulfur oxidation (S cycle)
Assimilation	Incorporation of inorganic compounds into metabolites (N and S cycle)
Bioenergy	ATP and other high energy releasing organic compounds
Central metabolism	Energy conversion encompassing the TCA cycle, carbon fixation and respiration
Carbon fixation	Incorporation of CO_2 into metabolites following diverse pathways
Respiratory chain	Series of redox enzymes that transport electrons from donors to acceptors
ISP or FeS	Iron-sulfur proteins of respiratory chains (and other biological pathways)
MQ or MK	Menaquinone, membrane lipidic redox carrier
Aerobic metabolism	Using oxygen as terminal electron acceptor for bioenergy producing reactions
Protonmotive	Reaction coupled to vectorial transport of protons or cations across the membrane
Terminal oxidases	Membrane enzymes that reduce oxygen in respiratory chains
HCO	Heme copper oxygen reductases
COX	Cytochrome *c* oxidase, the terminal oxidase of mammalian mitochondria
NOR	Nitric oxide reductase (N cycle)
Medical microbiology	Study of microbes affecting human and animal health

Table 1 contd. ...

...Table 1 contd.

Term	Functional significance
Environmental microbiology	Study of microbes living in any environment
Microbiome	Community of bacteria and other microbes living in a given environment
Metagenomics	Analysis of genetic material (DNA) directly obtained from the environment
MAG	Metagenomic assembled genome
Candidate phylum	MAG belonging to a new bacterial *phylum* with no cultivated representative
LGT	Lateral gene transfer between different organisms

found in hydrothermal vents, cold seeps, whale falls and wood falls in the ocean floor, forming microbial communities combining **chemolithoheterotrophs** with **chemolithoautotrophs** (Kalenitchenko et al. 2016).

Together with Martinus Bejerinck, Winogradsky pioneered **environmental microbiology** working on harmless bacteria that live in the environment, while the contemporary and much more famous Pasteur and Kock were leading **medical microbiology** on pathogenic bacteria to clarify the etiology of human and animal disease (Maczulak 2010). With the advent of **metagenomics**, contemporary microbiology studies have taken two main directions: one is biomedically related and investigates the **microbiome** of humans and animals (Nielsen et al. 2014, Anton et al. 2014, Degli Esposti and Martinez-Romero 2017), while the other expands the scope and breadth of **environmental microbiology** by exploring the **microbiome** of every known environment on earth (Eloe-Fadrosh et al. 2016, Louca et al. 2016, Anantharaman et al. 2016, Probst et al. 2017, Ji et al. 2017, Schulz et al. 2017, Tully et al. 2018). Recent studies have produced an ever increasing wealth of genomic information that enables the uncovering of previously unknown bacterial phyla (Hug et al. 2016, Anantharaman et al. 2016, Eloe-Fadrosh et al. 2016, Soo et al. 2017, Schulz et al. 2017, Parks et al. 2017, Becraft et al. 2017), as well as the origin of eukaryotic functional traits (Ravcheev and Thiele 2016, Degli Esposti et al. 2016, Spang et al. 2017, Degli Esposti 2017). Currently available genomic information has been exploited here to delineate the possible evolution of the major enzymes for **aerobic metabolism** presented at the end of the chapter.

The principal metabolic pathways by which prokaryotes obtain **bioenergy** are summarized in Fig. 1 (for extended reviews on the subject see: Canfield et al. 2006, Schoepp-Cothenet et al. 2013, Marreiros et al. 2016, Martin et al. 2016, Moore et al. 2017). As discussed in Chapter one, molecular hydrogen, H_2, constitutes the major electron donor for anaerobic respiratory chains, while elemental sulfur (S_0) and its derivatives such as polysulfides represent the oldest electron donors or acceptors for strictly anaerobic bacteria (Weiss et al. 2016, Martin et al. 2016, Moore et al. 2017). Hydrogen sulfide (H_2S) of geochemical origin was relatively abundant on early earth (Archean era—Canfield et al. 2006), and could function as electron donor to anaerobic respiratory chains, including the photosynthetic ones of green sulfur bacteria (Sakurai et al. 2010). Contrary to carbon monoxide (CO), which was certainly present in the Archaean (Canfield et al. 2006) and is still utilized by chemosynthetic organisms at its contemporary trace levels (Ji et al. 2017), nitrogen compounds such as nitrite or nitrate were formed (Canfield et al. 2006) but did not become abundant on early earth (Moore et al. 2017). Conversely, ammonia can be released by geochemical reactions and nitric oxide (NO) can be formed by lightening in the atmosphere (Canfield et al. 2006), therefore suggesting their presence on early earth (Ducluzeau et al. 2014). Whether these N compounds were relatively abundant in Archaean oceans so as to sustain specific chemosynthetic pathways is unclear from geochemical evidence (Moore et al. 2017), and controversial in biological terms (Vlaeminck et al. 2011, Ducluzeau et al. 2014). In any case, nitrate and nitrite, the major nitrogen compounds that are used as alternative electron acceptors of current anaerobic bacteria (Simon and Klotz 2013), became abundant in earth environments only after the GOE (Vlaeminck et al. 2011, Lyons et al. 2014, Moore et al. 2017). Consequently, the major aspects of N metabolism are linked to aerobic pathways in facultatively anaerobic bacteria, as discussed below.

Both sulfur (S) and nitrogen (N) produce environmental cycles in which abiotically formed intermediates are intermixed with biological reactions that have an enormous impact on earth geochemistry (Moore et al. 2017). The majority of known enzymes that participate in these cycles react with membrane quinones-menaquinone (MQ) and its derivatives in the earliest, predominantly anaerobic organisms (Schoepp-Cothenet et al. 2013, Marreiros et al. 2016)—see also Chapters three and four of this book. However, minimal protonmotive schemes that are present in some strict anaerobes drive ATP synthesis in the absence of membrane quinones (Fig. 2, cf. Degli Esposti et al. 2016, Martin et al. 2016; see also Chapter one). This evidence contradicts the hypothesis that the earliest life forms already had menaquinone and respiratory enzymes reacting with it (Schoepp-Cothenet et al. 2013, Ducluzeau et al. 2014).

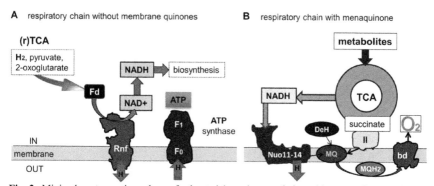

Fig. 2. Minimal protonmotive scheme for bacterial respiratory chains without membrane quinones (A) and with menaquinone (B). IN and OUT indicate the cytoplasmic and the periplasmic side of the inner membrane, respectively.

The Sulfur Cycle

Elemental sulfur reduction is considered among the oldest forms of metabolism of prokaryotes (Canfield et al. 2006), in particular of hyperthermophilic organisms that form deep-branching clades of sulfur reducers (Campbell et al. 2009, Imhoff 2016). The abundance of these organisms in oceanic hydrothermal environments suggests that sulfur reduction evolved in hydrothermal vents. Potential sulfur metabolism was present 3.8 billion years ago (Moore et al. 2017) and, among extant bacteria, it is concentrated in the epsilon and delta class of proteobacteria; it is also present in some Aquificae and Nitrospirae (Campbell et al. 2009; see Chapters three and four of the book).

Together with sulfur reduction, sulfate reduction is widespread in anoxic environments, especially in marine sediments, and is also thought to be a very old metabolism (Canfield et al. 2006, Zhou et al. 2011, Moore et al. 2017). However, elemental sulfur reduction most likely started as the dominant sulfur metabolism because sulfate was present at low concentrations in the Archean oceans and significantly increased only after the GOE (Moore et al. 2017). Sulfate reducing organisms are predominantly distributed among deltaproteobacteria (Chapter four) and other taxa of Gram negative bacteria, as well as some Gram positive organisms, in particular Clostridiales (Lens and Kuenen 2001, Muyzer et al. 2008, Zhou et al. 2011, Imhoff 2016).

Oxidation of sulfur compounds via electron transport to oxygen forms an important part of the contemporary sulfur cycle (Zhou et al. 2011, Dahl 2015, Imhoff 2016). The organisms that carry out this metabolism are called sulfur-oxidizers and are particularly abundant in one group of gammaproteobacteria that are either free living, such as *Beggiatoa*, or endosymbionts of metazoans that thrive in hydrothermal vents (see Chapter five and Dubilier et al. 2008). Even if these chemolithotrophic organisms evolved after oxygen became abundant in the oceans, ca. 500 million years ago (Lyons et al. 2014), the basic enzymes sustaining the metabolism of sulfur oxidation are much older, since they are in part shared with sulfur reducers. The pathways of sulfur oxidation are described in Chapter five; see Imhoff (2016) for a detailed review on the subject.

Sulfur metabolism encompasses the assimilation of inorganic sulfur to initiate the biosynthesis of **iron-sulfur proteins (ISP)**, the oldest and most widespread redox active proteins in biology (Chapter one). The biosynthetic pathway of ISP is shared by bacteria and all forms of mitochondria and related organelles (Lill 2009, Blanc et al. 2015), but is mentioned only briefly in this book. See Lill (2009) for a detailed review on this aspect of sulfur metabolism. ISP biosynthesis also requires the assimilation of iron (Fe), a metal which was abundant on early earth (Canfield et al. 2006, Anbar 2008). Fe assimilation is not mentioned further here, while aspects of its chemolithotrophy are presented in Chapter four, in particular its part regarding zetaproteobacteria.

The Nitrogen Cycle

The nitrogen cycle integrates redox modification of various nitrogen compounds carried out by different enzymes, often from bacteria associated in physical or metabolic consortia that utilizes parts of the cycle. An example of these nitrogen-utilizing consortia is ammonia-oxidizing (AO) organisms such as the betaproteobacterium *Nitrosomonas* and nitrite-oxidizing organisms such as *Nitrospira* along the depths of ocean layers (Lücker et al. 2010, Louca et al. 2016). The consecutive reactions of ammonia oxidation to nitrite and nitrite oxidation to nitrate are usually separated in different marine organisms, often belonging to diverse phyla as in the case of *Nitrosomonas* and *Nitrospira*. Recently, however, *Nitrospira* strains have been found to carry out both reactions (see Chapter three). Ammonia in the ocean may derive from abiotic reactions driven by volcanoes (Canfield et al. 2006), but most significantly originates from two biological sources: nitrogen fixation and ammonification. The latter is a recycling process of oxygenated nitrogen compounds that can follow different biochemical pathways (Vlaeminck et al. 2011, Simon and Klotz 2013), while N_2 fixation is one of the most important and ancient biological traits (Weiss et al. 2016, Moore et al. 2017). Indeed, an ancestral form of N_2 fixation seems to have been present already in the Last Universal Common Ancestor—LUCA (Chapter one, cf. Weiss et al. 2016). Transformation of atmospheric dinitrogen into ammonia, the only form of nitrogen that can be directly assimilated by all life forms, is a biochemical feat carried out by an enzyme called nitrogenase, which contains various **ISP** and either Molybdenum (Mo) or Vanadium (Vd) in its catalytic site. The reaction requires multiple molecules of ATP and is therefore very energy demanding; however, it has a useful byproduct, H_2, which can be exploited as electron donor for bioenergy production by most organisms adapted to anaerobiosis (Fig. 1). Of note is the structure of the catalytic subunit of nitrogenase which is homologous to that of [FeFe]-hydrogenases (Degli Esposti et al. 2016). Nitrogenase activity is generally very sensitive to low levels of oxygen; thus, nitrogen fixing bacteria need to have terminal oxidases with high affinity for oxygen to get rid of it around nitrogenase (Degli Esposti and Martinez Romero 2016). Nitrogenases are distributed in many bacterial phyla, as will be discussed in Chapters three to seven of the book. However, the best known N_2 fixing organisms are the nodulating symbionts of the Rhizobiaceae family of alphaproteobacteria (see dedicated section in Chapter seven).

The part of the N cycle just introduced, including production of ammonia from dinitrogen and its subsequent oxidation to form nitrite, is relatively straightforward. However, nitrite is at the center of a complex metabolic hub connecting other parts of the cycle and diverse respiratory chains via a series of overlapping or alternative biochemical reactions, which are difficult to summarize graphically (Simon and Klotz 2013, Sparacino-Watkins et al. 2014, Nelson et al. 2016). Let's start with the single reaction by which nitrite is oxidized to nitrate, usually referred to as nitrification. This reaction is carried by nitrite oxidase, Nxr, which is a particular version of the membrane-bound enzyme that carries out the opposite reaction, namely Nar nitrate reductase (Simon and Klotz 2013). Both enzymes contain three proteins: a transmembrane di-heme b subunit reacting with menaquinol—the electron donor substrate; a **ISP** subunit and a Molybdenum-containing large subunit exposed to the cytoplasmic side of the membrane, which directly reacts with NO_2 and NO_3. This enzyme architecture corresponds to an ancient and widespread module for membrane redox enzymes, the superfamily of Complex Iron-Sulfur Molybdoenzymes, usually abbreviated as CISM (Rothery et al. 2008, Simon and Klotz 2013, Marreiros et al. 2016). While membrane-bound nitrate reductases are widespread among diverse bacterial taxa (Louca et al. 2016, Marreiros et al. 2016, Nelson et al. 2016), Nxr enzymes are present in a few genera of alpha- and gamma-proteobacteria, plus *Nitrospina* and various members of the Nitrospirae *phylum* (Lücker et al. 2010, Vlaeminck et al. 2011, Louca et al. 2016, Anantharaman et al. 2016).

Nitrate represents the most oxygenated stage of the N cycle and can follow diverse pathways of reduction, either back to ammonia in assimilatory processes, or to dinitrogen along the dissimilatory process of denitrification. Nar reductase (menaquinol:nitrate reductase) functions as the first enzyme of the latter process and its product nitrite, once exported from the cytoplasm to the periplasm by a dedicated permease, can be further reduced to NO by either the cd1 nitrite reductase or the Cu-containing NirK, two soluble proteins with different evolutionary histories (Simon and Klotz 2013). NO is then reduced by membrane-bound NOR, a member of the HCO superfamily of terminal oxidases (see later the section on terminal oxidases), which can use either reduced cytochrome c (cNOR) or menaquinol (qNOR) as electron donor (Simon and Klotz 2013, Ducluzeau et al. 2104). The product of NOR is the stable gas N_2O, which can be fully reduced to N_2 by another membrane-bound enzyme, nitrous oxide reductase or Nos (Simon and Klotz 2013). This last enzyme of the pathway is much less common than the above mentioned nitrite reductases or NOR in bacteria and microbiomes (Louca et al. 2016, Nelson et al. 2016, Marreiros et al. 2016, Anantharaman et al. 2016), thereby suggesting that the complete pathway of denitrification has been acquired late along evolution (Ducluzeau et al. 2014, Louca et al. 2016).

Nitrite can be formed also by periplasmic enzymes of the NapA type (Sparacino-Watkins et al. 2014) and is alternatively reduced to ammonia via the following nitrite reductases (Simon and Klotz 2013): (1) soluble assimilatory nitrite reductase of the NirBD type, which uses NAD(P)H as electron donor in the cytosol; (2) membrane-bound NrfABCD complex, which uses menaquinol as electron donor, and: (3) periplasmic soluble octaheme-c cytochromes, with unknown electron donors, but related to other multiheme c cytochromes that are involved in the oxidoreduction of external metals such as iron (see Chapter four). Of note, NirBD is usually associated with a cytoplasmic NADH-dependent nitrate reductase and a specific nitrate/nitrite transporter called NarK, together forming a gene cluster that is essentially conserved from bacteria to eukaryotes (Degli Esposti et al. 2014). Cyanobacteria and diverse proteobacteria have another assimilatory nitrate reductase, NarA, which uses ferredoxin as electron donor and may well be the oldest enzyme for nitrate assimilation (Sparacino-Watkins et al. 2014). Conversely, periplasmic nitrate reductases of the NapA type have high affinity for nitrate and use membranous menaquinol as electron donor, similarly to their metabolic partners of Nrf nitrite reductases (Simon and Klotz 2013, Sparacino-Watkins et al. 2014). These enzymes have been discovered and well characterized in *E. coli* and later studied in other facultatively anaerobes of the gamma, delta- and epsilonproteobacteria (Simon and Klotz 2013, Sparacino-Watkins et al. 2014); their distribution in other bacterial groups has been recently uncovered by metagenomic studies (Louca et al. 2016, Nelson et al. 2016, Anantharaman et al. 2016). The same studies have also revealed the overlapping presence of multiple enzymes for the reduction and assimilation of nitrite and nitrate in related organisms (Sparacino-Watkins et al. 2014) and in microbial communities of marine and soil environments (Louca et al. 2016, Nelson et al. 2016, Anantharaman et al. 2016).

In evolutionary terms, N_2 fixation is a part of the N cycle for which there is strong geochemical evidence, since this trait is estimated to have evolved between 3.2 and 2.9 billion years ago (Moore et al. 2017). A primitive form of nitrogen fixation almost certainly appeared earlier than 3.2 billion years ago (Chapter one cf., Weiss et al. 2016), but isotopic signatures of specific nitrogenases remain ambiguous in older rocks, as reviewed by Moore et al. (2017). A phylogenetic tree of NapA type reductases has been reported (Sparacino-Watkins et al. 2014) and shows its deepest branches among proteins from deltaproteobacteria. Hence, periplasmic dissimilatory nitrate reduction is likely to have originated within ancient proteobacterial organisms related to extant deltaproteobacteria (see Chapter four for their phylogeny).

Nitrate and anaerobic iron oxidation

Nitrate can also function as electron acceptor for anaerobic oxidation of external Fe^{II} (Ilbert and Bonnefoy 2013, Carlson et al. 2013). The evolutionary significance of this oxidative pathway is unclear, since it is restricted to nitrate-respiring proteobacteria such as *Acidovorax* (Ilbert and Bonnefoy 2013, Carlson et al. 2013, Melton et al. 2014) and perhaps also in **candidate phylum** *Kryptonia* (Eloe-Fadrosh et al. 2016). The reaction pathway involves abiotic oxidation of periplasmic reduced *c* cytochromes by nitrate, a powerful oxidant (Carlson et al. 2013, Melton et al. 2014). This, together with the lack of genetic evidence for the existence of an enzymatic pathway of electron transport from Fe^{II} to nitrate (Melton et al. 2014), renders anaerobic oxidation of Fe^{II} a much less interesting metabolic trait than the micro-aerobic oxidation of the same metal (Fig. 1, bottom), which will be presented in Chapter four.

Methylotrophy and methanotrophy

Methylotrophy defines metabolic traits that are typically present in aerobic or facultatively anaerobic proteobacteria, even if some of their key enzymes have a much older history (Chistoserdova et al. 2009, Chistoserdova 2016). Methylotrophs can use one-carbon compounds such as methanol or methane as the sole carbon source for their growth. Hence, methanotrophs share the same overall pathway as methylotrophs, except for the signature enzyme methane monooxygenase, which produces methanol and defines methanotrophy, the capacity of living using methane as carbon and energy source (Chistoserdova et al. 2009, Ho et al. 2016). In turn, methane monooxygenase is a Cu-containing enzyme highly homologous of ammonia oxygenase, the enzyme responsible for the aerobic oxidation of ammonia in the nitrogen cycle which we have encountered before. The characteristic enzyme that distinguishes methylotrophs is methanol dehydrogenase (Chistoserdova et al. 2009), which contains pyrroloquinoline quinone (PQQ) and c-cytochromes, is membrane-bound and transfers electrons to ubiquinone (Keltjens et al. 2014). The final electron acceptor of methanol oxidation is oxygen, which is generally used also for methane oxidation by methanotrophs. Hence, these metabolic traits originated during or after the GOE (Battistuzzi et al. 2004), which explains their phylogenetic distribution among various families of alphaproteobacteria, one order of gammaproteobacteria (the Methylococcales) and a few genera of betaproteobacteria (Chistoserdova et al. 2009). Metagenomic data suggests the presence of signature enzymes of methylotrophy and methanotrophy also in some organisms outside the proteobacterial *phylum* (Louca et al. 2016, Anantharaman et al. 2016, Chistoserdova 2017), but phylogenetic significance of this distribution is presently unclear, given the likely contribution by lateral gene transfer (**LGT**). In any case, there is a strong link between methylotrophy and **aerobic metabolism**, as discussed earlier (Degli Esposti et al. 2014) and below.

The Central Role of Menaquinone in Bacterial Bioenergetics

As previously mentioned, the majority of the enzymes that form part of the S and N cycle, as well as those that reduce CO (Can et al. 2014, Ji et al. 2017), are bound to the cytoplasmic membranes of bacteria and use membrane quinones as electron acceptors or donors, the latter in their reduced quinol form (Marreiros et al. 2016). Menaquinone (**MQ**, 2-methyl-3-isoprenyl-naphthalene-1,4-dione) is the dominant membrane quinone in most facultative anaerobes and several strictly anaerobes (Zhi et al. 2014, Marreiros et al. 2016, Ravcheev and Thiele 2016). MQ has a redox potential considerably lower than that of ubiquinone (Simon and Klotz 2013) and is biosynthesized via two pathways that branch from the common precursor chorismate (Zhi et al. 2014, Ravcheev and Thiele 2016).

The classical pathway was elucidated in *E. coli* and involves a series of enzymes in a precise reaction sequence, starting with MenF (isochorismate synthase). Isochorismate is then transformed into o-succinylbenzoyl-CoA by the concerted reaction of MenD, MenH, MenC and MenE. The subsequent formation of the naphthalene ring is catalyzed by the MenB enzyme, followed by the attachment of a poyprenyl tail at position 3 of the ring by MenA. The demethylymenaquinone thus produced often remains as a substrate for membrane enzymes or is subsequently methylated by MenG or UbiE enzymes, which can use either the naphtalene or the quinone ring as substrates. MenG, and in part MenA also, catalyze reactions that are shared with the biosynthesis of ubiquinone (Ravcheev and Thiele 2016), but the biosynthetic pathways of the two quinones are clearly distinct (Degli Esposti 2017). The alternative biosynthesis of MQ follows the futalosine pathway (Zhi et al. 2014), which is present in facultatively anaerobic bacteria of the *phyla* Aquificae, Firmicutes and *Deinococcus-Thermus*, as well as in several deltaproteobacteria. At least one reaction of the futalosine pathway seems to be in common with the ubiquinone pathway (Ravcheev and Thiele 2016), in particular the ring decarboxylase. However, the decarboxylase enzyme associated with the futalosine pathway is different and longer than the one used in the biosynthesis of ubiquinone, UbiD (Degli Esposti 2017). Some bacterial groups such as Chlorobi and Aquificales possess distinctive derivatives of menaquinone as their major membrane quinones. Conversely, facultatively anaerobes of the proteobacterial *phylum* such as *E. coli* have both menaquinone and ubiquinone, switching to the latter when oxygen levels are abundant (Zhi et al. 2014, Marreiros et al. 2016, Ravcheev and Thiele 2016).

Although the presence of membrane quinones is not absolutely required to drive protonmotive respiratory chains in anaerobic bacteria, as shown in Fig. 2A, membrane enzymes of anaerobic respiratory chains predominantly use **MQ** and its derivatives as their substrates (Fig. 2B, cf. Marreiros et al. 2016). However, proton-pumping NADH oxidoreductase, also known as respiratory complex I, equally reacts with MQ and ubiquinone (Baradaran et al. 2013). The evolutionary history of this enzyme complex, the largest and most efficient protonmotive machine in biology (Sazanov 2015), is of interest in the context of the present chapter because it encompasses both anaerobic and aerobic organisms. Indeed, most bacterial phyla possess one or more forms of respiratory complex I (Spero et al. 2015, Degli Esposti 2015). The most ancestral form of the enzyme is found in organisms that thrive in oceanic hydrothermal vents such as epsilonproteobacteria of the Nautiliales order (Campbell et al. 2009) and the deep branching *Desulfurobacterium* and *Thermovibrio* of the *phylum* Aquificae (Giovannelli et al. 2012). The ancestral form of the enzyme contains 11 subunits and is clearly related to membrane-bound NiFe hydrogenases (Coppi 2006, Marreiros et al. 2013, Spero et al. 2015); it transfers electrons from reduced ferredoxin to NAD(P)+, like the Rnf complex but, most likely, without proton pumping (Degli Esposti 2015). The MQ-oxidoreductase form of complex I (Fig. 2B) evolved in photosynthetic Chlorobi and various anaerobic bacteria, for instance Clostridiales, using ferredoxin or other low potential proteins as donor substrates. The next stage in complex I evolution appears to have occurred in deltaproteobacteria and led to the incorporation of NADH-reacting subunits from soluble formate dehydrogenases (Degli Esposti 2015). Indeed, the same NADH-oxidizing subunits are retained as a separate module in hydrogenosomes and other organelles derived from mitochondria, which have lost respiratory complex I inherited from proto-mitochondria (Müller et al. 2012, Atteia et al. 2013, Degli Esposti et al. 2016). Finally, the 14-subunit NADH-ubiquinone reductase of mitochondrial type emerged in facultatively anaerobes of the alphaproteobacterial class such as *Magnetococcus* (Degli Esposti 2015) and then spread to multiple bacterial lineages possessing aerobic metabolism, presumably by lateral gene transfer (Spero et al. 2015). For example, aerobic Aquificales have a complex I enzyme that is very similar to that of aerobic proteobacteria. Hence, the last evolutionary stage of complex I, before being inherited by eukaryotes (Chapter eight), is intimately connected with the development and spread of **aerobic metabolism** in bacteria, the subject of next section of this chapter.

Aerobic Metabolism in Bacteria

The enormous increase in atmospheric oxygen that occurred around 2.3 billion years ago has changed the way life forms adapt to the environment and extract bioenergy from it. Before this dramatic event, life on earth was dominated by anaerobic organisms (Chapter one), even if some of them had acquired aerotolerance by intermittent exposure to abiotically produced O_2 in some environmental niches (Anbar et al. 2007). Indeed, it has been frequently assumed that cytochrome *c* oxidase (COX), the major enzyme that utilizes oxygen in eukaryotes, is older than GOE (Catresana and Saraste 1995, Schäfer et al. 1996). This apparently counter-intuitive concept derived from the comparative analysis of a limited set of cytochrome oxidase proteins that were available twenty years ago (Catresana and Saraste 1995, Pereira et al. 2001); nevertheless, it turns out to be at least partially correct (Han et al. 2011, Ducluzeau et al. 2014). Recently, metagenomic information has clearly indicated that cyanobacterial organisms responsible for the GOE acquired their COX enzymes from other organisms, which were, presumably, facultatively anaerobes related to extant proteobacteria (Soo et al. 2017, Martin et al. 2018).

The above paragraph on **aerobic metabolism** has already introduced a number of terms that are commonly used in microbiology and bioenergetics to define, sometimes confusingly, the metabolic relationships with oxygen. Such terms are listed in Table 2 to provide the reader with basic knowledge on the fundamental terminology, traditional and modern, surrounding the aerobic metabolism of microbes (see also Morris and Schmidt 2013, Ducluzeau et al. 2014, Martin 2017). COX is only one family of various enzymes that reduce oxygen in the electron transport chains of prokaryotes, collectively called **terminal oxidases** (Garcia-Horsman et al. 1994, Schäfer et al. 1996, Brocher-Armamet et al. 2009, Morris and Schmidt 2013, Ducluzeau et al. 2014). The simplest and, in all likelihood, oldest of these terminal oxidases is the bd ubiquinol oxidase, which contains only *b* hemes for reducing oxygen (Borisov et al. 2011, Degli Esposti et al. 2015, Forte et al. 2017). This ubiquinol oxidase has high affinity for O_2 and is widespread among facultatively anaerobes of nearly all bacterial *phyla* (Brocher-Armamet et al. 2009), also in aerotolerant organisms usually considered to be anaerobes, for instance members of the *Bacteroidetes* (Baughn and Malamy 2004) and Nitrospirae (see Chapter three). Evidently, the majority of bacteria had to come to terms with the pervasive presence of O_2 in the environment after the GOE, and bd oxidase

Table 2. Definition of aerobic metabolism in bacteria.

References	Definition of bacteria with respect to their aerobic/anaerobic metabolism		
	Madigan et al. (2010), Baughn and Malamy (2004)	Morris and Shmidt (2013)	This book
Growth requirement for microbes	traditional definition	new definition	alternative definition
Require normal oxygen levels to grow optimally	obligate aerobes	aerobes	strictly aerobes, with only low affinity terminal oxidases
Grow optimally at oxygen levels well below normal	microaerophiles	microaerobes, with high affinity terminal oxidases, either alone or in combination with other oxidases	
Grow by using either oxygen or alternative electron acceptors, or even by fermentation	facultatively anaerobic		facultatively anaerobic, with high affinity terminal oxidases in any combination with other terminal oxidases
Grow best under anoxic conditions but can use very low levels of oxygen for energy conservation	nanoaerobes		
Grow optimally without oxygen but can tolerate its presence, without using it for energy conservation	acrotolerant anaerobes	anaerobes	
Grow under anoxic conditions	obligate anaerobes		strictly anaerobes, without any terminal oxidase

provided the simplest solution (Baughn and Malamy 2004), given its structural compactness and biochemical versatility (Borisov et al. 2011, Forte et al. 2016, 2017). The original function of this enzyme might have been to detoxify oxygen or other gaseous compounds (Forte et al. 2017) that would interfere with redox reactions of bioenergetic importance, for instance nitrogen fixation—a pathway that is normally sensitive to even low levels of oxygen (Simon and Klotz 2013, Moore et al. 2017).

However, the majority of **terminal oxidases** belong to the superfamily of Heme Copper Oxygen reductases, **HCO** (Fig. 3). These enzymes contain an O_2-reducing center formed by a high-spin heme and a Cu atom ligated to the protein via a complex, conserved geometry (Pereira et al. 2001, Han et al. 2011, Sousa et al. 2012, Ducluzeau et al. 2014, Sharma and Wikström 2014). Distant members of this superfamily contain a Fe atom instead and predominantly reduce NO—the Nitric Oxide Reductases or NOR (Sharma and Wikström 2014, Ducluzeau et al. 2014) of the nitrogen cycle encountered before. Although several members of the HCO superfamily oxidize quinols like the bd oxidase, many other members use

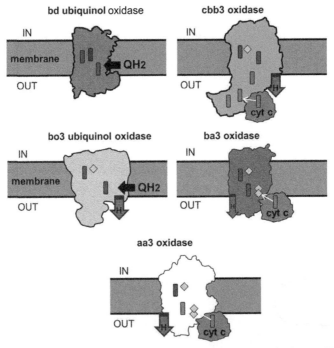

Fig. 3. Bacterial terminal oxidases. The silhouettes are drawn to match the known 3D structure of the membrane oxygen reductases from the following references: Buschmann et al. (2010) for *Pseudomonas* cbb3 oxidase (C family of HCO); Abramson et al. (2000) for *E. coli* bo3 ubiquinol oxidase; Tiefenbrunn et al. (2011) for the ba3 oxidase of *Thermus* and; Iwata (1998) for the aa3 cytochrome c oxidase of *Paracoccus*. The arrows with H inside indicate proton pumping. Note the smaller arrow for the ba3 oxidase which lacks one proton pumping channel (see text).

water-soluble proteins with higher midpoint potential as electron donors, typically cytochrome *c* as in mitochondria. These soluble electron carriers link aerobic metabolism with other metabolic pathways, such as those belonging to the nitrogen cycle (see the part of Chapter four on epsilonproteobacteria) and chemolithotrophic Fe^{II} oxidation (see Chapter four, part on zetaproteobacteria). HCO enzymes are divided in three families: A, including mitochondrial COX; B including the ba3 oxidases, and C, including cbb3 oxidases (Pereira et al. 2001, Han et al. 2011, Sousa et al. 2012, Ducluzeau et al. 2014, Sharma and Wikström 2014). NCBI resources list a fourth family of HCO, containing bo3 ubiquinol oxidases (https://www.ncbi. nlm.nih.gov/Structure/cdd/cddsrv.cgi?uid=cl00275, accessed 5 December 2017), which actually constitutes a subtype of family A (Pereira et al. 2001, Han et al. 2011), as discussed below.

The recent definition of microaerobes is based upon the presence of high affinity terminal oxidases such as bd ubiquinol oxidase (Morris and Schmidt 2013). This modern concept (Table 2) can be refined further to deduce the kind of aerobic metabolism that a bacterial organism is likely to have on the basis of its genomic properties regarding **terminal oxidases**, which come in different flavors and colors, as described below.

Terminal Oxidases in Bacteria

Prokaryotes have six types of O$_2$-reducing terminal oxidases which differ in their subunit composition and oxygen affinities (Sousa et al. 2012, Degli Esposti et al. 2014, Sharma and Wikström 2014). Only two of these oxidases are present in eukaryotic mitochondria: the proton pumping cytochrome aa3 oxidase (here called COX) and the alternative oxidase that oxidizes ubiquinol (called AOX-Pennisi et al. 2016). The latter is not considered further here due to its scant distribution among bacteria Chapter seven. All other terminal oxidases are named on the basis of their typical cytochrome type and have diverse affinities for oxygen, in any case well below the normal concentration of O$_2$ in the ocean, which is around 250 μM (Morris and Schmidt 2013). The already introduced bd ubiquinol oxidase is formed by two similar transmembrane proteins which bind two heme *b* and one heme *d*; it oxidizes membrane quinols without pumping protons across the membrane (Borisov et al. 2011). The enzyme comes in two forms, the bd-I typical of *E. coli* and the cyanide insensitive oxidase (Cio) typical of *Pseudomonas* (Cunningham et al. 1997, Degli Esposti et al. 2015). The former has high affinity for oxygen, with nanomolar K$_m$ values (D'Mello et al. 1996), while the latter has a broader affinity for oxygen extending to the sub-micromolar range (Arai et al. 2014).

The cbb3 oxidase is probably the oldest among proton pumping HCO (Sharma and Wikström 2014, Ducluzeau et al. 2014). It was originally discovered in nitrogen fixing bacterial symbionts and its genes are usually named *fixNOP* (for subunits 1, 2 and 3, respectively; Pitcher and Watmough 2004). Its subunits 2 and 3 contain one or two *c*-type cytochromes but no Cu (Pereira et al. 2001, Pitcher and Watmough 2004). The binuclear center of these oxidases is bound to subunit

1, which is homologous to COX1 proteins of heme *a*-containing oxidases (Pereira et al. 2001). Cbb3 oxidases are known for their strong affinity for oxygen, with K_m values in the nanomolar range (Pitcher and Watmough 2004, Buschmann et al. 2010, Arai et al. 2014).

The bo3 ubiquinol oxidases have been classified within the A1 type of HCO (Pereira et al. 2001, Sousa et al. 2012), even if they lack the CuA center in subunit 2 (Abramson et al. 2000, Han et al. 2011). These oxidases derive their common name from the characteristic presence of cytochrome *o*, which was originally discovered by Keilin after reaction of *E. coli* with CO. This reaction alters the absorbance spectrum of the high-spin *o* heme that, compared with the *b* heme, has a farnesyl group attached to the protoporphyrin IX ring. The heme *b* to heme *o* reaction is catalyzed by the membrane-bound assembly subunit CtaB, which corresponds to eukaryotic COX10 and is related to the prenyl-transferase MenA of MQ biosynthesis encountered before. The bo3 oxidase of *E. coli* has an affinity for oxygen that is intermediate between that of cbb3 and that of aa3 oxidases (D'Mello et al. 1995, Abramson et al. 2000). However, homologous bo3 oxidases from alphaproteobacteria display low oxygen affinity, essentially superimposable to that of COX (Richhardt et al. 2013). Heme *o* is an obligate intermediate in the biosynthesis of heme *a*, which is catalyzed by the membrane protein CtaA or heme A synthase, corresponding to eukaryotic COX15 (Hederstedt 2012).

The *ctaA* and *ctaB* genes are often part of gene clusters of heme *a*-containing oxidases, which have been divided in family A and B (Pereira et al. 2001). Family B is the simplest of all HCO and comprises the ba3 oxidase originally found in *Thermus* (Zimmermann et al. 1988)—of note is the term ba3 that has been used for other enzymes also that we now know belong to subtypes of the A family. At difference with A family oxidases, which only have hemes of type *a*, classical ba3 oxidases have a cytochrome *b* as the low-spin heme connecting the CuA center in subunit 2 with the O_2-reacting binuclear center (Radzi Noor and Soulimane 2013). These enzymes are common in Archaea (Scharf et al. 1997, Ducluzeau et al. 2014) and also in many more bacterial *phyla* than previously considered (Pereira et al. 2001, Sousa et al. 2012). See additional discussion on this topic in Chapter seven. Most recent studies on ba3 oxidases have focused on their diminished capacity of proton pumping (Han et al. 2011), especially in the light of their different three-dimensional structure around the proton channels (Pereira et al. 2001, Tiefenbrunn et al. 2011, Radzi Noor and Soulimane 2013). The available structural information supports the concept that ba3 oxidases have an intermediate affinity for oxygen (Radzi Noor and Soulimane 2013), as it has been commonly assumed (Han et al. 2011, Ducluzeau et al. 2014) even if detailed biochemical evidence is scant. Notably, ba3 oxidases have a simple gene cluster with a distinctively short COX2 subunit, which normally has a single transmembrane helix at the N terminus and a cupredoxin fold at the C-terminus for binding CuA. COX2 proteins of other heme *a*-containing oxidases characteristically contain two transmembrane helices towards the N terminus (Tsukihara et al. 1996, Pereira et al. 2011, Ducluzeau et al. 2014); in several ba3 oxidases, the second transmembrane helix is provided instead by a short hydrophobic subunit that is associated with the gene cluster (Prunetti et al. 2011).

Conversely, many organisms contain COX2 variants having a cytochrome *c* domain at the C-terminus, thereby inspiring the name of caa3 oxidases such as those characteristic of Bacilli (Quirck et al. 1993, Degli Esposti et al. 2014). The three dimensional structure of *Thermus* caa3 oxidase has been reported (Lyons et al. 2012, Radzi Noor and Soulimane 2013) and belongs to the A2 subdivision of the vast A family of aa3 oxidases (Pereira et al. 2001, Sousa et al. 2012). The oxygen affinity for these A2 type oxidases is generally higher than that of A1 type oxidases (Sone and Fujiwara 1991, Garcia Horsman et al. 1991, Ramel et al. 2013). Four variants of the *Bacillus* caa3 operon have been found in alphaproteobacteria (Degli Esposti et al. 2014) and their COX1 subunit generally belong to the A1 type, as defined by the protein signatures of CuB binding and proton channels (Pereira et al. 2001, M.D.E. unpublished).

The aa3 oxidases of A1 type include mitochondrial cytochrome *c* oxidase, the most studied respiratory enzyme now known at the atomic level in its whole complexity (Tsukihara et al. 1996). The homologous bacterial enzyme is coded by an operon that is specific to alphaproteobacteria, comprising various subunits required for the assembly of the redox cofactors, apparently related to the *ctaABCDEFG* operon originally described in *Bacillus* (Quirck et al. 1993, Degli Esposti et al. 2014). The oxygen affinity of mitochondrial cytochrome *c* oxidase varies depending upon the energetic state of the organelle, but the apparent K_m values for oxygen are usually in the micromolar range (Krab et al. 2011), higher than those reported for A2 type oxidases, that is, mitochondrial COX is a low affinity terminal oxidase, a property that can be extended to all its bacterial homologs of A1 type (Iwata 1998).

On the Evolution of COX and Related Terminal Oxidases

Given the characteristics described above, the most conservative scheme for the evolution of the HCO superfamily would follow the progressively more complex types of hemes they contain. So, NOR and cbb3 oxidases, which contain only *b* hemes, would be the most ancient while the aa3 oxidases of the A family would be the most recent (Sharma and Wikström 2014), with B family oxidases in an intermediate position. Surprisingly, this scheme has been considered only in a couple of reviews on the subject (Ducluzeau et al. 2014, Sharma and Wikström 2014). In contrast, Brochier-Armanet et al. (2009) have proposed a diametrically opposite scheme, in which A family oxidases would be the oldest and C family oxidases the most recent, with the additional hypothesis that B family oxidases originated in Archaea, from which they spread to bacteria. This hypothesis echoes earlier concepts (Catresana and Saraste 1995) and has been viewed with favor by experts in bacterial bioenergetics (Han et al. 2011, Simon and Klotz 2013). However, mounting evidence indicates that the genes of ba3 and other terminal oxidases have been laterally transferred the opposite way, from bacteria to Archaea (Sousa et al. 2012, Nelson-Sathi et al. 2012, Dibrova et al. 2014). Such contrasting views on HCO evolution underscore the problem of considering archaean and bacterial COX1 sequences together to infer the phylogeny of the oxidase families, a problem that has remained in the fields of bioenergetics and molecular evolution since the very first analysis of COX proteins (Catresana and Saraste 1995).

Recently, it has been shown that all oxidases of Archaea have been acquired by LGT from either Gram positive bacteria or proteobacteria (Sousa et al. 2012, Nelson-Sathi et al. 2012, 2015, Dibrova et al. 2014, Garushyants et al. 2015, Gupta 2016). Consequently, it is biologically misleading to include proteins from Archaea to obtain phylogenetic trees of HCO, which are most frequently based upon the largest catalytic subunit that is shared by all families (Brochier-Armanet et al. 2009, Sousa et al. 2012, Ducluzeau et al. 2014). Such trees have been considered to have no discernable root due to the enormous sequence divergence of COX1 homologs and the apparent lack of rooting relatives (Sousa et al. 2012). However, the currently available wealth of HCO proteins derived from metagenomic assembled genomes (**MAG**—Anantharaman et al. 2016, Soo et al. 2017, Becraft et al. 2017, Parks et al. 2017) could provide new opportunities to revisit the evolution of HCO oxidases—without the confusing and often distorting signals of the laterally acquired proteins from Archaea. Efforts are being implemented to exploit these opportunities.

Acknowledgements

I thank Lars Hederstedt (University of Lund, Sweden) for critically reading the manuscript and useful discussion. I also thank Bill Martin (University of Dusseldorf, Germany) for discussion.

References

Abramson, J., S. Riistama, G. Larsson, A. Jasaitis, M. Svensson-Ek, L. Laakkonen et al. 2000. The structure of the ubiquinol oxidase from Escherichia coli and its ubiquinone binding site. Nat. Struct. Biol. 7: 910–917.

Anbar, A.D., Y. Duan, T.W. Lyons, G.L. Arnold, B. Kendall, R.A. Creaser et al. 2007. A whiff of oxygen before the great oxidation event? Science 317: 1903–1906.

Anbar, A.D. 2008. Oceans. Elements and evolution. Science 322: 1481–1483.

Anantharaman, K., C.T. Brown, L.A. Hug, I. Sharon, C.J. Castelle, A.J. Probst et al. 2016. Thousands of microbial genomes shed light on interconnected biogeochemical processes in an aquifer system. Nat. Commun. 7: 13219. doi:10.1038/ncomms13219.

Anton, B.P., S. Kasif, R.J. Roberts and M. Steffen. 2014. Objective: biochemical function. Front. Genet. 5: 210. doi:10.3389/fgene.2014.00210.

Arai, H., T. Kawakami, T. Osamura, T. Hirai, Y. Sakai and M. Ishii. 2014. Enzymatic characterization and *in vivo* function of five terminal oxidases in Pseudomonas aeruginosa. J. Bacteriol. 196: 4206–4215. doi:10.1128/JB.02176-14.

Atteia, A., R. van Lis, A.G. Tielens and W.F. Martin. 2013. Anaerobic energy metabolism in unicellular photosynthetic eukaryotes. Biochim. Biophys. Acta. 1827: 210–223.

Baughn, A.D. and M.H. Malamy. 2004. The strict anaerobe Bacteroides fragilis grows in and benefits from nanomolar concentrations of oxygen. Nature 427: 441–444.

Baradaran, R., J.M. Berrisford, G.S. Minhas and L.A. Sazanov. 2013. Crystal structure of the entire respiratory complex I. Nature 494: 443–448.

Battistuzzi, F.U., A. Feijao and S.B. Hedges. 2004. A genomic timescale of prokaryote evolution: insights into the origin of methanogenesis, phototrophy, and the colonization of land. BMC Evol. Biol. 4: 44.

Becraft, E.D., T. Woyke, J. Jarett, N. Ivanova, F. Godoy-Vitorino, N. Poulton et al. 2017. Rokubacteria: Genomic giants among the uncultured bacterial phyla. Front. Microbiol. 8: 2264. doi:10.3389/fmicb.2017.02264.

Blanc, B., C. Gerez and S. Ollagnier de Choudens. 2015. Assembly of Fe/S proteins in bacterial systems: Biochemistry of the bacterial ISC system. Biochim. Biophys. Acta. 1853: 1436–1447. doi:10.1016/j.bbamcr.2014.12.009.

Borisov, V.B., R.B. Gennis, J. Hemp and M.I. Verkhovsky. 2011. The cytochrome bd respiratory oxygen reductases. Biochim. Biophys. Acta. 1807: 1398–1413. doi:10.1016/j.bbabio.2011.06.016.

Brochier-Armanet, C., E. Talla and S. Gribaldo. 2009. The multiple evolutionary histories of dioxygen reductases: Implications for the origin and evolution of aerobic respiration. Mol. Biol. Evol. 26: 285–297.

Buchanan, B.B. and D.I. Arnon. 1990. A reverse KREBS cycle in photosynthesis: Consensus at last. Photosynth Res. 24: 47–53.

Buschmann, S., E. Warkentin, H. Xie, J.D. Langer, U. Ermler and H. Michel. 2010. The structure of cbb3 cytochrome oxidase provides insights into proton pumping. Science 329: 327–330. doi:10.1126/science.1187303.

Campbell, B.J. and S.C. Cary. 2004. Abundance of reverse tricarboxylic acid cycle genes in free-living microorganisms at deep-sea hydrothermal vents. Appl. Environ. Microbiol. 70: 6282–6289.

Campbell, B.J., J.L. Smith, T.E. Hanson, M.G. Klotz, L.Y. Stein et al. 2009. Adaptations to submarine hydrothermal environments exemplified by the genome of Nautilia profundicola. PLoS Genet. 5: e1000362.

Can, M., F.A. Armstrong and S.W. Ragsdale. 2014. Structure, function, and mechanism of the nickel metalloenzymes, CO dehydrogenase, and acetyl-CoA synthase. Chemical Rev. 114: 4149–4174.

Canfield, D.E., M.T. Rosing and C. Bjerrum. 2006. Early anaerobic metabolisms. Philos. Trans. R. Soc. Lond. B Biol. Sci. 361: 1819–1834. doi:10.1098/rstb.2006.1906.

Carlson, H.K., I.C. Clark, S.J. Blazewicz, A.T. Iavarone and J.D. Coates. 2013. Fe(II) oxidation is an innate capability of nitrate-reducing bacteria that involves abiotic and biotic reactions. J. Bacteriol. 195: 3260–3268. doi:10.1128/JB.00058-13.

Castresana, J. and M. Saraste. 1995. Evolution of energetic metabolism: the respiration-early hypothesis. Trends Biochem. Sci. 20: 443–448.

Chistoserdova, L., M.G. Kalyuzhnaya and M.E. Lidstrom. 2009. The expanding world of methylotrophic metabolism. Annu. Rev. Microbiol. 63: 477–499. doi:10.1146/annurev.micro.091208.073600.

Chistoserdova, L. 2016. Wide distribution of genes for tetrahydromethanopterin/methanofuran-linked C1 transfer reactions argues for their presence in the common ancestor of bacteria and archaea. Front. Microbiol. 13: 1425. doi:10.3389/fmicb.2016.01425.

Chistoserdova, L. 2017. Application of Omics Approaches to Studying Methylotrophs and Methylotroph Comunities. Curr. Issues Mol. Biol. 24: 119–142. doi:10.21775/cimb.024.119.

Coppi, M.V. 2005. The hydrogenases of geobacter sulfurreducens: a comparative genomic perspective. Microbiology 151: 1239–1254.

Cunningham, L., M. Pitt and H.D. Williams. 1997. The cioAB genes from Pseudomonas aeruginosa code for a novel cyanide-insensitive terminal oxidase related to the cytochrome bd quinol oxidases. Mol. Microbiol. 24: 579–591.

David, L.A. and E.J. Alm. 2011. Rapid evolutionary innovation during an Archaean genetic expansion. Nature 469: 93–96. doi:10.1038/nature09649.

Degli Esposti, M., B. Chouaia, F. Comandatore, E. Crotti, D. Sassera, P.M. Lievens et al. 2014. Evolution of mitochondria reconstructed from the energy metabolism of living bacteria. PLoS One 9: e96566. doi:10.1371/journal.pone.0096566.

Degli Esposti, M. 2015. Genome analysis of structure-function relationships in respiratory complex I, an ancient bioenergetic enzyme. Genome Biol. Evol. 8: 126–147. doi:10.1093/gbe/evv239.

Degli Esposti, M., T. Rosas-Pérez, L.E. Servín-Garcidueñas, L.M. Bolaños, M. Rosenblueth and E. Martínez-Romero. 2015. Molecular evolution of cytochrome bd oxidases across proteobacterial genomes. Genome Biol. Evol. 7: 801–820. doi:10.1093/gbe/evv032.

Degli Esposti, M. and E. Martinez Romero. 2016. A survey of the energy metabolism of nodulating symbionts reveals a new form of respiratory complex I. FEMS Microbiol. Ecol. 92:fiw084. doi:10.1093/femsec/fiw084.

Degli Esposti, M., D. Cortez, L. Lozano, S. Rasmussen, H.B. Nielsen and E. Martinez Romero. 2016. Alpha proteobacterial ancestry of the [Fe-Fe]-hydrogenases in anaerobic eukaryotes. Biol. Direct. 11: 34. doi: 10.1186/s13062-016-0136-3.

Degli Esposti, M. and E. Martinez Romero. 2017. The functional microbiome of arthropods. PLoS One. 12: e0176573. doi:10.1371/journal.pone.0176573.

Degli Esposti, M. 2017. A journey across genomes uncovers the origin of ubiquinone in cyanobacteria. Genome Biol. Evol. 9: 3039–3053. doi:10.1093/gbe/evx225.

Dibrova, D.V., M.Y. Galperin and A.Y. Mulkidjanian. 2014. Phylogenomic reconstruction of archaeal fatty acid metabolism. Environ. Microbiol. 16: 907–918. doi:10.1111/1462-2920.12359.

D'Mello, R., S. Hill and R.K. Poole. 1995. The oxygen affinity of cytochrome bo' in Escherichia coli determined by the deoxygenation of oxyleghemoglobin and oxymyoglobin: Km values for oxygen are in the submicromolar range. J. Bacteriol. 177(3): 867–70.

D'Mello, R., S. Hill and R.K. Poole. 1996. The cytochrome bd quinol oxidase in Escherichia coli has an extremely high oxygen affinity and two oxygen-binding haems: Implications for regulation of activity *in vivo* by oxygen inhibition. Microbiology 142: 755–763.

Dubilier, N., C. Bergin and C. Lott. 2008. Symbiotic diversity in marine animals: the art of harnessing chemosynthesis. Nat. Rev. Microbiol. 6: 725–740. doi:10.1038/nrmicro1992.

Ducluzeau, A.L., B. Schoepp-Cothenet, R. van Lis, F. Baymann, M.J. Russell and W. Nitschke. 2014. The evolution of respiratory O2/NO reductases: an out-of-the-phylogenetic-box perspective. J. R. Soc. Interface 11: 20140196. doi:10.1098/rsif.2014.0196.

Dworkin, M. 2012. Sergei Winogradsky: a founder of modern microbiology and the first microbial ecologist. FEMS Microbiol. Rev. 36: 364–379. doi:10.1111/j.1574-6976.2011.00299.x.

Eloe-Fadrosh, E.A., D. Paez-Espino, J. Jarett, P.F. Dunfield, B.P. Hedlund, A.E. Dekas et al. 2016. Global metagenomic survey reveals a new bacterial candidate phylum in geothermal springs. Nat. Commun. 7: 10476. doi:10.1038/ncomms10476.

Forte, E., V.B. Borisov, M. Falabella, H.G. Colaço, M. Tinajero-Trejo, R.K. Poole et al. 2016. The terminal oxidase cytochrome bd promotes sulfide-resistant bacterial respiration and growth. Sci. Rep. 6: 23788. doi:10.1038/srep23788.

Forte, E., V.B. Borisov, J.B. Vicente and A. Giuffrè. 2017. Cytochrome bd and Gaseous Ligands in Bacterial Physiology. Adv. Microb. Physiol. 71: 171–234. doi:10.1016/bs.ampbs.2017.05.002.

Garcia-Horsman, J.A., B. Barquera and J.E. Escamilla. 1991. Two different aa3-type cytochromes can be purified from the bacterium Bacillus cereus. Febs. J. 199: 61–68.

García-Horsman, J.A., B. Barquera, J. Rumbley, J. Ma and R.B. Gennis. 1994. The superfamily of heme-copper respiratory oxidases. J. Bacteriol. 176: 5587–5600.

Garushyants, S.K., M.D. Kazanov and M.S. Gelfand. 2015. Horizontal gene transfer and genome evolution in Methanosarcina. BMC Evol. Biol. 15: 102. doi:10.1186/s12862-015-0393-2.

Giovannelli, D., J. Ricci, I. Pérez-Rodríguez, M. Hügler, C. O'Brien, R. Keddis et al. 2012. Complete genome sequence of Thermovibrio ammonificans HB-1(T), a thermophilic, chemolithoautotrophic bacterium isolated from a deep-sea hydrothermal vent. Stand. Genomic Sci. 7: 82–90.

Gupta, R.S. 2016. Impact of genomics on the understanding of microbial evolution and classification: the importance of Darwin's views on classification. FEMS Microbiol. Rev. 40: b520–553. doi:10.1093/femsre/fuw011.

Han, H., J. Hemp, L.A. Pace, H. Ouyang, K. Ganesan, J.H. Roh et al. 2011. Adaptation of aerobic respiration to low O2 environments. Proc. Natl. Acad. Sci. USA 108: 14109–14114. doi:10.1073/pnas.1018958108.

Hederstedt, L. 2012. Heme A biosynthesis. Biochim. Biophys. Acta. 1817: 920–927. doi:10.1016/j.bbabio.2012.03.025.

Ho, A., R. Angel, A.J. Veraart, A. Daebeler, Z. Jia, S.Y. Kim et al. 2016. Biotic interactions in microbial communities as modulators of biogeochemical processes: Methanotrophy as a model system. Front. Microbiol. 7: 1285. doi:10.3389/fmicb.2016.01285.

Holland, H.D. 2006. The oxygenation of the atmosphere and oceans. Philos. Trans. R. Soc. Lond. B Biol. Sci. 361: 903–915.

Hug, L.A., B.J. Baker, K. Anantharaman, C.T. Brown, A.J. Probst, C.J. Castelle et al. 2016. A new view of the tree of life. Nat. Microbiol. 1: 16048. doi: 10.1038/nmicrobiol.2016.48.

Ilbert, M. and V. Bonnefoy. 2013. Insight into the evolution of the iron oxidation pathways. Biochim. Biophys. Acta. 1827: 161–175. doi:10.1016/j.bbabio.2012.10.001.

Iwata, S. 1998. Structure and function of bacterial cytochrome c oxidase. J. Biochem. 123: 369–375.

Ji, M., C. Greening, I. Vanwonterghem, C.R. Carere, S.K. Bay and J.A. Steen. 2017. Atmospheric trace gases support primary production in Antarctic desert surface soil. Nature 552: 400–403. doi:10.1038/nature25014.

Jormakka, M., K. Yokoyama, T. Yano, M. Tamakoshi, S. Akimoto, T. Shimamura et al. 2008. Molecular mechanism of energy conservation in polysulfide respiration. Nat. Struct. Mol. Biol. 15: 730–737. doi:10.1038/nsmb.1434.

Kalenitchenko, D., M. Dupraz, N. Le Bris, C. Petetin, C. Rose, N.J. West and P.E. Galand. 2016. Ecological succession leads to chemosynthesis in mats colonizing wood in sea water. ISME J. 10: 2246–2258. doi:10.1038/ismej.2016.12.

Keltjens, J.T., A. Pol, J. Reimann and H.J. Op den Camp. 2014. PQQ-dependent methanol dehydrogenases: rare-earth elements make a difference. Appl. Microbiol. Biotechnol. 98: 6163–6183. doi:10.1007/s00253-014-5766-8.

Krab, K., H. Kempe and M. Wikström. 2011. Explaining the enigmatic K(M) for oxygen in cytochrome c oxidase: a kinetic model. Biochim. Biophys. Acta. 1807: 348–358. doi:10.1016/j.bbabio.2010.12.015.

Lens, P.N. and J.G. Kuenen. 2001. The biological sulfur cycle: novel opportunities for environmental biotechnology. Water Sci. Technol. 44: 57–66.

Lill, R. 2009. Function and biogenesis of iron-sulphur proteins. Nature 460: 831–838. doi:10.1038/nature08301.

Louca, S., L.W. Parfrey and M. Doebeli. 2016. Decoupling function and taxonomy in the global ocean microbiome. Science 353: 1272–1277. doi:10.1126/science.aaf4507.

Lyons, J.A., D. Aragão, O. Slattery, A.V. Pisliakov, T. Soulimane and M. Caffrey. 2012. Structural insights into electron transfer in caa3-type cytochrome oxidase. Nature 487: b514–518. doi:10.1038/nature11182.

Lyons, T.W., C.T. Reinhard and N.J. Planavsky. 2014. The rise of oxygen in Earth's early ocean and atmosphere. Nature 506: 307–315. doi:10.1038/nature13068.

Lwoff, A., C.B. van Niel, P.J. Ryan and E.L. Tatum. 1946. Nomenclature of nutritional types of microorganisms. Cold Spring Harbor Symposia on Quantitative Biology (5th edn.), Vol. XI, The Biological Laboratory, Cold Spring Harbor, NY, pp. 302–303.

Maczulak, A. 2010. Allies and enemies: how the world depends on bacteria. FT Press, Upper Saddle River NJ, USA.

Marreiros, B.C., A.P. Batista, A.M. Duarte and M.M. Pereira. 2013. A missing link between complex I and group 4 membrane-bound [NiFe] hydrogenases. Biochim. Biophys. Acta. 1827: 198–209.

Marreiros, B.C., F. Calisto, P.J. Castro, A.M. Duarte, F.V. Sena, A.F. Silva et al. 2016. Exploring membrane respiratory chains. Biochim. Biophys. Acta. 1857: 1039–1067. doi:10.1016/j.bbabio.2016.

Martin, W.F., M.C. Weiss, S. Neukirchen, S. Nelson-Sathi and F.L. Sousa. 2016. Physiology, phylogeny, and LUCA. Microb. Cell. 3: 582–587. doi:10.15698/mic2016.12.545.

Martin, W.F. 2017. Unmiraculous facultative anaerobes (comment on DOI 10.1002/bies.201600174). Bioessays 39. doi:10.1002/bies.201700041.

Martin, W.F., D.A. Bryant and J.T. Beatty. 2018. A Physiological Perspective on the Origin and Evolution of Photosynthesis 42: 205–231. doi:10.1093/femsre/fux056.

Melton, E.D., E.D. Swanner, S. Behrens, C. Schmidt and A. Kappler. 2014. The interplay of microbially mediated and abiotic reactions in the biogeochemical Fe cycle. Nat. Rev. Microbiol. 12: 797–808. doi:10.1038/nrmicro3347.

Moore, E.K., B.I. Jelen, D. Giovannelli, H. Raanan and P.G. Falkowski. 2017. Metal availability and the expanding network of microbial metabolisms in the Archaean eon. Nat. Geo. 10: 629–636.

Morris, R.L. and T.M. Schmidt. 2013. Shallow breathing: bacterial life at low O(2). Nat. Rev. Microbiol. 11: 205–212. doi:10.1038/nrmicro2970.

Müller, M., M. Mentel, J.J. van Hellemond, K. Henze, C. Woehle, S.B. Gould et al. 2012. Biochemistry and evolution of anaerobic energy metabolism in eukaryotes. Microbiol. Mol. Biol. Rev. 76: 444–495.

Muyzer, G. and A.J. Stams. 2008. The ecology and biotechnology of sulphate-reducing bacteria. Nat. Rev. Microbiol. 6: 441–454. doi:10.1038/nrmicro1892.

Nelson, M.B., A.C. Martiny and J.B. Martiny. 2016. Global biogeography of microbial nitrogen-cycling traits in soil. Proc. Natl. Acad. Sci. USA 113: 8033–8040. doi:10.1073/pnas.1601070113.

Nelson-Sathi, S., T. Dagan, G. Landan, A. Janssen, M. Steel, J.O. McInerney et al. 2012. Acquisition of 1,000 eubacterial genes physiologically transformed a methanogen at the origin of Haloarchaea. Proc. Natl. Acad. Sci. USA 109: 20537–20542. doi:10.1073/pnas.1209119109.

Nelson-Sathi, S., F.L. Sousa, M. Roettger, N. Lozada-Chávez, T. Thiergart, A. Janssen et al. 2015. Origins of major archaeal clades correspond to gene acquisitions from bacteria. Nature 517: 77–80. doi:10.1038/nature13805.

Nielsen, H.B., M. Almeida, A.S. Juncker, S. Rasmussen, J. Li, S. Sunagawa et al. 2014. Identification and assembly of genomes and genetic elements in complex metagenomic samples without using reference genomes. Nat. Biotechnol. 32: 822–828. doi:10.1038/nbt.2939.

Parks, D.H., C. Rinke, M. Chuvochina, P.A. Chaumeil, B.J. Woodcroft, P.N. Evans et al. 2017. Recovery of nearly 8,000 metagenome-assembled genomes substantially expands the tree of life. Nature Microbiol. 2: 1533–1542. doi:10.1038/s41564-017-0012-71533.

Pennisi, R., D. Salvi, V. Brandi, R. Angelini, P. Ascenzi and F. Polticelli. 2016. Molecular evolution of alternative oxidase proteins: a phylogenetic and structure modeling approach. J. Mol. Evol. 82: 207–218. doi:10.1007/s00239-016-9738-8.

Pereira, M.M., M. Santana and M. Teixeira. 2001. A novel scenario for the evolution of haem-copper oxygen reductases. Biochim. Biophys. Acta 1505: 185–208.

Pitcher, R.S. and N.J. Watmough. 2004. The bacterial cytochrome cbb3 oxidases. Biochim. Biophys. Acta. 1655: 388–399.

Probst, A.J., C.J. Castelle, A. Singh, C.T. Brown, K. Anantharaman, I. Sharon et al. 2017. Genomic resolution of a cold subsurface aquifer community provides metabolic insights for novel microbes adapted to high CO(2) concentrations. Environ. Microbiol. 19: 459–474. doi:10.1111/1462-2920.13362.

Prunetti, L., M. Brugna, R. Lebrun, M.T. Giudici-Orticoni and M. Guiral. 2011. The elusive third subunit IIa of the bacterial B-type oxidases: the enzyme from the hyperthermophile Aquifex aeolicus. PLoS One 6: e21616.

Quirk, P.G., D.B. Hicks and T.A. Krulwich. 1993. Cloning of the cta operon from alkaliphilic Bacillus firmus OF4 and characterization of the pH-regulated cytochrome caa3 oxidase it encodes. J. Biol. Chem. 268: 678–685.

Radzi Noor, M. and T. Soulimane. 2012. Bioenergetics at extreme temperature: Thermus thermophilus ba(3)- and caa(3)-type cytochrome c oxidases. Biochim. Biophys. Acta. 1817: 638–49. doi:10.1016/j.bbabio.2011.08.004.

Ramel, F., A. Amrani, L. Pieulle, O. Lamrabet, G. Voordouw, N. Seddiki et al. 2013. Membrane-bound oxygen reductases of the anaerobic sulfate-reducing Desulfovibrio vulgaris Hildenborough: roles in oxygen defence and electron link with periplasmic hydrogen oxidation. Microbiology 159: 2663–2673.

Ravcheev, D.A. and I. Thiele. 2016. Genomic Analysis of the Human Gut Microbiome Suggests Novel Enzymes Involved in Quinone Biosynthesis. Front. Microbiol. 7: 128. doi:10.3389/fmicb.2016.00128.

Richhardt, J., B. Luchterhand, S. Bringer, J. Büchs and M. Bott. 2013. Evidence for a key role of cytochrome bo3 oxidase in respiratory energy metabolism of *Gluconobacter oxydans*. J. Bacteriol. 195: 4210–4220.

Rothery, R.A., G.J. Workun and J.H. Weiner. 2008. The prokaryotic complex iron-sulfur molybdoenzyme family. Biochim. Biophys. Acta. 1778: 1897–1929.

Sakurai, H., T. Ogawa, M. Shiga and K. Inouc. 2010. Inorganic sulfur oxidizing system in green sulfur bacteria. Photosynth. Res. 104: 163–176. doi:10.1007/s11120-010-9531-2.

Sazanov, L.A. 2015. A giant molecular proton pump: Structure and mechanism of respiratory complex I. Nat. Rev. Mol. Cell. Biol. 16: 375–388.

Schäfer, G., W. Purschke and C.L. Schmidt. 1996. On the origin of respiration: Electron transport proteins from archaea to man. FEMS Microbiol. Rev. 18: 173–188.

Scharf, B., R. Wittenberg and M. Engelhard. 1997. Electron transfer proteins from the haloalkaliphilic archaeon Natronobacterium pharaonis: Possible components of the respiratory chain include cytochrome bc and a terminal oxidase cytochrome ba3. Biochemistry 36: 4471–4479.

Schoepp-Cothenet, B., R. van Lis, A. Atteia, F. Baymann, L. Capowiez, A.L. Ducluzeau et al. 2013. On the universal core of bioenergetics. Biochim. Biophys. Acta. 1827: 79–93. doi:10.1016/j.bbabio.2012.09.005.

Schübbe, S., T.J. Williams, G. Xie, H.E. Kiss, T.S. Brettin, D. Martinez et al. 2009. Complete genome sequence of the chemolithoautotrophic marine magnetotactic coccus strain MC-1. Appl. Environ. Microbiol. 75: 4835–4852.

Schulz, F., E.A. Eloe-Fadrosh, R.M. Bowers, J. Jarett, T. Nielsen, N.N. Ivanova et al.2017. Towards a balanced view of the bacterial tree of life. Microbiome. 5: 140. doi:10.1186/s40168-017-0360-9.

Segata, N., D. Börnigen, X.C. Morgan and C. Huttenhower. 2013. PhyloPhlAn is a new method for improved phylogenetic and taxonomic placement of microbes. Nat. Commun. 4: 2304.

Sharma, V. and M. Wikström. 2014. FEBS Lett. 588: 3787–3792. doi:10.1016/j.febslet.2014.09.020.

Shih, P.M., L.M. Ward and W.W. Fischer. 2017. Evolution of the 3-hydroxypropionate bicycle and recent transfer of anoxygenic photosynthesis into the Chloroflexi. Proc. Natl. Acad. Sci. USA 114: 10749–10754. doi:10.1073/pnas.1710798114.

Sone, N. and Y. Fujiwara. 1991. Haem O can replace haem A in the active site of cytochrome c oxidase from thermophilic bacterium PS3. FEBS Letters 288: 154–158

Soo, R.M., J. Hemp, D.H. Parks, W.W. Fischer and P. Hugenholtz. 2017. On the origins of oxygenic photosynthesis and aerobic respiration in Cyanobacteria. Science 355: 1436–1440.

Sousa, F.L., R.J. Alves, M.A. Ribeiro, J.B. Pereira-Leal, M. Teixeira and M.M. Pereira. 2012. The superfamily of heme-copper oxygen reductases: types and evolutionary considerations. Biochim. Biophys. Acta. 1817: 629–637. doi:10.1016/j.bbabio.2011.09.020.

Spang, A., E.F. Caceres and T.J.G. Ettema. 2017. Genomic exploration of the diversity, ecology, and evolution of the archaeal domain of life. Science 357. pii: eaaf3883. doi:10.1126/science.aaf3883.

Sparacino-Watkins, C., J.F. Stolz and P. Basu. 2014. Nitrate and periplasmic nitrate reductases. Chem. Soc. Rev. 43: 676–706. doi:10.1039/c3cs60249d.

Spero, M.A., F.O. Aylward, C.R. Currie and T.J. Donohue. 2015. Phylogenomic analysis and predicted physiological role of the proton-translocating NADH: Quinone oxidoreductase (complex I) across bacteria. MBio. 6: ve00389–15.

Stolper, D.A., N.P. Revsbech and D.E. Canfield. 2010. Aerobic growth at nanomolar oxygen concentrations. Pro. Natl. Acad. Sci. USA 107: 18755–18760. doi:10.1073/pnas.1013435107.

Thiergart, T., G. Landan and W.F. Martin. 2014. Concatenated alignments and the case of the disappearing tree. BMC Evol. Biol. 14: 266. doi:10.1186/s12862-014-0266-0.

Tiefenbrunn, T., W. Liu, Y. Chen, V. Katritch, C.D. Stout, J.A. Fee et al. 2011. High resolution structure of the ba3 cytochrome c oxidase from Thermus thermophilus in a lipidic environment. PLoS One 6: e22348. doi:10.1371/journal.pone.0022348.

Tsukihara, T., H. Aoyama, E. Yamashita, T. Tomizaki, H. Yamaguchi, K. Shinzawa-Itoh et al. 1996. The whole structure of the 13-subunit oxidized cytochrome c oxidase at 2.8 A. Science 272: 1136–1144.

Tully, B.J., C.G. Wheat, B.T. Glazer and J.A. Huber. 2018. A dynamic microbial community with high functional redundancy inhabits the cold, oxic subseafloor aquifer. ISME J. 12: 1–16. doi:10.1038/ismej.2017.187.

Vlaeminck, S.E., A.G. Hay, L. Maignien and W. Verstraete. 2011. In quest of the nitrogen oxidizing prokaryotes of the early Earth. Environ. Microbiol. 13: 283–295. doi:10.1111/j.1462-2920.2010.02345.x

Waksman, S.A. 1953. Sergei Nikolaevitch Winogradsky: 1856–1953. Science 118: 36–37.

Weiss, M.C., F.L. Sousa, N. Mrnjavac, S. Neukirchen, M. Roettger, S. Nelson-Sathi et al. 2016. The physiology and habitat of the last universal common ancestor. Nat. Microbiol. 1: 16116. doi:10.1038/nmicrobiol.2016.116.

Woese, C.R. 1987. Bacterial evolution. Microbiol. Rev. 51: 221–271.

Zhi, X.Y., J.C. Yao, S.K. Tang, Y. Huang, H.W. Li and W.J. Li. 2014. The futalosine pathway played an important role in menaquinone biosynthesis during early prokaryote evolution. Genome Biol. Evol. 6: 149–160. doi:10.1093/gbe/evu007.

Zimmermann, B.H., C.I. Nitsche, J.A. Fee, F. Rusnak and E. Münck. 1988. Properties of a copper-containing cytochrome ba3: a second terminal oxidase from the extreme thermophile Thermus thermophilus. Proc. Natl. Acad. Sci. USA 85: 5779–5783.

3

Gram Negative Bacteria Related to Proteobacteria

Mauro Degli Esposti

Introduction

The definition of Gram negative (diderms, with two membranes) vs. Gram positive (monoderms, single membrane) bacteria is ancient in microbiology, but its phylogenetic value is limited. Recently, Gribaldo and coworkers (Antunes et al. 2016) examined the phylogeny of a major group of Gram positive bacteria, the Firmicutes (including Bacilli), and concluded that their reactivity to the Gram dye is a derived feature of their envelope, which ancestrally was structured with a lipopolysaccharide (LPS) containing outer membrane as in classical Gram negative bacteria (Antunes et al. 2016). Hence, the presence of a LPS-containing outer membrane producing negative staining with the Gram dye is likely to be a common and ancestral feature of bacteria. In the context of this book, we shall consider only a subset of Gram negative bacteria that are phylogenetically intermediate between the deepest branching group, hence LUCA (Chapter one), and proteobacteria (next Chapter four) which constitute the *phylum* of most interest in relation to the evolution of mitochondria (Chapter seven). This subset includes the Bacteroidetes (with only a brief mention of their relatives Chlorobi) and the Nitrospirae, representative proteins of which are either in sister (Antunes et al. 2016, Degli Esposti et al. 2015, Lin et al. 2017) or in close proximity to their proteobacterial homologs in most phylogenetic trees (Lücker et al. 2013, Anantharaman et al. 2016).

Center for Genomic Sciences, UNAM Campus de Cuernavaca, Cuernavaca, 62130 Morelos, Mexico.
Email: mauro1italia@gmail.com

The Bacteroidetes

The *phylum* Bacteroidetes includes five major classes and representative orders (Krieg et al. 2010): Bacteroidia (Bacteroidales); Flavobacteriia (Flavobacteriales); Cytophagia (Cytophagales); Sphingobacteriia (Sphingobacteriales) and; Chitinophagia (Chitinophagales). Bacteroidetes also include thermophilic *Rhodothermus* and related organisms of the family Rhodothermaceae, which are classified as Bacteroidetes *Incertae sedis* (https://www.ncbi.nlm.nih.gov/Taxonomy/Browser/wwwtax.cgi?mode=Undef&id=1100069&lvl=3&lin=f&keep=1&srch mode=1&unlock, accessed March 20, 2018). The taxonomic classification of the diverse bacteria belonging to the Bacteroidetes *phylum* has changed over time (Krieg et al. 2010, Thomas et al. 2011, Hahnke et al. 2016), but their traditional definition as CFB (Cytophaga-Flavobacteria-Bacteroidetes—Woese 1987, Gupta 2004) persists, also in NCBI (National Center for Biotechnology Information) web resources. Nowadays, the *phylum* Bacteroidetes is associated with the *phyla* of Chlorobi and Fibrobacteres, as well as to the recently introduced *phylum* of Ignavibacteria, together forming the FCB group or superphylum (Gupta 2004, Kadnikov et al. 2013, Hamilton et al. 2016). https://www.ncbi.nlm.nih.gov/Taxonomy/Browser/wwwtax.cgi?mode=Undef&id=1783270&lvl=3&lin=f&kee p=1&srchmode=1&unlock.

Bacteroidetes (CFB) include ecologically and physiologically diverse Gram negative, non-motile or gliding bacteria that inhabit most ecosystems of our planet. The gliding behavior does not involve flagella and has only recently been understood (Shrivastava and Berg 2015). While Bacteroidales are (facultatively) anaerobes and form the largest group of Gram negative bacteria in animal microbiota (Martens et al. 2009, Thomas et al. 2011, Karlsson et al. 2011), Flavobacteriales constitute a large and continuously expanding group of predominantly aerobic taxa, which are present in diverse environmental niches, especially aquatic but also as endosymbionts of insects (see Chapter six). Cytophagales include a relatively small number of environmental aerobic heterotrophs, including endosymbionts of protists; Sphingobacteriales are also aerobic heterotrophs and characteristically contain sphingolipids in their membranes (Kämpfer 2015, Geiger et al. 2018). Cytophagales and Chitinophagales are known for their capacity of degrading various recalcitrant organic macromolecules such as chitin, pectin, and cellulose. Similar degradative properties are widespread among other lineages of Bacteroidetes, some of which (e.g., the Flavobacteriales *Capnocytophaga*) can even degrade hydroxyapatite, the basis of dentin (Peros and Gibbons 1982, Piau et al. 2013). *Capnocytophaga*, *Porphyromonas* and other Flavobacteriales inhabiting the oral microbiota can become opportunistic pathogens and are responsible for periodontitis and other conditions affecting human health (Meuric et al. 2010, Costalonga and Herzberg 2014, Nibali 2015). Some Flavobacteriales are pathogenic to mammals, freshwater fish or marine fish (Thomas et al. 2011, Hahnke et al. 2016). However, a great majority of Bacteroidetes organisms, including their growing number discovered by metagenomic analysis (Anantharaman et al. 2016), are heterotrophic commensals or environmental degraders of complex biopolymers.

Functional Properties of Bacteroidetes

Central energy metabolism and protonmotive systems

Unlike the related Chlorobi, which are autotrophic green sulfur bacteria (Buchanan and Arnon 1990), Bacteroidetes are metabolically diverse chemoorganotrophs that are able to grow on a variety of complex biopolymers, such as cellulose, chitin and agar (Thomas et al. 2011). The majority of these organisms appear to be facultatively anaerobes, even if Flavobacteriales, Cytophagales and Sphingobacteriales are usually considered to be aerobic (Thomas et al. 2011, Hahnke et al. 2016). Their energy metabolism is centered on the TCA cycle, which can also go partially in reverse in several anaerobic Bacteroidetes, similar to the rTCA of Chlorobi (Buchanan and Arnon 1990) (Fig. 1). This is due to the presence of enzymes such as 2-oxoglutarate:ferredoxin oxidoreductase which bypass the cycle reducing ferredoxin instead of NAD+ (Buchanan and Arnon 1990, Campbell and Cary 2004, Degli Esposti and Martinez-Romero 2016). Reduced ferredoxin is then re-oxidized in the cytosol by either a [FeFe]-hydrogenase or nitrogenase (Degli Esposti and Martinez-Romero 2016). About one-half of anaerobic Bacteroidales do express [FeFe]-hydrogenases (Wolf et al. 2016, Greening et al. 2016, Anantharaman et al. 2016). Alternatively, reduced ferredoxin is re-oxidized at the internal surface of the cytoplasmic membrane by the ionmotive Rnf complex (see Chapter two), which is widespread among anaerobic Bacteroidales and is also sporadically present in other CFB organisms (Fig. 1, cf. Meuric et al. 2010, Biegel et al. 2011).

Because biochemical analysis of the bioenergetic properties of Bacteroidetes is scant, most information on their bioenergy capacity derives from genetic data and comparative genomics (Nelson et al. 2003, Hongoh et al. 2008, Meuric et al. 2010, Kimura et al. 2011, Kadnikov et al. 2013, Wilkins et al. 2014). The respiratory chain producing most of the ATP required for living is significantly different in anaerobic Bacteroidetes than in predominantly aerobic Flavobacteriales, as illustrated in Fig. 1. Compared with the more familiar structure of the respiratory chain of alphaproteobacteria (Degli Esposti 2014, Degli Esposti and Martinez-Romero 2016), the respiratory chain of Bacteroidetes often lacks complex I (Spero et al. 2015), containing instead alternative NADH dehydrogenase systems such as the Na-pumping NADH-MQ oxidoreductase (NQR) and its related Rnf complex (Fig. 1, cf. Meuric et al. 2010, Kadnikov et al. 2013). The same respiratory chain always lacks the cytochrome bc_1 complex or its relatives, which is sometimes compensated by the non ortholog Alternative Complex III (Act–Fig. 1, cf. González et al. 2013). Act, which appears to combine the function of menaquinol (MQH_2)-nitrite reductase with that of MQH_2-cytochrome c reductase (Refojo et al. 2010, 2013), was indeed discovered in the Bacteroidetes organism, *Rhodothermus* (Refojo et al. 2010). However, the Act complex is very rare among

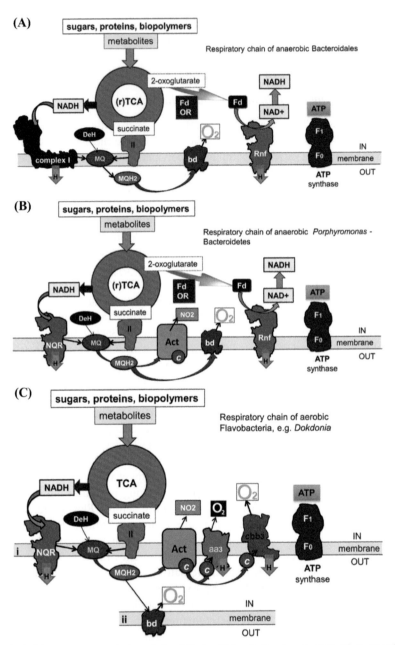

Fig. 1. A. Most common respiratory chain of Bacteroidales. B. Respiratory chain of an anaerobic Bacteroidales, *Porphyromonas gengivalies*. C. Respiratory chain of aerobic Flavobacteriales, e.g., *Dokdonia*.

anaerobic Bacteroidales, in which menaquinol is oxidised by a ubiquitous bd-I type ubiquinol oxidase (Fig. 1A, cf. Degli Esposti et al. 2015, Leclerc et al. 2015), while it is relatively common in Flavobacteriia. The main reason for the diverse distribution of the Act complex among the various classes of Bacteroidetes is that its gene cluster is generally associated with the operon of an aa3 terminal oxidase (Lamrabet et al. 2011), as discussed below and later in Chapter four.

Aerobic Bacteroidetes regularly have another terminal oxidase, the cbb3 cytochrome *c* oxidase (Fig. 1c–Meuric et al. 2010, Wilkins et al. 2014), which has a peculiar structure of its catalytic proteins (Ducluzeau et al. 2008). Whether CFB possesses a true aa3 cytochrome *c* oxidase has been considered unlikely, owing to the presence of evident cases of Lateral Gene Transfer (LGT) from other bacterial phyla (Kadnikov et al. 2013) and the previous belief that complete COX operons may be present only in Planctomycetes and proteobacteria (Degli Esposti 2014). However, detailed genomic analysis has revealed the presence of a split but complete operon for aa3 oxidase in Flavobacteriales. It appears to combine the properties of COX operon type a-I, characterized by its combination with the Act gene cluster, and those of COX operon type a-III, which contains a duplicate of the COX3 subunit (cf. Degli Esposti 2014). The catalytic subunit COX1 of this cluster generally conforms to type A1 (Pereira et al. 2001). Further details of this novel type of COX operon in Flavobacteriales will be discussed in subsequent Chapter four. Leaving aside this A family terminal oxidase, the respiratory chains of Bacteroidetes generally have two proton pumping redox systems that can build effective protonmotive force to drive ATP synthesis via the FoF1 ATP synthase (Fig. 1), the ancient enzyme defined as rotor stator ATPase in Chapter one. Additional contributions to the protonmotive force can be provided by sodium pumping membrane decarboxylases, as in anaerobic *Porphyromonas* (Meuric et al. 2010), or by light-driven proton pumping via proteorhodopsin, as in the marine flavobacterium *Dokdonia* (Kimura et al. 2011, Riedel et al. 2013).

Nitrogen metabolism

Bacteroidetes have a versatile set of systems associated with nitrogen metabolism. A recent survey of oceanic microbiomes has indicated that the CFB group expresses, besides the characteristic traits of organic compound degradation such as chitinolysis and ureolysis, the following traits of nitrogen metabolism: nitrate reduction and respiration, nitrite respiration and denitrification (Louca et al. 2016). These traits are also present among 35 CFB taxa found in a groundwater metagenome (Anantharaman et al. 2016), even if the last enzyme for denitrification is absent in most of these taxa, as in other metagenomic surveys (Fan et al. 2012, Louca et al. 2016—see also the section on the N cycle in Chapter two). The same traits, especially denitrification, are present in Bacteroidetes thriving in other environments (Johansen and Binnerup 2002, Mohan et al. 2004, Mann et al. 2013), including animal microbiota (Martens et al. 2009, Sun et al. 2010, Wolf et al. 2016).

Interestingly, some Bacteroidetes appear to fix nitrogen, as in the case of *Flavobacterium nitrogenifigens* (Kämpfer et al. 2015) and *Bacteroidetes* strains found in the guts of termites (Sakamoto and Ohkuma 2013). A survey of the distribution of genes for the subunits of the nitrogenase enzyme has indicated the presence of *nif* genes in 0.5% of the genomes of Bacteroidales, but only in seven organisms these genes group together in a conventional nitrogenase operon, therefore presumably coding for fully functional proteins (Inoue et al. 2015). This suggests that the ancient metabolic trait of N_2 fixation (Chapter one) has been initially inherited by ancestral Bacteroidales and subsequently lost in the great majority of current members of the *phylum* Bacteroidetes (Inoue et al. 2015), which sometimes just retain genomic vestiges of the diazotrophic trait (Meuric et al. 2010, Inoue et al. 2015).

Membrane lipids in Bacteroidetes and immunomodulation

Bacteroidetes include one of the few bacterial groups that synthesize sphingolipids (Sohlenkamp and Geiger 2016, Geiger et al. 2018). Some Gram-negative bacteria lack LPS and seem to have sphingolipids instead in the outer leaflet of their outer membrane (OM). Such bacteria include various members of the alpha, beta, gamma and delta class of proteobacteria and many Bacteroidetes (Geiger et al. 2018). The whole *phylum* of Bacteroidetes contains two types of sphingolipids, ceramide phosphoethanolamine and ceramide phosphoglycerol, which are generally intermixed with LPS (Kato et al. 1995, Geiger et al. 2018). Besides Sphingobacteriales such as *Pedobacter*, CFB genera with abundant presence of these lipids include: *Bacteroides*, *Prevotella* and *Mucilaginibacter* among Bacteroidales; *Flectobacyllus* and *Cytophaga* among Cytophagales, and *Porphyromonas* among Flavobacteriales (Geiger et al. 2018). Intriguingly, sphingolipids are also common in Myxococcales (deltaproteobacteria), which share with Bacteroidetes the capacity of gliding movements (Geiger et al. 2018). The likely occurrence of sphingolipids in the inner (cytoplasmic) membrane of the same organisms has not been specifically documented, so far (O. Geiger, personal communication). The bacterial biosynthesis of sphingolipids is poorly known with respect to that occurring in eukaryotes; most information concerns the first enzyme of the biosynthetic pathway, serine palmitoyltransferase (Geiger et al. 2018). Intriguingly, serine palmitoyltransferase proteins of predatory deltaproteobacteria cluster with their homologs of Bacteroidetes (Wieland Brown et al. 2013, Geiger et al. 2018), indicating a possible vertical transmission from the latter *phylum* to proteobacteria.

Functional aspects of bacterial sphingolipids have emerged from recent studies on *Bacteroides* species which are common in human gut microbiota, where they appear to modulate the innate immune response towards prokaryotic symbionts (Wieland Brown et al. 2013—reviewed by Geiger et al. 2018). Very recently, the same *Bacteroides* species forming part of children gut microbiota have been found to alter the immune response, predisposing towards autoimmune disease such as type 1 diabetes (Davis-Richardson et al. 2014, Vatanen et al. 2016). However, in this case, it is the LPS component and not the sphingolipid part of Bacteroidetes in children microbiota that is implicated in immunomodulation. This *Bacteroides*-specific LPS is significantly different from *E. coli* LPS and inhibits innate immune signaling as well as endotoxin tolerance. Moreover, the LPS of the common human symbiont/commensal, *Bacteroidetes dorei*, does not decrease the incidence of autoimmune diabetes in laboratory diabetic mice, contrary to LPS molecules from other common bacteria of human microbiota such as *E. coli* (Vatanen et al. 2016).

Notably, membrane lipids in Bacteroidetes usually comprise of significant amounts of amide-linked lipids (aminolipids) rather than ester-linked polar lipids (Sohlenkamp and Geiger 2016). Recently, ornithine lipids have been inferred to be common in Bacteroidetes, since homologs of the biosynthetic key enzyme OlsF are widely distributed among gamma-, delta- and epsilonproteobacteria, and also in the CFB group (Vences-Guzmán et al. 2015). Presumably, these amide-linked lipids are predominant in the cytoplasmic membrane of the bacteria which synthesize them. Finally, characteristic fatty acids of Bacteroidetes are 2-hydroxy and 3-hydroxy modified acids, predominantly iso-C17:0 3-OH, iso-C15:0 3-OH and iso-C16:0 3-OH (Krieg et al. 2010), whereas strains of the separate groups, which sometimes are defined as *phyla,* of Balneolaeota and Rhodothermaeota do not possess such 2-hydroxy and 3-hydroxy fatty acids (Hahnke et al. 2016).

Menaquinone (MQ) biosynthesis in Bacteroidetes proceeds via the classical as well as the futalosine pathway (Zhi et al. 2014)—see also the part of Chapter two dedicated to MQ biosynthesis.

Phylogeny of Bacteroidetes

A direct connection between the bioenergetic properties of Bacteroidetes and their phylogeny is represented by the unusual structure of catalytic subunit I of their cbb3 type cytochrome *c* oxidase (coded by the *fixN* gene), which is fused at the C terminus with subunit II (coded by the *fixO* gene) of the enzyme operon (Ducluzeau et al. 2008, Wilkins et al. 2014). Such a fusion is present also in Planctomycetes, Verrucomicrobia and aerobic predatory deltaproteobacteria, the Myxococcales (Ducluzeau et al. 2008), as will be further discussed in Chapter four. Phylogenetic trees of the fused subunit I-II of cbb3 oxidase show that this version of the enzyme complex is ancient and originated in Planctomycetes, from which it appears to have been vertically transmitted to CFB organisms (M.D.E. unpublished results). The orthologs of the same fused protein that are present in Myxococcales and Verrucomicrobia follow in two sister clades that do not match the known phylogeny of these diverse bacteria. This tree topology may suggest pervasive LGT phenomena for the gene cluster of cbb3 oxidase, which is also present in plasmids or other movable genetic elements in proteobacteria (Ducluzeau et al. 2008, Degli Esposti and Martinez-Romero 2017). However, the evolutionary history of the cbb3 oxidase is rather complex, weakening its value as a phylogenetic marker (Ducluzeau et al. 2008). Recent genome-based analysis has shown that Chitinophagia and Cytophagia are the earliest branching classes, while Flavobacteria is the latest branching class

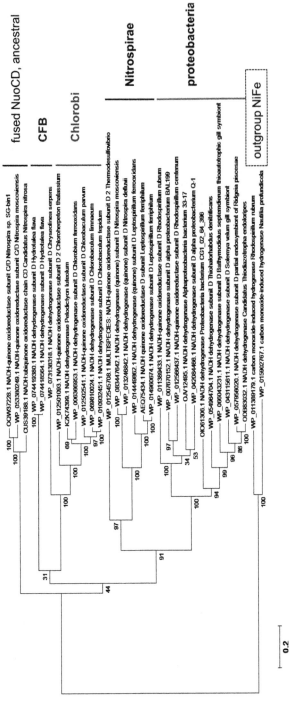

Fig. 2. Phylogenetic tree of the NuoD subunit of complex I from CFB, Chlorobi, Nitrospirae and proteobacteria. The tree was obtained with the Maximal Likelihood method using the program MEGA5 (Tamura et al. 2011) with 500 bootstraps, the percentage value of which is annotated by each node.

of Bacteroidetes (Hahnke et al. 2016). Chlorobi are often intermixed with the early branching lineages of Bacteroidetes, as shown in the representative phylogenetic tree of Fig. 2.

Figure 2 shows the Maximum likelihood (ML) tree of the NuoD subunit of complex I, which has a strong phylogenetic signal across all bacterial taxa (Degli Esposti 2015), to show the relationship between ancestral CFB and the sister *phyla* of Nitrospirae and proteobacteria. CFB organisms are clearly older than Nitrospirales and proteobacteria, clustering with photosynthetic green-sulfur bacteria (Chlorobi) as expected from taxonomic considerations (Gupta 2004, Gupta and Lorenzini 2007, Hiras et al. 2016). However, this phylogenetic trend can be seen only using appropriate CFB taxa, especially those in deep branching groups such as *Hydrotalea*, a member of the Cytophagales, because of the widespread presence of polyphyletic genera among Bacteroidetes (Hahnke et al. 2016). Indeed, the majority of Flavobacteriales and other common Bacteroidetes such as polyphyletic *Prevotella* are fast evolving (Hahnke et al. 2016) and thus do not fit the basal phylogenetic sequence that can be envisaged using a conserved protein such as the NuoD subunit of complex I, as shown in Fig. 2.

The *phylum* Nitrospirae and Related Nitrospinae

Nitrospirae is a recently introduced *phylum* of Gram negative, metabolically versatile bacteria that predominantly live in aquatic environments and are responsible for a great proportion of the nitrification process on the whole planet (Lücker et al. 2010). It contains only one class, Nitrospira, with the single order of Nitrospirales. Besides the type genus *Nitrospira*, Nitrospirae include aerobic chemolitothrophs such as the acidophilic Fe^{II}-oxidizer *Leptospirillum* and anaerobic, thermophilic sulfate reducers such as *Thermodesulfovibrio* (Garrity and Holt 2001). These species, despite their diverse physiology, are all related to *Nitrospina*, the type genus of the newly proposed *phylum* Nitrospinae (Lücker et al. 2013, Spieck et al. 2014). Whether Nitrospinae constitute a *bona fide* separate *phylum* or just a branch of the Nitrospirae remains unclear.

The first *Nitrospira* organism was isolated and characterized for its capacity of carrying out the second step of nitrification, the oxidation of nitrite to nitrate (Watson et al. 1986). The same capacity was later associated with Nitrospinae (Lücker et al. 2013, Spieck et al. 2014). To date, the biochemical capacity of nitrite oxidation is restricted to Nitrospirae, Nitrospinae and two genera of proteobacteria: the alphaproteobacterium *Nitrobacter* and the gammaproteobacterium *Nitrococcus* (Vlaeminck et al. 2011). Cumulatively, these organisms are called NOB, for nitrite-oxidizing bacteria (Vlaeminck et al. 2011). By far, Nitrospirales constitute the most abundant and biotechnologically relevant NOB for they are the predominant nitrite oxidizers in wastewater treatment plants also (Lücker et al. 2010, Lau et al. 2014). Contrary to *Nitrobacter*, *Nitrospira* organisms are the dominant NOB under oxygen-limited conditions (Vlaeminck et al. 2011, Lau et al. 2014, Koch et al. 2015) because they have a versatile metabolism capable of producing bioenergy at low levels of ambient oxygen, namely under micro-oxic conditions (Lücker et al. 2010, Koch et al. 2015), as discussed below.

Functional Properties of Nitrospirae and Nitrospinae

Respiratory chain and energy metabolism

The respiratory chain of Nitrospirae is complex, reflecting the metabolic versatility of the *phylum* and the adaptation of its species to a variety of ecological niches with different levels of oxygen (Fig. 3A, cf. Lücker et al. 2010). Effectively, various Nitrospirae can be considered facultatively anaerobic chemolithotrophs, since they contain only one type of terminal oxidases, the bd ubiquinol oxidase (Lücker et al. 2010), which is probably ancestral to the CIO (Cyanide Insensitive Oxidase) type of the same enzyme that is present in proteobacteria (Degli Esposti et al. 2015). Blast (Basic Local Alignment Search Tool) searches have identified the presence of the catalytic subunit of the cbb3 cytochrome *c* oxidase, another terminal oxidase with high affinity for oxygen (Chapter two), in a few metagenomic Nitrospirae

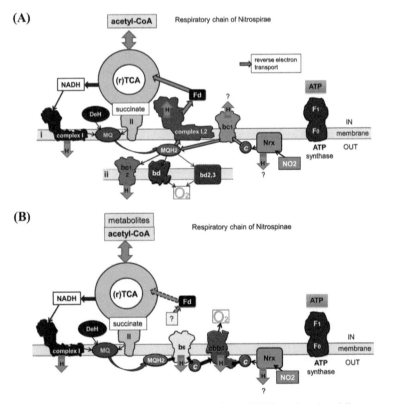

Fig. 3. Respiratory chain of Nitrospirae and Nitrospinae. (A) Nitrospirae. i and ii represent two sections of the same cytoplasmic membrane. (B) Nitrospinae.

such as Nitrospirae bacterium GWD2_57_9 from a subsurface metagenome (Anantharaman et al. 2016). The genes for these proteins are related to a previously reported protein of *Nitrospina gracilis* containing subunit I, I and III fused together (*fixNOP* of cbb3 oxidase–Lücker et al. 2013), which resembles the *fixN-O* fused genes of Bacteroidetes already discussed in this chapter (cf. Ducluzeau et al. 2008). Similar trifunctional fused proteins are found in very few proteobacteria, such as Gallionellales bacterium GWA2_59_43 (subsurface metagenome) and *Candidatus* Tenderia electrophaga, an uncultivated gammaproteobacterium isolated from a biocathode enrichment. Even if the scattered distribution of these fused cbb3 oxidases suggests possible instances of LGT, *Nitrospina* has been reported to constitutively harbor a functional cbb3 oxidase (Lücker et al. 2013), thereby introducing a major difference from the respiratory chain of Nitrospirae (Fig. 3), which fundamentally has only terminal oxidases of the bd type (Lücker et al. 2010). In *Nitrospina*, the cbb3 complex re-oxidizes reduced *c* cytochromes associated with the enzyme for nitrite oxidation, Nxr (Lücker et al. 2013). The same reaction cannot be carried out by *Nitrospira*, given the above mentioned absence of cytochrome *c* oxidases and the unlikely possibility that one of its four bd-type oxidases may function as a cytochrome *c* oxidase too (Lücker et al. 2010)—a possibility that has neither structural nor functional significance, considering the recent three-dimensional structure of the bd oxidase (Safarian et al. 2016). Intriguingly, there is no functional bd ubiquinol oxidase in *Nitrospina* (Fig. 3B), since subunit II of the enzyme complex is missing from its genome. However, magnetotactic Nitrospirae do have cytochrome *c* oxidases of the A and B family, as will be discussed in Chapter seven.

An alternative route for the re-oxidation of the *c* cytochromes accepting electrons from nitrite oxidation is reverse electron transport via the bc1 complex, which is coded by two different operons in the genome of *Nitrospira* (Lücker et al. 2010). One of these operons has the classical sequence Rieske iron-sulfur protein (ISP)–cytochrome *b*–cytochrome *c*1–as the *petabc* operon of photosynthetic alphaproteobacteria (Degli Esposti 2014—see also Chapter four). Its cytochrome *b* contains a fused *c* cytochrome at its C-terminus, an ancestral condition that probably originated in Planctomycetes (Degli Esposti 2014). The other operon has a different gene sequence and a short cytochrome *b* protein that may well be responsible for reverse electron transport to menaquinone (MQ), thereby producing menaquinol (MQH$_2$) which could be re-oxidized by one of the two types of complex I that are present in the genome of many Nitrospirae, working in reverse too, as hypothesized earlier (Lücker et al. 2010). The nuo13 type of complex I is characterized by the fusion of the NuoC with the NuoD subunit and an additional iron sulfur center, an ancestral situation (see Fig. 2) which renders the complex more apt to work in reverse electron transport under anaerobic conditions (Degli Esposti 2015, Spero et al. 2015). Of note is the fact that the genome of *Nitrospina* contains instead a paralogue of the bc1 complex, which is structurally equivalent to the b6f complex of plastids, and one complete operon for nuo14 complex I (Lücker et al. 2013, cf. Spero et al. 2015). Another operon for complex I is present in the genome of *Nitrospina* but is split and lacks the FMN binding site; it has nevertheless been considered to carry out reverse electron transport from menaquinol to ferredoxin (Lücker et al. 2013), like in ancestral forms of complex I (Degli Esposti 2015, Degli Esposti and Martinez-Romero 2016). This hypothetical function could help producing reduced ferredoxin to drive CO_2 fixation under anaerobic conditions via the rTCA cycle, which is present in *Nitrospina* (Lücker et al. 2013) as well as in *Nitrospira* (Lücker et al. 2010, Kolinko et al. 2016). Indeed, the genome of *Nitrospira* codes for all the elements of the rTCA cycle, which is considered to be the main route of CO_2 fixation in this organism (Lücker et al. 2010), similar to Chlorobi (Buchanan and Arnon 1990).

The enzymes responsible for producing reduced ferredoxin to drive the reductive action of the rTCA cycle have not been defined in *Nitrospira* (Lücker et al. 2010) but were later found as part of the rTCA in magnetotactic Nitrospirae (Kolinko et al. 2016). One of the two types of complex I present in *Nitrospira* may be able to reduce ferredoxin via reverse electron transfer, as suggested for *Nitrospina* (Lücker et al. 2013), or directly from NADH as postulated for so-called green complex I in Rhizobi (Degli Esposti and Martinez-Romero 2016). Ferredoxin can be oxidized, or reduced at the expense of protonmotive force, via another NADH-reacting enzyme, the Rnf complex (Biegel et al. 2011), which is not present in cultivated strains of *Nitrospira* (Kits et al. 2017). However, recent metagenomic studies have uncovered the presence of this complex in several unclassified Nitrospirae (Baker et al. 2015), plus strains of the magnetotactic *Ca.* Magnetobacterium (M.D.E. unpublished results), thereby expanding the bioenergetic portfolio of Nitrospirae. Figure 3 illustrates the deduced respiratory chain of *Nitrospira* and *Nitrospina* and their protonmotive steps. See Chapter seven for additional information regarding the terminal oxidases of magnetotactic Nitrospirae in relation to those of ancestral alphaproteobacteria.

Other metabolic traits of Nitrospirae

The metabolic capacity of Nitrospirae has been extended further by the recent finding that marine *Nitrospira* organisms can carry out complete nitrification, namely ammonia oxidation to nitrite coupled with nitrite oxidation to nitrate (Daims et al. 2015, Kits et al. 2017). Furthermore, metabolic analysis of metagenomic Nitrospirae has shown the presence of various traits associated with the sulfur cycle, in particular reduction of polysulfides to H_2S (Anantharaman et al. 2016, Frank et al. 2016) and sulfide oxidation (Kolinko et al. 2016)—see more on this topic in chapter four. The magnetotactic properties of some Nitrospirae are also well known (Jogler et al. 2011, Kolinko et al. 2016, Lin et al. 2017), while ureolysis has only been recently uncovered in a ubiquitous *Nitrospira* organism which is also capable of aerobic oxidation of H_2 (Koch et al. 2015). The latter capacity produces additional protonmotive reactions that can help balancing the energy demand for reverse electron transport (cf. Fig. 3A). The genome of various genera of Nitrospirae, including *Thermodesulfovibrio, Leptospirillum, Ca.* Magnetoovum and unclassified organisms have *nifD* and other genes for the subunits of nitrogenase, thereby potentially being able to fix N_2 (Frank et al. 2016, Kolinko et al. 2016, Lin et al. 2017). The majority of these organisms do not live on nitrite oxidation but have other chemolithotrophic traits to sustain their bioenergy production (Frank et al. 2016, Kolinko et al. 2016). However, there is no biochemical characterization of N_2 fixation for any member of the Nitrospirae, so far. On a final note, the respiratory chain of the sulfate reducer *Thermodesulfovibrio* is much more similar to physiologically equivalent deltaproteobacteria such as *Desulfovibrio* than to that of *Nitrospira* (Fig. 3A), as will be further discussed in Chapter four.

Phylogeny of Nitrospirae and Nitrospinae

The sister *phyla* of Nitrospirae and Nitrospinae are mainly based on phylogenetic considerations (Garrity and Holt 2001, Lücker et al. 2013, Spieck et al. 2014). However, bioenergetic systems of these bacteria appear to be intermediate between those of CFB and those of proteobacteria (cf. Fig. 2); their proteins also appear to be ancestral to their homologs from proteobacteria. For instance, Nitrospirae contain the first operon for the bc1 complex outside proteobacteria, with a cytochrome *b* subunit that lies in the precursor position of phylogenetic trees (Degli Esposti 2014). Interestingly, the respiratory chain of *Nitrospira* as depicted in Fig. 3A closely resembles that of nodulating symbionts of alphaproteobacterial Rhizobi, with green complex I supplying reduced ferredoxin for N_2 fixation (Degli Esposti and Martinez-Romero 2016).

Systematic phylogenetic analysis of all proteins coded by *Nitrospina* has shown a mosaic profile of its genome; many proteins branch with deltaproteobacteria, followed by Nitrospirae, gammaproteobacteria and Planctomycetes (Lücker et al. 2013). A similar chimeric situation has been reported for the genome of a *Nitrospira* (Koch et al. 2015) and is reminiscent of that found in deep branching Magnetococcales of the alphaproteobacteria class (Ji et al. 2017—see also Chapter seven). Magnetotactic traits are indeed shared by some Nitrospirae and alphaproteobacteria of the Magnetococcales and Rhodospirillales order (Ji et al. 2017, Lin et al. 2017), while the trait of nitrite oxidation is shared between Nitrospirae, Nitrospinae and two genera of proteobacteria, as already mentioned (Vlaeminck et al. 2011). The other trait of nitrification, ammonia oxidation to nitrite, is shared instead by some Nitrospirae (Daims et al. 2015) and a few proteobacteria of the gamma and beta class (Vlaeminck et al. 2011). Moreover, the rare trait of chemolithotrophic iron oxidation is shared by *Leptospirillum* of Nitrospirae, zetaproteobacteria and the Gallionellales order of betaproteobacteria (Ilbert and Bonnefoy 2013). All these shared traits render Nitrospirae the closest *phylum* to proteobacteria, as also indicated by various phylogenetic trees (Fig. 2, cf. Degli Esposti et al. 2015, Lin et al. 2017).

To conclude, the bacteria examined in this chapter contain ancestral features in their respiratory chain and central metabolism that overall constitute a preliminary step along the evolution of prokaryotes from LUCA (chapter one) to proto-mitochondria (Chapter seven). Basically, all the metabolic traits present in CFB and Nitrospirae will be found later in the bacteria described in subsequent chapters of the book. The beginning of aerobic metabolism can be seen in some classes of the CFB *phylum* (Fig. 1) and groups of the Nitrospira *phylum* (Fig. 3), paving the way for the subsequent evolution of the respiratory chain of mitochondrial ancestors.

Acknowledgements

I thank various colleagues at CCG (Cuernavaca, Mexico) for discussion and support. Among these colleagues, Otto Geiger is especially thanked for his willingness to discuss at length various issues on bacterial lipids.

References

Anantharaman, K., C.T. Brown, L.A. Hug, I. Sharon, C.J. Castelle, A.J. Probst et al. 2016. Thousands of microbial genomes shed light on interconnected biogeochemical processes in an aquifer system. Nat. Commun. 7: 13219. doi:10.1038/ncomms13219.

Antunes, L.C., D. Poppleton, A. Klingl, A. Criscuolo, B. Dupuy, C. Brochier-Armanet et al. 2016. Phylogenomic analysis supports the ancestral presence of LPS-outer membranes in the Firmicutes. Elife. 5. pii: e14589. doi:10.7554/eLife.14589.

Baker, B.J., C.S. Lazar, A.P. Teske and G.J. Dick. 2015. Genomic resolution of linkages in carbon, nitrogen, and sulfur cycling among widespread estuary sediment bacteria. Microbiome 3: 14. doi:10.1186/s40168-015-0077-6.

Biegel, E., S. Schmidt, J.M. González and V. Müller. 2011. Biochemistry, evolution and physiological function of the Rnf complex, a novel ion-motive electron transport complex in prokaryotes. Cell. Mol. Life Sci. 68: 613–634. doi:10.1007/s00018-010-0555-8.

Buchanan, B.B. and D.I. Arnon. 1990. A reverse KREBS cycle in photosynthesis: Consensus at last. Photosynth. Res. 24: 47–53.

Campbell, B.J. and S.C. Cary. 2004. Abundance of reverse tricarboxylic acid cycle genes in free-living microorganisms at deep-sea hydrothermal vents. Appl. Environ. Microbiol. 70: 6282–6289.

Costalonga, M. and M.C. Herzberg. 2014. The oral microbiome and the immunobiology of periodontal disease and caries. Immunol. Lett. 162: 22–38. doi:10.1016/j.imlet.2014.08.017.

Daims, H., E.V. Lebedeva, P. Pjevac, P. Han, C. Herbold, M. Albertsen et al. 2015. Complete nitrification by Nitrospira bacteria. Nature 528: 504–509. doi:10.1038/nature16461.

Davis-Richardson, A.G., A.N. Ardissone, R. Dias, V. Simell, M.T. Leonard, K.M. Kemppainen et al. 2014. *Bacteroides dorei* dominates gut microbiome prior to autoimmunity in Finnish children at high risk for type 1 diabetes. Front. Microbiol. 5: 678. doi:10.3389/fmicb.2014.00678.

Degli Esposti, M. 2014. Bioenergetic evolution in proteobacteria and mitochondria. Genome Biol. Evol. 6: 3238–3251. doi:10.1093/gbe/evu257.

Degli Esposti, M., T. Rosas-Pérez, L.E. Servín-Garcidueñas, L.M. Bolaños, M. Rosenblueth and E. Martínez-Romero. 2015. Molecular evolution of cytochrome bd oxidases across proteobacterial genomes. Genome Biol. Evol. 7: 801–820. doi:10.1093/gbe/evv032.

Degli Esposti, M. 2015. Genome analysis of structure-function relationships in respiratory complex I, an ancient bioenergetic enzyme. Genome Biol. Evol. 8: 126–147. doi:10.1093/gbe/evv239.

Degli Esposti, M. and E. Martinez Romero. 2016. A survey of the energy metabolism of nodulating symbionts reveals a new form of respiratory complex I. FEMS Microbiol. Ecol. 92: fiw084. doi:10.1093/femsec/fiw084.

Ducluzeau, A.L., S. Ouchane and W. Nitschke. 2008. The cbb3 oxidases are an ancient innovation of the domain bacteria. Mol. Biol. Evol. 25: 1158–1166. doi: 10.1093/molbev/msn062.

Fan, L., D. Reynolds, M. Liu, M. Stark, S. Kjelleberg, N.S. Webster et al. 2012. Functional equivalence and evolutionary convergence in complex communities of microbial sponge symbionts. Proc. Natl. Acad. Sci. USA 109: E1878–E1887. doi:10.1073/pnas.1203287109.

Frank, Y.A., V.V. Kadnikov, A.P. Lukina, D. Banks, A.V. Beletsky, A.V. Mardanov et al. 2016. Characterization and genome analysis of the first facultatively Alkaliphilic *Thermodesulfovibrio* isolated from the deep terrestrial subsurface. Front. Microbiol. 7: 2000. doi:10.3389/fmicb.2016.02000.

Garrity, G.M. and J.G. Holt. 2001. Phylum BVIII. Nitrospirae phy. nov. pp. 451–464. *In*: Bergey's Manual of Systematic Bacteriology. Springer, New York.

Geiger, O., J. Padilla-Gómez and I.M. López-Lara. 2018. Bacterial sphingolipids and sulfonolipids. *In*: Geiger, O. (ed.). Biogenesis of Fatty Acids, Lipids and Membranes. Springer International Publishing AG 2018. doi:10.1007/978-3-319-43676-0_12-1.

González, J.M., J. Pinhassi, B. Fernández-Gómez, M. Coll-Lladó, M. González-Velázquez, P. Puigbò et al. 2011. Genomics of the proteorhodopsin-containing marine flavobacterium *Dokdonia* sp. strain MED134. Appl. Environ. Microbiol. 77(24): 8676–8686. doi:10.1128/AEM.06152-11.

Greening, C., A. Biswas, C.R. Carere, C.J. Jackson, M.C. Taylor, M.B. Stott et al. 2016. Genomic and metagenomic surveys of hydrogenase distribution indicate H2 is a widely utilised energy source for microbial growth and survival. ISME J. 10: 761–777. doi:10.1038/ismej.2015.153.

Gupta, R.S. 2004. The phylogeny and signature sequences characteristics of Fibrobacteres, Chlorobi, and Bacteroidetes. Crit. Rev. Microbiol. 30: 123–143.

Gupta, R.S. and E. Lorenzini. 2007. Phylogeny and molecular signatures (conserved proteins and indels) that are specific for the Bacteroidetes and Chlorobi species. BMC Evol. Biol. 7: 71.

Hahnke, R.L., J.P. Meier-Kolthoff, M. García-López, S. Mukherjee, M. Huntemann, N.N. Ivanova et al. 2016. Genome-based taxonomic classification of bacteroidetes. Front. Microbiol. 7: 2003. doi:10.3389/fmicb.2016.02003.

Hamilton, T.L., R.J. Bovee, S.R. Sattin, W. Mohr, W.P. Gilhooly, 3rd, T.W. Lyons et al. 2016. Carbon and Sulfur Cycling below the Chemocline in a Meromictic Lake and the Identification of a Novel Taxonomic Lineage in the FCB Superphylum, Candidatus Aegiribacteria. Front. Microbiol. 7: 598. doi:10.3389/fmicb.2016.00598.

Hiras, J., Y.W. Wu, S.A. Eichorst, B.A. Simmons and S.W. Singer. 2016. Refining the phylum Chlorobi by resolving the phylogeny and metabolic potential of the representative of a deeply branching, uncultivated lineage. ISME J. 10: 833–845. doi:10.1038/ismej.2015.158.

Hongoh, Y., V.K. Sharma, T. Prakash, S. Noda, H. Toh, T.D. Taylor et al. 2008. Genome of an endosymbiont coupling N2 fixation to cellulolysis within protist cells in termite gut. Science 322: 1108–1109. doi:10.1126/science.1165578.

Ilbert, M. and V. Bonnefoy. 2013. Insight into the evolution of the iron oxidation pathways. Biochim. Biophys. Acta. 1827: 161–175. doi:10.1016/j.bbabio.2012.10.001.

Inoue, J., K. Oshima, W. Suda, M. Sakamoto, T. Iino, S. Noda et al. 2015. Distribution and evolution of nitrogen fixation genes in the phylum Bacteroidetes. Microbes Environ. 30: 44–50. doi:10.1264/jsme2.ME14142.

Ji, B., S.D. Zhang, W.J. Zhang, Z. Rouy, F. Alberto, C.L. Santini et al. 2017. The chimeric nature of the genomes of marine magnetotactic coccoid-ovoid bacteria defines a novel group of Proteobacteria. Environ. Microbiol. 19: 1103–1119. doi:10.1111/1462-2920.13637.

Jogler, C., G. Wanner, S. Kolinko, M. Niebler, R. Amann, N. Petersen et al. 2011. Conservation of proteobacterial magnetosome genes and structures in an uncultivated member of the deep-branching Nitrospira phylum. Proc. Natl. Acad. Sci. USA 108: 1134–1139. doi:10.1073/pnas.1012694108.

Johansen, J.E. and S.J. Binnerup. 2002. Contribution of cytophaga-like bacteria to the potential of turnover of carbon, nitrogen, and phosphorus by bacteria in the rhizosphere of barley (*Hordeum vulgare* L.). Microb. Ecol. 43: 298–306.

Kadnikov, V.V., A.V. Mardanov, O.A. Podosokorskaya, S.N. Gavrilov, I.V. Kublanov, A.V. Beletsky et al. 2013. Genomic analysis of *Melioribacter roseus*, facultatively anaerobic organotrophic bacterium representing a novel deep lineage within Bacteriodetes/Chlorobi group. PLoS One 8: e53047. doi:10.1371/journal.pone.0053047.

Kämpfer, P. 2015. Sphingobacteriales ord. nov. Bergey's manual of systematics of archaea and bacteria. 1. John Wiley & Sons, Inc., in association with Bergey's Manual Trust.

Kämpfer, P., H.J. Busse, J.A. McInroy, J. Xu and S.P. Glaeser. 2015. *Flavobacterium nitrogenifigens* sp. nov., isolated from switchgrass (*Panicum virgatum*). Int. J. Syst. Evol. Microbiol. 65: 2803–2809. doi:10.1099/ijs.0.000330.

Kato, M., Y. Muto, K. Tanaka-Bandoh, K. Watanabe and K. Ueno. 1995. Sphingolipid composition in Bacteroides species. Anaerobe 1: 135–139.

Karlsson, F.H., D.W. Ussery, J. Nielsen and I. Nookaew. 2011. A closer look at bacteroides: phylogenetic relationship and genomic implications of a life in the human gut. Microb. Ecol. 61: 473–485. doi:10.1007/s00248-010-9796-1.

Kimura, H., C.R. Young, A. Martinez and E.F. Delong. 2011, Light-induced transcriptional responses associated with proteorhodopsin-enhanced growth in a marine flavobacterium. ISME J. 5: 1641–1651. doi:10.1038/ismej.2011.36.

Kits, K.D., C.J. Sedlacek, E.V. Lebedeva, P. Han, A. Bulaev, P. Pjevac et al. 2017. Kinetic analysis of a complete nitrifier reveals an oligotrophic lifestyle. Nature 549: 269–272. doi:10.1038/nature23679.

Koch, H., S. Lücker, M. Albertsen, K. Kitzinger, C. Herbold, E. Spieck et al. 2015. Expanded metabolic versatility of ubiquitous nitrite-oxidizing bacteria from the genus Nitrospira. Proc. Natl. Acad. Sci. USA 112: 11371–11376. doi:10.1073/pnas.1506533112.

Kolinko, S., M. Richter, F.O. Glöckner, A. Brachmann and D. Schüler. 2016. Single-cell genomics of uncultivated deep-branching magnetotactic bacteria reveals a conserved set of magnetosome genes. Environ. Microbiol. 18: 21–37. doi:10.1111/1462-2920.12907.

Krieg, N.R., W. Ludwig, J. Euzeby and W.B. Whitman. 2010. Phylum XIV. Bacteroidetes phyl. nov. pp. 25–470. *In*: Boone, D.R. and R.W. Castenholz (eds.). Bergey's Manual of Systematic Bacteriology, 2nd edition. Vol. 4. Springer, New York.

Lamrabet, O., L. Pieulle, C. Aubert, F. Mouhamar, P. Stocker, A. Dolla and G. Brasseur. 2011. Oxygen reduction in the strict anaerobe Desulfovibrio vulgaris Hildenborough: characterization of two membrane-bound oxygen reductases. Microbiology 157: 2720–2732. doi:10.1099/mic.0.049171-0.

Lau, M.C., C. Cameron, C. Magnabosco, C.T. Brown, F. Schilkey, S. Grim et al. 2014. Phylogeny and phylogeography of functional genes shared among seven terrestrial subsurface metagenomes reveal N-cycling and microbial evolutionary relationships. Front. Microbiol. 5: 531. doi:10.3389/fmicb.2014.00531.

Leclerc, J., E. Rosenfeld, M. Trainini, B. Martin, V. Meuric, M. Bonnaure-Mallet et al. 2015. The Cytochrome bd oxidase of porphyromonas gingivalis contributes to oxidative stress resistance and dioxygen tolerance. PLoS One 10: e0143808. doi:10.1371/journal.pone.0143808.

Lin, W., G.A. Paterson, Q. Zhu, Y. Wang, E. Kopylova, Y. Li et al. 2017. Origin of microbial biomineralization and magnetotaxis during the Archean. Proc. Natl. Acad. Sci. USA 114: 2171–2176. doi:10.1073/pnas.1614654114.

Louca, S., L.W. Parfrey and M. Doebeli. 2016. Decoupling function and taxonomy in the global ocean microbiome. Science 353: 1272–1277. doi:10.1126/science.aaf4507.

Lücker, S., M. Wagner, F. Maixner, E. Pelletier, H. Koch, B. Vacherie et al. 2010. A Nitrospira metagenome illuminates the physiology and evolution of globally important nitrite-oxidizing bacteria. Proc. Natl. Acad. Sci. USA 107: 13479–13484. doi:10.1073/pnas.1003860107.

Lücker, S., B. Nowka, T. Rattei, E. Spieck and H. Daims. 2013. The genome of *Nitrospina gracilis* illuminates the metabolism and evolution of the major marine nitrite oxidizer. Front. Microbiol. 4: 27. doi:0.3389/fmicb.2013.00027.

Mann, A.J., R.L. Hahnke, S. Huang, J. Werner, P. Xing, T. Barbeyron et al. 2013. The genome of the alga-associated marine flavobacterium *Formosa agariphila* KMM 3901T reveals a broad potential for degradation of algal polysaccharides. Appl. Environ. Microbiol. 79: 6813–6822. doi:10.1128/AEM.01937-13.

Martens, E.C., N.M. Koropatkin, T.J. Smith and J.I. Gordon. 2009. Complex glycan catabolism by the human gut microbiota: the Bacteroidetes Sus-like paradigm. J. Biol. Chem. 284: 24673–24677. doi:10.1074/jbc.R109.022848.

Meuric, V., A. Rouillon, F. Chandad and M. Bonnaure-Mallet. 2010. Putative respiratory chain of *Porphyromonas gingivalis*. Future Microbiol. 5: 717–734. doi:10.2217/fmb.10.32.

Mohan, S.B., M. Schmid, M. Jetten and J. Cole. 2004. Detection and widespread distribution of the nrfA gene encoding nitrite reduction to ammonia, a short circuit in the biological nitrogen cycle that competes with denitrification. FEMS Microbiol. Ecol. 49: 433–443. doi:10.1016/j.femsec.2004.04.012.

Nelson, K.E., R.D. Fleischmann, R.T. DeBoy, I.T. Paulsen, D.E. Fouts, J.A. Eisen et al. 2003. Complete genome sequence of the oral pathogenic Bacterium porphyromonas gingivalis strain W83. Bacteriol. 185(18): 5591–601.

Nibali, L. 2015. Aggressive Periodontitis: microbes and host response, who to blame? Virulence. 6: 223–228. doi:10.4161/21505594.2014.986407.

Pereira, M.M., M. Santana and M. Teixeira. 2001. A novel scenario for the evolution of haem-copper oxygen reductases. Biochim. Biophys. Acta 1505: 185–208.

Peros, W.J. and R.J. Gibbons. 1982. Influence of sublethal antibiotic concentrations on bacterial adherence to saliva-treated hydroxyapatite. Infect. Immun. 35: 326–334.

Piau, C., C. Arvieux, M. Bonnaure-Mallet and A. Jolivet-Gougeon. 2013. *Capnocytophaga* spp. involvement in bone infections: a review. Int. J. Antimicrob. Agents 41: 509–515. doi:10.1016/j.ijantimicag.2013.03.001.

Refojo, P.N., F.L. Sousa, M. Teixeira and M.M. Pereira. 2010. The alternative complex III: a different architecture using known building modules. Biochim. Biophys. Acta. 1797: 1869–1876.

Refojo, P.N., M.A. Ribeiro, F. Calisto, M. Teixeira and M.M. Pereira. 2013. Structural composition of alternative complex III: variations on the same theme. Biochim. Biophys. Acta. 1827: 1378–1382. doi:10.1016/j.bbabio.2013.01.001.

Riedel, T., L. Gómez-Consarnau, J. Tomasch, M. Martin, M. Jarek, J.M. González et al. 2013. Genomics and physiology of a marine flavobacterium encoding a proteorhodopsin and a xanthorhodopsin-like protein. PLoS One 8: e57487. doi:10.1371/journal.pone.0057487.

Safarian, S., C. Rajendran, H. Müller, J. Preu, J.D. Langer, S. Ovchinnikov et al. 2016. Structure of a bd oxidase indicates similar mechanisms for membrane-integrated oxygen reductases. Science 352: 583–586. doi:10.1126/science.aaf2477.

Sakamoto, M. and M. Ohkuma. 2013. *Bacteroides reticulotermitis* sp. nov., isolated from the gut of a subterranean termite (Reticulitermes speratus). Int. J. Syst. Evol. Microbiol. 63: 691–695. doi:10.1099/ijs.0.040931-0.

Shrivastava, A. and H.C. Berg. 2015. Towards a model for Flavobacterium gliding. Curr. Opin. Microbiol. 28: 93–97. doi:10.1016/j.mib.2015.07.018.

Sohlenkamp, C. and O. Geiger. 2016. Bacterial membrane lipids: diversity in structures and pathways. FEMS Microbiol. Rev. 40: 133–159. doi:10.1093/femsre/fuv008.

Spero, M.A., F.O. Aylward, C.R. Currie and T.J. Donohue. 2015. Phylogenomic analysis and predicted physiological role of the proton-translocating NADH:quinone oxidoreductase (complex I) across bacteria. MBio. 6: pii: e00389–15. doi:10.1128/mBio.00389-15.

Spieck, E., S. Keuter, T. Wenzel, E. Bock and W. Ludwig. 2014. Characterization of a new marine nitrite oxidizing bacterium, Nitrospina watsonii sp. nov., a member of the newly proposed phylum "Nitrospinae". Syst. Appl. Microbiol. 37: 170–176. doi:10.1016/j.syapm.2013.12.005.

Sun, W., J. Wu, L. Lin, Y. Huang, Q. Chen and Y. Ji. 2010. *Porphyromonas gingivalis* stimulates the release of nitric oxide by inducing expression of inducible nitric oxide synthases and inhibiting endothelial nitric oxide synthases. J. Periodontal. Res. 45: 381–388. doi:10.1111/j.1600-0765.2009.01249.x.

Tamura, K., D. Peterson, N. Peterson, G. Stecher, M. Nei and S. Kumar. 2011. MEGA5: Molecular evolutionary genetics analysis using maximum likelihood, evolutionary distance, and maximum parsimony methods. Mol. Biol. Evol. 28: 2731–2739. doi:10.1093/molbev/msr121.

Thomas, F., J.H. Hehemann, E. Rebuffet, M. Czjzek and G. Michel. 2011. Environmental and gut bacteroidetes: the food connection. Front. Microbiol. 2: 93. doi:10.3389/fmicb.2011.00093.

Vatanen, T., A.D. Kostic, E. d'Hennezel, H. Siljander, E.A. Franzosa, M. Yassour et al. 2016. Variation in microbiome LPS immunogenicity contributes to autoimmunity in humans. Cell. 165: 842–853. doi:10.1016/j.cell.2016.04.007.

Vences-Guzmán, M.Á., Z. Guan, W.I. Escobedo-Hinojosa, J.R. Bermúdez-Barrientos, O. Geiger and C. Sohlenkamp. 2015. Discovery of a bifunctional acyltransferase responsible for ornithine lipid synthesis in *Serratia proteamaculans*. Environ. Microbiol. 17: 1487–1496. doi:10.1111/1462-2920.12562.

Vlaeminck, S.E., A.G. Hay, L. Maignien and W. Verstraete. 2011. In quest of the nitrogen oxidizing prokaryotes of the early Earth. Environ. Microbiol. 13: 283–295. doi:10.1111/j.1462-2920.2010.02345.x.

Watson, S.W., E. Bock, F.W. Valois, J.B. Waterbury and U. Schlosser. 1986. *Nitrospira marina* gen. nov. sp. nov.: a chemolithotrophic nitrite-oxidizing bacterium. Arch. Microbiol. 144: 1–7. doi:10.1007/BF00454947.

Wieland Brown, L.C., C. Penaranda, P.C. Kashyap, B.B. Williams, J. Clardy, M. Kronenberg et al. 2013. Production of α-galactosyl ceramide by a prominent member of the human gut microbiota. PLoS Biology 11: e1001610. doi:10.1371/journal.pbio.1001610.

Wilkins, M.J., D.W. Kennedy, C.J. Castelle, E.K. Field, R. Stepanauskas, J.K. Fredrickson et al. 2014. Single-cell genomics reveals metabolic strategies for microbial growth and survival in an oligotrophic aquifer. Microbiology 160: 362–372. doi:10.1099/mic.0.073965-0.

Woese, C.R. 1987. Bacterial evolution. Microbiol. Rev. 51: 221–271.

Wolf, P.G., A. Biswas, S.E. Morales, C. Greening and H.R. Gaskins. 2016. H2 metabolism is widespread and diverse among human colonic microbes. Gut. Microbes. 7: 235–245. doi:10.1080/19490976.2016.1182288.

Zhi, X.Y., J.C. Yao, S.K. Tang, Y. Huang, H.W. Li and W.J. Li. 2014. The futalosine pathway played an important role in menaquinone biosynthesis during early prokaryote evolution. Genome Biol. Evol. 6: 149–160. doi:10.1093/gbe/evu007.

4

Proteobacteria: From Anaerobic to Aerobic Organisms

Mauro Degli Esposti

Proteobacteria, an Introduction

This chapter will present the proteobacteria, a vast *phylum* of Gram negative prokaryotes from which proto-mitochondria originated. The term proteobacteria was introduced by Woese (1987) and then defined taxonomically by Stackebrandt et al. (1988). Previously, the diverse Gram negative bacteria, now classified as proteobacteria, were known as 'purple bacteria and their relatives' (Woese 1987, Stackebrandt et al. 1988, Gupta 2000), for they were traditionally considered to derive from, or be related to, photosynthetic purple bacteria such as *Rhodobacter* (see Chapter seven). The name 'proteobacteria' stands for 'protean group of bacteria of diverse properties', from the mythological god Proteus, capable of assuming many different shapes (Stackebrandt et al. 1988). Initially, proteobacteria comprised the four subdivisions of alpha, beta, gamma and delta (Woese 1987, Stackebrandt et al. 1988). Subsequently, epsilonproteobacteria (Tenover et al. 1992, Campbell et al. 2006) and zetaproteobacteria (Emerson et al. 2007) were added to the *phylum*. More recently, two additional classes have been proposed within proteobacteria: Acidithiobacillia, derived from acidophiles previously classified as gammaproteobacteria (Williams and Kelly 2013; see Chapter five), and Oligoflexia, comprising environmental organisms together with predatory bacteria previously classified in the delta class (Nakai et al. 2016, Hahn et al. 2017). Presently, the

Center for Genomic Sciences, UNAM Campus de Cuernavaca, Cuernavaca, 62130 Morelos, Mexico.
Email: mauro1italia@gmail.com

consistency and phylogenetic robustness of the last proposed classes is uncertain; hence, the *phylum* proteobacteria can be conservatively considered to contain the six classes listed in Table 1.

At the time of their definition, three decades ago, proteobacteria were the largest division within prokaryotes and included many of the Gram-negative bacteria then known (Woese 1987, Stackebrandt et al. 1988, Gupta 2000). Today, thanks to the explosion of metagenomic sequences, new bacterial organisms are constantly reported, so that the phylogenetic diversity of prokaryotes has increased enormously (Hug et al. 2016, Schulz et al. 2017). Yet, the cumulative number of proteobacterial genomes is by far the largest of any prokaryotic *phylum* (Rinke et al. 2013, Land et al. 2015, Schulz et al. 2017—Table 1). Indeed, proteobacteria remain among the most common prokaryotes in the majority of earth habitats, except for the microbiota of humans and vertebrates, which are dominated by Bacteroidetes, Firmicutes and Actinomycetes (Quin et al. 2010, Karlsson et al. 2011, Colston and Jackson 2016). They dominate, however, the microbiota of arthropods and marine invertebrates (Degli Esposti and Martinez-Romero 2017) as well as oceanic microbiomes (Louca et al. 2016, Schulz et al. 2017).

Traditionally, different taxa have been placed among proteobacteria based on 16S rRNA sequencing, phylogenetic analysis of universal proteins and other molecular techniques (Gupta 2000, Hug et al. 2016). In recent metagenomic studies, however, the taxonomic affiliation of new organisms has often been based upon phylogenetic trees of concatenated ribosomal proteins (Ciccarelli et al. 2006, Williams and Kelly 2013, Anantharaman et al. 2016, Hug et al. 2016, Probst et al. 2017). Although this approach is considered to have superior resolution to 16S rRNA analysis, the benchmark of bacterial classification (Woese 1987, Yarza et al. 2014, Schulz et al. 2017), it has failed to properly classify a number of organisms, often generated from metagenomic data (Tully et al. 2016, 2017,

Table 1. Number of genomes for the various classes of proteobacteria and some *phyla* of Gram negative bacteria as reported in NCBI web resources, accessed at the indicated dates. Taxonomic richness is taken from Schulz et al. (2017) and computed from the number of operational taxonomic units (OTU) defined by 97% identity in rRNA.

Class or phylum	Genomes 10 Aug 2017	Genomes 21 Mar 2018	Taxonomic richness 17 Oct 2017*
alphaproteobacteria	1450	2031	1113
betaproteobacteria	717	943	509
gammaproteobacteria	1855	2485	1269
deltaproteobacteria	430	649	865
epsilonproteobacteria	108	126	97
zetaproteobacteria	17	42	8
phylum Proteobacteria	**4678**	**6562**	**3764**
phylum Bacteroidetes	**1174**	**2438**	**1892**
Phylum Nitrospirae	**81**	**127**	**69**

*Day of publication of the paper by Schulz et al. (2017).

Devers-Lamrani et al. 2016, Probst et al. 2017, Slaby et al. 2017). Consequently, at least 120 genomes are currently listed under the broad definition of 'unclassified proteobacteria' [https://www.ncbi.nlm.nih.gov/Taxonomy/Browser/wwwtax.cgi, accessed 3 Jan 2018]. Recently, wide searches of several key proteins involved in central metabolism have been found to improve the taxonomic assignment of some unclassified proteobacteria (Degli Esposti 2017b). This central chapter of the book presents a compendium of the functional properties and phylogeny of the *phylum* proteobacteria as a whole and then summarizes current knowledge on the epsilon, delta and zeta classes. The other major classes of proteobacteria will be presented in the following chapters, with Chapter five dedicated to gammaproteobacteria, Chapter six to betaproteobacteria and Chapter seven to alphaproteobacteria.

The plethora of functional properties of proteobacteria

Proteobacteria constitute the phenotypically most diversified *phylum* of prokaryotes. Indeed, metabolically proteobacteria have it all! The only prokaryotic traits that are not represented in proteobacteria are methanogenesis, which is exclusively found in a large group of Archaea (Chapter one), and anammox (anaerobic oxidation of ammonia), which is restricted to a small group of Planctomycetes (Vlaeminck et al. 2011, Simon and Klotz 2013). Moreover, proteobacteria share with eukaryotes various metabolic traits and biochemical signatures that are not present in other bacteria, starting with the membrane quinones, ubiquinone (Q) and rhodoquinone (RQ); the biosynthesis of these quinones is summarized at the end of this chapter. Recent surveys (Marreiros et al. 2016, Louca et al. 2016, Anantharaman et al. 2016, Chistoserdova 2017, Haase et al. 2017, Waite et al. 2017) have indicated that the following functional traits are exclusively present in proteobacteria, among all bacteria: sorbitol:Q reductase, alternative quinol oxidase (AOX), (mena) quinol:tetrathionate and thiosulfate reductase, menaquinol:nitrate reductase of Nrf and NapA type, menaquinol:hydroxylamine oxidoreductase εHao, plus Qrc and Nhc of sulfate reducers. Conversely, traits of aerobic metabolism such as alcohol:Q reductase (Marreiros et al. 2016) and the *mxaF* gene for methanol dehydrogenase (Lau et al. 2013, Chistoserdova 2017) are predominantly present in alpha-, beta- and gammaproteobacteria, defining phenotypic characters that are routinely tested in microbiology research (see Chapter two). These and other metabolic traits fundamentally shared by gamma-, beta- and alphaproteobacteria will be discussed in the following chapters dedicated to such classes. Below I will present the traits of electron transport and energy metabolism that are commonly found in epsilon- and deltaproteobacteria, which thrive in habitats containing high levels of sulfide that normally are poisonous to other proteobacteria, as well as to eukaryotes.

Epsilonproteobacteria

The epsilon class is the second smallest of the *phylum* proteobacteria (Table 1), but comprises ecophysiologically diverse organisms varying from human pathogens to dominant microbes of hydrothermal vents in the oceans (Campbell et al. 2006, Waite et al. 2017). The class includes only two valid orders, the Campylobacterales, comprising the microaerophilic pathogens *Helicobacter* and *Campylobacter*, and the Nautiliales, comprising strictly anaerobic chemoautotrophs such as *Nautilia* and *Lebetimonas* living in hydrothermal vents (Campbell et al. 2006, 2009, Meyer and Huber 2014). Several other organisms that also inhabit hydrothermal vents cannot be consistently grouped with either of the above orders, and are therefore considered 'unclassified epsilonproteobacteria' (Campbell et al. 2006, Waite et al. 2017). These include the genera of *Thiovulum*, *Sulfurovum* and *Thioreductor*, which physiologically oxidize sulfide (see next Chapter five for the pathways of sulfur oxidation), while Nautiliales thrive on the complementary physiology of sulfur reduction, which is shared by the unclassified *Sulfurospirillum* genus too (Campbell et al. 2006, Stolz et al. 2015—see also the section of Chapter two on the sulfur cycle). Sulfur reducers are strict anaerobes that utilize elemental sulfur and its polysulfide derivatives as final electron acceptors of their electron transport chain linked to carbon fixation via reductive TCA, rTCA (Campbell et al. 2006, 2009). rTCA is present in most epsilonproteobacteria, even those with microaerophilic ecophysiology, as the fundamental route of carbon fixation (Campbell et al. 2006, Waite et al. 2017). Interestingly, epsilonproteobacteria possess a large repertoire of enzymes and systems associated with the nitrogen cycle (Campbell et al. 2006, 2009, Kern and Simon 2009, Meyer and Huber 2014, Marreiros et al. 2016, Haase et al. 2017, Waite et al. 2017), which is considered part of the ancestral core physiology of the class (Waite et al. 2017). Indeed, many epsilonproteobacterial organisms can grow on nitrate as electron acceptor, using either hydrogen or sulfide as electron donor (Campbell et al. 2006).

Functional properties of epsilonproteobacteria

Epsilonproteobacteria share with gammaproteobacteria many traits of the nitrogen cycle, but overall they contain a less versatile electron transport chain. They lack the following bioenergetic systems of other proteobacteria (Marreiros et al. 2016): alcohol, lactate and sorbitol:Q reductase; high potential soluble iron-sulfur carrier (Hipip); sodium-pumping NADH-quinone reductase, NQR; Alternative complex three, Act (or AcIII); the Rnf complex; Quinol:electron acceptor oxidoreductase, Qmo; and, except for *Lebetimonas* (Meyer and Huber 2014), Nar nitrate reductase. Moreover, epsilonproteobacteria do not have homologues of the short, Cyanide Insensitive Oxidase (CIO) type of bd ubiquinol oxidase (Degli Esposti et al. 2015). Conversely, these bacteria are enriched in systems of the nitrogen cycle such as menaquinol:nitrate reductase of NapA and Nrf types (Marreiros et al. 2016) and the menaquinol:hydroxylamine oxidoreductase εHao (Haase et al. 2017), consistent with the pervasive and ancestral properties of this cycle in the epsilon class (Waite et al. 2017, Haase et al. 2017). *Wolinella succinogenes*, a symbiont of cattle rumen, is the best known epsilonbacterium with regard to the electron transport chain and associated energy metabolism (Kröger et al. 2002, Kern and

Simon 2009). Fumarate reductase (Fdr) and polysulfide reductase (Psr) of *Wolinella* are actually the reference enzymes of their superfamilies (Kern and Simon 2009, Simon and Klotz 2013). However, *Wolinella* has an anaerobic physiology, contrary to most of its relatives of the Campylobacterales order, which are microaerophilic (Campbell et al. 2006, Hofreuter 2014). Therefore, *Wolinella* cannot be considered a valid reference organism for the respiratory chain of epsilonproteobacteria. To this end, *Campylobacter jejunii* is a much better example (Hofreuter 2014), also because it contains the bc1 complex which is commonly present in microaerophilic epsilonproteobacteria (Marreiros et al. 2016), but not in *Wolinella* (Kröger et al. 2002, Kern and Simon 2009—Fig. 1A). Genomic analysis shows that the operon of the bc1 complex in *Campylobacter jejuni* has the sequence cyt *b*-cyt *c*1-Rieske iron-sulfur protein (ISP), which is similar to that present in alphaproteobacteria from the Acetobacteraceae family. However, this gene sequence does not conform to the *petabc* operon of the bc1 complex present in most proteobacteria, including other epsilonproteobacteria such as *Helicobacter*. Moreover, the ISP of all epsilonproteobacteria lacks a central alpha helix and the protein turn just before the ISP cluster, similar to the same protein of deltaproteobacteria (Degli Esposti et al. 2014, M.D.E. unpublished data). In *Campylobacter jejuni*, the soluble cytochrome *c* that is reduced by the bc1 complex is re-oxidized by a cbb3 oxidase (Hofreuter 2014), which is widespread in Campilobacterales but not present in Nautiliales or other epsilonproteobacteria (Degli Esposti and Martinez-Romero 2017—Fig. 1).

The respiratory chain of *Campylobacter* (Fig. 1A) is the most complex of those discussed in this book so far, reflecting the accumulation of electron transport systems acquired for sustaining microaerophilic heterotrophic physiology, which is associated with the colonization of animal and human gastro-intestinal apparata (Hofreuter 2014), plus the complete denitrification process (Kern and Simon 2009). In *Campylobacter*, at least nine different systems re-oxidize menaquinol with diverse electron acceptors such as oxygen and polysulfide, but only one of these is protonmotive—the bc1 complex (Fig. 1A). Comparison with the electron transport chain of anaerobic *Nautilia* (Fig. 1B, cf. Campbell et al. 2009) highlights the ancestral core of the respiratory chain of epsilonproteobacteria, which includes: group 1 MQ-reducing Ni-Fe hydrogenase (Hup) and formate:Q reductase (Fdh), both of which are protonmotive (Kern and Simon 2009); menaquinol:fumarate reductase, associated with rTCA (Campbell et al. 2006); NapA menaquinol:nitrate reductase (Kern and Simon 2009); Psr menaquinol:polysulfide reductase (Jormakka et al. 2008); and the bd-I type ubiquinol oxidase (Degli Esposti et al. 2015). Of note is that the fumarate reductase of *Campylobacter* is very similar to that of *Wolinella*, which contains a single transmembrane subunit binding a diheme cytochrome *b* (Kröger et al. 2002), but is rather different from that of *Nautilia*, which does not have transmembrane subunits and is linked to a membrane-associated protein homologous of the HrdB subunit of heterosulfide reductase (cf. Marreiros et al. 2016). In the respiratory chain of epsilonproteobacteria, the bioenergy-producing protonmotive reactions are concentrated in the enzyme complexes that reduce MQ, in particular Group 1 hydrogenase, given that complex I is present with a reduced

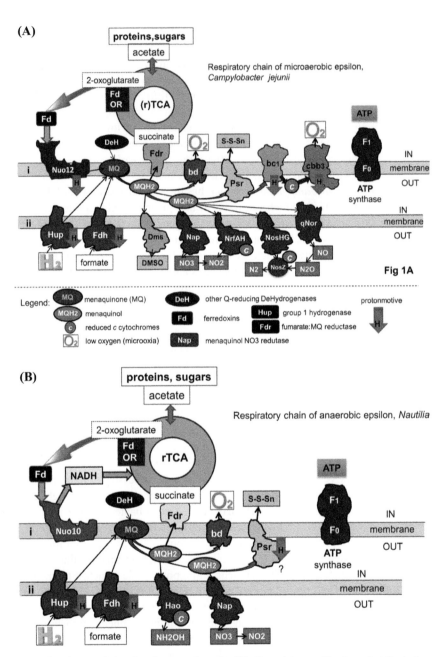

Fig. 1. Respiratory chain of epsilonproteobacteria. (A) *Campylobacter*. The legend at the bottom illustrates the most common graphical symbols used in this and all subsequent illustrations of respiratory chains in the book. (B) *Nautilia*.

number of subunits and has limited protonmotive capacity, if any (Weerakoon and Olson 2008). The nuo10 complex I of Nautiliales (Fig. 1B) is equivalent to the 11-subunit ancestral version of the enzyme described in Chapter two but with NuoC and NuoD subunits fused together; it re-oxidizes ferredoxin (or flavodoxin) reduced by the rTCA to produce NAD(P)H for biosynthetic reactions (Degli Esposti 2015), without generation of protonmotive force. *Nautilia* genome contains several [NiFe]-hydrogenases, including the membrane-bound Ech (Energy conserving hydrogenase) and the CO-dependent one (Vignais and Billoud 2007, Campbell et al. 2009, Greening et al. 2016).

In sum, a good proportion of the respiratory chain of epsilonproteobacteria is contributed by enzymes that participate in the process of denitrification (Fig. 1), which seems to be complete and widespread in this class (Hofreuter 2014, Marreiros et al. 2016, Waite et al. 2017). The opposite trait of nitrogen fixation is present in some environmental epsilonproteobacteria, since *nif* gene clusters for nitrogenase are found in *Lebetimonas* (Meyer and Huber 2014), *Wolinella* (Kern and Simon 2009), *Sulfuricurvum*, *Arcobacter* and *Sulfurospirillum* (Waite et al. 2017), as well as in three metagenomic *Sulfurimonas* (Anantharaman et al. 2016). Although the true capacity of N_2 fixation has been established biochemically only in *Wolinella* (Kern and Simon 2009), environmental epsilonproteobacteria such as *Lebetimonas* are likely to have a functional nitrogenase too, given that they often have two *nif* operons (Meyer and Huber 2014).

Phylogenetic features of epsilonproteobacteria

Despite their physiological and ecological diversity, epsilonproteobacteria usually form a coherent, monophyletic group in phylogenetic analyses, with the order Nautiliales at its base (Campbell et al. 2006, 2009). However, the phylogeny of the epsilon class, which was originally considered to be at the root of the *phylum* proteobacteria (Tenover et al. 1992, Gupta 2000), has been increasingly questioned as 16S rRNA gene repositories have expanded, together with the complementary analysis of protein markers (Gupta and Sneath 2007, Williams and Kelly 2013, Yarza et al. 2014, Zhang and Sievert 2014, Hug et al. 2016, Waite et al. 2017, Schulz et al. 2017). Recently, Waite et al. (2017) have proposed that the organisms classified within the epsilon class may form a separate *phylum* from that containing other proteobacteria, except for the small order of Desulfurellales from the class of deltaproteobacteria. Desulfurellales contain only two genera, *Desulfurella* and *Hippea*, of moderately thermophilic, anaerobic sulfur-reducing bacteria (Rainey et al. 1993, Miroshnichenko et al. 1999), characters that are shared with the Nautiliales (Campbell 2009, Florentino et al. 2017, Waite et al. 2017). The new proposed *phylum* of 'Epsilonbacteraeota' would cluster with the *phylum* Aquificae, while the *phylum* of (remaining) proteobacteria would cluster with Nitrospirae (Waite et al. 2017), as previously reported (see Chapter three), and also Acidobacteria (cf. Yarza et al. 2014, Hug et al. 2016, Hahn et al. 2017). This re-classification of Gram negative bacteria derives from the detailed examination of the large number of genomes that are currently available, even if the actual 'phylogenomic' analysis has been based upon trees of about 100 concatenated marker proteins (Ormerod et al. 2016, Waite et al. 2017, Parks et al. 2017), as in previous studies (Williams and Kelly 2013). Such phylogenetic markers are overwhelmingly represented by proteins reacting or associated with polynucleotides, especially RNA (Ciccarelli et al. 2006, Williams and Kelly 2013, Hug et al. 2016, Probst et al. 2017, Parks et al. 2017), thereby providing congruent results, in general, with those obtained by phylogenetic analysis of RNA genes (Yarza et al. 2014, Waite et al. 2017, Schulz et al. 2017). For instance, 58% of the housekeeping marker proteins used by Williams and Kelly (2013) are ribosomal and only 2% are enzymes that do not react with poly- or mono-nucleotides. However, both approaches suffer from the fundamental problem that RNA phylogenies constitute poor proxies for functional genomics (Ku et al. 2015, Imhoff 2016, Degli Esposti and Marinez-Romero 2017). The drawback is well summarized by Imhoff (2016): '...[RNA] is not related to metabolic properties *per se*, but only through phylogenetic observations and assumptions'. Moreover, the use of concatenated protein alignments is prone to technical problems that hamper resolution of the deep branches in phylogenetic trees (Thiergart et al. 2014, cf. Chapter one).

The above considerations prompted an evaluation of the proposed affiliation of Desulfurellales with epsilonproteobacteria (Zhang and Sievert 2014, Waite et al. 2017, Hahn et al. 2017) using two important bioenergetic systems shared by these prokaryotes: the bd type ubiquinol oxidase (Degli Esposti et al. 2015) and complex I (Spero et al. 2015, Degli Esposti 2015). As noted earlier in this chapter, epsilonproteobacteria characteristically have the long, bd-I form of bd ubiquinol oxidase, as *E. coli*. Conversely, the majority of deltaproteobacteria have the short, CIO form of the same ubiquinol oxidase (Degli Esposti et al. 2015) and this applies to the bd oxidases of Desulfurellales too, which, thus, do not share such a character with epsilonproteobacteria. Regarding complex I, both epsilon- and deltaproteobacteria have a wide variety of *nuo* gene clusters, often present in multiple forms within the same genome (Marreiros et al. 2013, Spero et al. 2015, Degli Esposti 2015). However, the conserved catalytic NuoD subunits of Nautiliales and a few other epsilonproteobacteria such as *Sulfurovum* show a rare molecular signature at their C-terminus, which is a relic from the cysteine motif binding to the [NiFe] cluster in their ancestral hydrogenase relatives (Degli Esposti 2015). Alignment of the C-terminal part of the NuoD subunit of Desulfurellales does not show this vestigial signature but another Cys residue aligning instead with the NuoD subunit from other deltaproteobacteria and Nitrospirae (Fig. 2, cf. Degli Esposti 2015). Hence, it appears unlikely that Nautiliales and Desulfurellales have a common sulfur-reducing ancestor, as recently claimed (Waite et al. 2017), since they would be expected to share ancestral traits of key bioenergetic systems such as complex I, but they are not. Clearly, further studies are necessary to resolve the phylogeny of epsilonproteobacteria, which firmly remain within the *phylum* proteobacteria in this book.

accession	position	sequence	aa	organism	enzyme type	Class
WP_011389179	328	MNWIDALNVMMAGARISDIPLIVNSIDPCISCTER	361	R.rubrum	Co-induced Ech	α
WP_012625292	327	ANVPPLLIMLPG·KLPDVPVIVLSIDPCISCTER	360	Desulfovibrio desulfuricans	Ech	δ
WP_040306309	339	-GLLLAQKHLPKMKIDVAAWWGSLAICPPDIDK	371	Caloramator australicus	Mbx-like cluster	Firmicutes
WP_005963985	348	-AVQVLEKLAVGSRIEDDVAQTMFSLDACPPEVDR	380	endosymb Riftia pachyptila	Mbx-like cluster	γ
WP_013638772	357	--LWAMPEMIKGVKLADVPVIFASLYMCHGDIDR	388	Desulfurobacterium thermo	NUO11 cluster	Aquificae
WP_013538242	358	--LWAMPEMMKNVKLADVPVIFASLYICHGDIDR	389	Thermovibrio ammonificans	NUO11 cluster	Aquificae
WP_015902513	518	--TMMLDKLLRGKTLSDIPLLYGSMHICOGDLDR	549	Lebetimonas	NUO10 cluster	ε Nautiliales
WP_007474985	526	--TMMLEKLLPGHTLSDIPLLYGSMHICOGDLDR	557	Nautilia profundicola	NUO10 cluster	ε Nautiliales
WP_008350072	502	--TMMVEKLLEGNTISDIPITVFGSMYICQGDLDR	533	Thiovulum sp. ES	NUO10 cluster	ε
WP_019980537	516	--TMMLNHLLAGETLSDVPLVFGSLYVCQGDLDR	547	Sulfurovum sp. NBC37-1	NUO10 cluster	ε
YP_007451141	331	-LQILPHILKGVKVADIMALLGSIDIIMGSVDR	394	Synechocystis sp. PCC 6803	NDB-1 plastid	Cyanobacteria
YP_002600964	328	-LQILPNIVQGMKLADIMTTLGSVDIIMGECDR	391	Pyramimonas parkeae	NDB-1 plastid	green algae
EEF60813	392	-LSILPHLLMGHHMDTVALLGSLLDFVMGECDR	423	Pedosphaera	NUO14 cluster	Verrucomicrobia
WP_013044308	378	-LCVLPEIAPGHHMTDITVLLGSLLDFVMGECDR	409	Coraliomargarita akajim.	NUO14 cluster	Verrucomicrobia
WP_013247211	520	-MGAFDHMWARGYLLSDIITIFGTKDIVMGECDR	583	Nitrospira defluvii	NUO13 cluster	Nitrospirae
CBK40015	384	-MGAFDHMSKGYMIADAVTIFGTKDIVMGECDR	415	Nitrospira defluvii	NUO12 - NuoH	Nitrospirae
WP_018048422	363	-TGALPELCKGMLADIVPTFDMLNMIGGECDR	396	Nitrospina sp. AB-629-B18	fragmented NUO14	Nitrospinae
YP_002248750	366	-AGVLPKLCEGSLVADVIANIGSIDIVLGECDR	397	Thermodesulfovibrio yellows.	NUO11 cluster	Nitrospirae
WP_013908230	379	-ISAIPDLCEGQMVADIIAVIGSIDIVLGECDR	410	Thermodesulfator indicus	NUO11 cluster	Thermodesulfobacteria
WP_025391350	366	-LSAMPVMLKGHIYIADIISVTGSLDFVGECDR	398	Desulfurella acetivorans	NUO11 cluster	δ
WP_013681256	366	-ISAMPKMVQGGLIADIISVIGSLDFVFGECDR	398	Hippea maritima	NUO11 cluster	δ
WP_012829609	352	-LQTVSEIIRGAFISDIVPIFGMINMIGGECDK	366	Haliangium ochraceum	fragmented NUO14	δ
AGP32092	373	-TQCLSQLITGLMIFDVVFTFGSLNMIGGECDH	404	Sorangium cellulosum	fragmented NUO14	δ
WP_006748238	376	-MGGLHRLLEGYLQLADVVSTFGTVNMIGGECDR	410	Thioalkalivibrio thiocy.	NUO14 cluster	γ
ADV45433	377	-TGILQDILPGTYIFDVVTIIGSTNIVFGEVDR	409	Nitratifractor salsuginis	derived NUO14	ε
ADN09876	392	-TGILTDLLPGHYIFDVVSIIGTTNIVFGEVDR	413	Sulfurimonas autotrophica	incomplete NUO14	ε
WP_011552796	384	-LAAVPHIIEGKMLADLLIPTFDTNMIGEVEQ	415	Myxococcus xanthus	fragmented NUO14	δ
WP_006366953	369	-LSAMKDLSKGQLIFDLVATIGSIDMVIPEIDR	400	Chlorobium ferrooxidans	NUO11 cluster	Chlorobi
WP_007524325	375	-LAIFAECARGTLLADAVAILGSLDMVIPEIDR	406	Desulfovibrio sp. A2	fragmented NUO14	δ
WP_010967799	373	-LQALFGVTNARYLADMIAVLGSLDPVMAEVDK	404	Sinorhizobium meliloti	NUO11 cluster	α
WP_018992401	373	-LQALAVLARG·VLADVIATIGSLDPVIAEVDR	404	Azoarcus toluclasticus	Green CI	β
AAA97941	378	-LQSLPTACKGFQVFDMVAIIASLDPVMGDVDR	409	Thermus thermophilus	NUO14 cluster	Thermus/Deinococcus

Fig. 2. Alignment of the conserved C-terminal part of the NuoD subunit of bacterial complex I. Conserved residues are in bold, while cysteine residues are highlighted in yellow. The two Desulfurrellales organisms are also highlighted in yellow. Unusual amino acid substitutions are in red and highlighted in gray (Degli Esposti 2015). The various types of gene clusters for complex I have been defined as reported earlier (Spero et al. 2015, Degli Esposti 2015, Degli Esposti and Martinez-Romero 2016). The orange arrow indicates the Cys residue that is a ligand to the [NiFe] cluster in hydrogenases and is present as a vestigial feature of such a ligand in ancestral NuoD subunits (Degli Esposti 2015).

Deltaproteobacteria

Epsilonproteobacteria and deltaproteobacteria are often considered together as the oldest group of proteobacteria, since they include obligate anaerobes which utilize sulfur compounds for their energy metabolism (Campbell et al. 2006, Barton and Fauque 2009, Degli Esposti 2016). However, the class of deltaproteobacteria has been formally defined as follows: 'bacteria having 16S rRNA gene sequences related to those of the members of the order Myxococcales' (Kuever et al. 2005). Myxococcales, also known as Myxobacteria, are aerobic predatory bacteria long studied for their capacity of social behavior, forming morphologically bizarre fruiting bodies (Kaiser et al. 2010), a property that originally led to their classification among fungi (Thaxter 1892). After the advent of 16S rRNA sequencing for bacterial classification, Myxococcales were found to share a specific relationship with bacteria having a completely different physiology, the anaerobic sulfate-reducers represented by *Desulfovibrio desulfuricans* and its relatives (Oyaizu and Woese 1985, Woese 1987, Kuever et al. 2005). The majority of currently known sulfate reducing bacteria (SRB) still belongs to deltaproteobacteria (Muyzer and Stams 2008, Barton and Fauque 2009, Zhou et al. 2011, Imhoff 2016), which also include sulfur-reducing *Desulfuromonas* and *Desulfurella* (see above, section on epsilonproteobacteria phylogeny), as well as chemolithotrophic organisms reacting with heavy metals such as *Geobacter*.

Deltaproteobacteria additionally included the *Bdellovibrio* group of predatory bacteria (Sockett 2009), which share many properties with unclassified alphaproteobacteria of the genus *Micavibrio* (Davidov et al. 2006). These taxa have been recently proposed by Hahn et al. (2017) to form part of a novel proteobacterial class, the Oligoflexia (Nakai et al. 2014)—see below section on deltaproteobacteria phylogeny. Even without *Bdellovibrio* and *Bactereiovorax*, the class of deltaproteobacteria contains a combination of aerobic predatory and facultatively anaerobic sulfur and metal reacting organisms which probably shows the widest phylogenetic breadth among proteobacteria (Hahn et al. 2017). Nevertheless, deltaproteobacterial SRB have enzymes and physiological properties that are shared only with members of the *phylum* Nitrospirae (McInerney et al. 2007, Zhou et al. 2011, Imhoff 2016, Marreiros et al. 2016, Frank et al. 2016). Of note is that SRB are also present among Gram positive bacteria and Archaea (Zhou et al. 2011, Imhoff 2016, Marreiros et al. 2016)—prokaryotes which are not considered in this book.

While Myxococcales are predominantly soil inhabitants (Reichenbach 1984, Goldman et al. 2007, Kaiser et al. 2010), sulfate- and sulfur-reducing deltaproteobacteria are common in sediments and stratified layers of aquatic environments that become anoxic due to processes of oxygen-consuming biodegradation (Muyzer and Stams 2008, Louca et al. 2016). They also inhabit hydrothermal vents and soil habitats, globally shaping the sulfur cycle and significantly contributing to carbon recycling (Muyzer and Stams 2008, Zhou et al. 2011). Some deltaproteobacteria inhabit human and animal microbiota (Quin et al. 2010, Nielsen et al. 2014, Degli Esposti and Martinez-Romero 2017), while others are engaged in syntrophic association with different prokaryotes to complete anaerobic degradation of biological material, with important biotechnological applications (McInerney et al. 2007, Muyzer and Stams 2008, Barton and Fauque 2009, Zhou et al. 2011). Conversely, predatory deltaproteobacteria play important roles in soil habitats; most Myxococcales are also economically important for their production of chemically different antibiotics (Reichenbach 1984, Herrmann et al. 2017). More recently, *Geobacter* and related deltaproteobacteria have received great attention for their biotechnological applications in electrochemical biofuel cells (Ishii et al. 2013, White et al. 2016). This chapter will focus on the functional properties of anaerobic deltaproteobacteria that define their energy metabolism and are linked to their phylogenesis.

Traditionally, deltaproteobacterial SRB have been divided between those that cannot oxidize organic compounds beyond acetate, such as *Desulfovibrio* (the type genus of the order Desulfovibrionales), and those that can undertake complete oxidation of fatty acids, including acetate (Muyzer and Stams 2008, Barton and Fauque 2009). *Desulfobacter*, *Desulfobacterium* and *Desulfomaculum* are prominent organisms of the latter group and all belong to the order Desulfobacterales. While *Desulfobacter* appears to have a complete TCA cycle to degrade acetate, *Desulfobacterium* and *Desulfomaculum* oxidize acetate via the carbon monoxide dehydrogenase system (Barton and Fauque 2009). *Desulfobulbus* also belongs to Desulfobacterales, but is an atypical SRB because it can disproportionate elemental sulfur to sulfate and sulfide, without being capable of oxidizing acetate (Pagani et al. 2011). Desulfuromonadales form another important order of anaerobic deltaproteobacteria, which includes sulfur-reducing taxa such as the type genus *Desulfuromonas* and various organisms using Fe or other heavy metals as final acceptors for their chemolithotrophic lifestyle, with *Geobacter* as their most prominent *genus* (Barton and Fauque 2009). The last major order of anaerobic deltaproteobacteria is represented by the Synthrophobacterales, which includes the syntrophic, fat-eating *Synthrophus* (McIrney et al. 2007) and a variety of other organisms with diverse physiology, such as *Desulfobacca* and *Smithella.*

Some deltaproteobacteria have evolved an endosymbiotic lifestyle, which is particularly interesting in the recently described endosymbionts of anaerobic flagellates living in insect guts (Ikeda-Ohtsubo et al. 2016). *Candidatus* Adiutrix intracellularis, for example, is an intracellular symbiont of the flagellate *Trichonympha* that represents a novel clade of uncultured deltaproteobacteria distributed in the guts of termites and cockroaches (Ikeda-Ohtsubo et al. 2016). These bacteria are strictly anaerobes and rely on [FeFe]-hydrogenases for re-oxidizing ferredoxins reduced by [NiFe]-hydrogenases and fermentative enzymes, similar to alphaproteobacterial anaerobes found in human gut microbiota (Degli Esposti et al. 2016).

Intriguingly, deltaproteobacterial SRB are also present in what would normally be atypical environments, considering their strictly anaerobic metabolism defined in microbiology studies (Barton and Fauque 2009, Lamrabet et al. 2011). Examples of such environments are aerobic wastewater biofilms and sea grass rhizosphere sediments—see Ramel et al. (2015) and references therein. These habitats can be temporarily exposed to elevated oxygen concentrations, thereby imposing significant aerotolerance to ecologically successful SRB. Such a tolerance is particularly strong in *Desulfovibrio* organisms (Lamrabet et al. 2011, Ramel et al. 2015) due to the presence of oxygen consuming systems that are discussed in detail below.

Functional properties of facultatively anaerobic deltaproteobacteria

Biochemically, deltaproteobacteria are well known for their anaerobic electron transport systems and capacity of using sulfate as terminal electron acceptor for their respiratory chain (Barton and Fauque 2009, Zhou et al. 2011, Marreiros et al. 2016). Indeed, *Desulfovibrio* and *Desulfobacter* are probably the best known SRB (Barton and Fauque 2009), even if some of their unique bioenergetic systems of the sulfur cycle remain poorly known (Marreiros et al. 2016). This is particularly the case for dissimilatory sulfite reductase, the Dsr system, which works in the reductive direction for SRB, namely from sulfite to sulfide. Alternatively, it can also work in the oxidative direction, from sulfide to sulfite, in sulfur-oxidizing bacteria (Grein et al. 2010, Müller et al. 2015, Dahl 2015, Imhoff 2016). Sulfur-oxidizing bacteria are rare among deltaproteobacteria, but widespread in epsilonproteobacteria and other classes of the *phylum* (Imhoff 2016)—see subsequent Chapter five for a description of the pathways of sulfur oxidation.

Most anaerobic deltaproteobacteria actually are facultatively anaerobes (see Chapter two for the modern definition of bacterial aerobic metabolism) due to a complex respiratory chain similar to that illustrated in Fig. 3A, which specifically represents the bioenergetic systems of *Desulfovibrio* spp. The respiratory chain is dominated by various MQ-reducing systems, the most prominent of which are membrane-bound [NiFe]-hydrogenases of Group 1 (Hup–Vignais and Billoud 2007) and Group 4 (Ech–Coppi 2005). *Desulfovibrio* and other deltaproteobacterial SRB contain several other types of hydrogenases (Coppi 2005, Barton and Fauque 2009), including cytosolic [FeFe]-hydrogenases that reoxidize the ferredoxin reduced by enzymes bypassing the TCA cycle, normally associated with fermentation of pyruvate (Vignais and Billoud 2007, Hug et al. 2010, Ramel et al. 2015, Greening et al. 2016), as well as periplasmic [NiFe]-hydrogenases which reduce multiheme *c* cytochromes (Barton and Fauque 2009, Da Silva et al. 2012). Indeed, H_2 is the major electron donor to environmental SRB, together with organic compounds (Barton and Fauque 2009—see Fig. 1 of Chapter two). Protonmotive hydrogenases are considered indispensable for supplying ATP to the energy-demanding reaction of sulfate activation into adenosine-phosphosulfate (APS), a reaction occurring in the cytosol after uptake of environmental sulfate by a specific transporter (Barton and Fauque 2009, Zhou et al. 2011). This reaction, catalysed by the enzyme ATP sulfurylase, overall requires 2 ATP per molecule of activated sulfate and is necessary

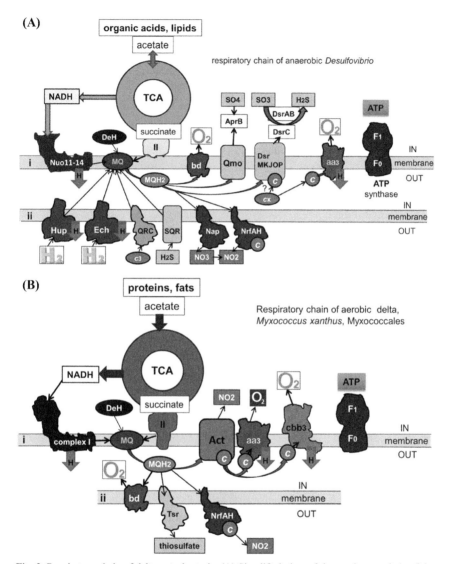

Fig. 3. Respiratory chain of deltaproteobacteria. (A) Simplified view of the respiratory chain of the SRB representative *Desulfovibrio*. The Rnf relay of ferredoxin is not shown. (B) Detailed view of the respiratory chain of *Myxococcus xanthus*, representative of Myxococcales.

for utilizing sulfate as terminal electron acceptor in SRB metabolism (Muyzer and Stams 2008, Barton and Fauque 2009, Zhou et al. 2011).

Two protonmotive systems drive additional ATP production along the respiratory chain of *Desulfovibrio* and related organisms (Fig. 3A): (i) NADH-MQ reductase, which is the main function of the different forms of complex I present in deltaproteobacteria (Spero et al. 2015, Degli Esposti 2015); (ii) aa3 cytochrome *c* oxidase (Lobo et al. 2008, Lamrabet et al. 2011, Ramel et al. 2015). This oxidase has been variably defined as cc(b/o)o3 cytochrome oxidase (Lamrabet et al. 2011, Ramel et al. 2015) or caa3 oxidase (Lobo et al. 2008), even if it squarely falls in the A2 type of the A family of heme copper oxygen reductases, HCO (Sousa et al. 2012). Genomically, the oxidase has a peculiar gene sequence that is apparently unique to deltaproteobacteria (Lamrabet et al. 2011, Degli Esposti 2014) and soluble cytochrome c-553 is considered to be its principal substrate (Lobo et al. 2008, Lamrabet et al. 2011). This cytochrome may be reduced by the Drs complex or other systems associated with periplasmic c-type cytochromes (indicated by the symbol cx in Fig. 3A), such as soluble formate dehydrogenases (Da Silva et al. 2012). In *Desulfovibrio* organisms, periplasmic formate dehydrogenases transfer electrons to a variety of periplasmic c cytochromes, in contrast to enzymes carrying out formate oxidation in other bacteria (Da Silva et al. 2012). One or more of these cytochromes can thus link formate oxidation to protonmotive oxygen reduction by the aa3 oxidase, which is likely to have moderate to intermediate affinity for oxygen (see below).

It is plausible that the cycle of MQ oxidoreduction that seems to occur within the Dsr complex (Grein et al. 2010) can contribute additional protonmotive force to *Desulfovibrio* organisms, considering that the complex contains a transmembrane diheme b cytochrome which can carry electrogenic reactions as in Nar nitrate reductase or the bc1 complex (Marreiros et al. 2016). The respiratory chain of *Desulfovibrio* contains two additional MQ reductases associated with the metabolism of sulfur: the sulfide-(mena)quinone reductase SQR and the Cytochrome c3: (mena)quinone oxidoreductase, QRC (Fig. 3A, cf. Marreiros et al. 2016). Both systems are restricted to deltaproteobacteria, while the menaquinol oxidoreductase Qmo is also present in functionally related Nitrospirae organisms (Marreiros et al. 2016, Frank et al. 2016). The Qmo complex transfers electrons from menaquinol to cytosolic AprAB (Adenosine Phosphosulfate Reductase), the enzyme which reduces APS to sulfite (Muyzer and Stams 2008, Barton and Fauque 2009). Moreover, *Desulfovibrio* has the capacity of growing on nitrate (Barton and Fauque 2009) due to the expression of NapA- and Nrf-type membrane complexes that oxidize menaquinol to reduce either nitrate or nitrite (Fig. 3A), similar to the denitrification systems of epsilonproteobacteria (Fig. 1). Finally, menaquinol can be re-oxidized by the above mentioned bd oxidase of CIO type, which is ubiquitous among deltaproteobacteria and contributes to the high aerotolerance of *Desulfovibrio vulgaris* and related organisms (Fig. 3—cf. Lamrabet et al. 2011, Ramel et al. 2015, Degli Esposti et al. 2015, Degli Esposti and Martinez-Romero 2017).

Marreiros et al. (2016) have reported that about 30% of the deltaproteobacteria they have examined contain two additional quinol:electron carrier oxidoreductases, a bc1 complex-like and the so-called alternative complex III, which is abbreviated here as Act (Singer et al. 2011, Degli Esposti et al. 2014)—we have already encountered Act in Chapter three. However, these systems are not present in SRB such as *Desulfovibrio*, which instead has the Rnf complex, an ancient protonmotive system linked to the ferredoxin cycle stemming from rTCA reactions (see previous Chapters two and three of this book). Overall, the respiratory chain of *Desulfovibrio* has at least six MQ-reducing systems, including a peculiar form of succinate dehydrogenase, and five MQH_2-oxidizing systems, if one considers the Dsr complex primarily as a quinol re-oxidizing enzyme (Grein et al. 2010, Marreiros et al. 2016). The redox balance thus appears to favor upstream reduction of MQ, a situation similar to that existing in nodulating symbionts living in environments poor in oxygen but rich in reducing equivalents to feed the nitrogenase reaction (Degli Esposti and Martinez-Romero 2016). Indeed, N_2 fixation is a common trait among SRB and other deltaproteobacteria (Barton and Fauque 2009, Degli Esposti and Martinez-Romero 2017).

Additional functional properties of deltaproteobacteria are present in organisms of the *Geobacter* genus that reduce external heavy metals such as Fe, Mn and uranium (Lovley et al. 2004, Feist et al. 2014). *Geobacter* genomes code for a large number of diverse *c*-type cytochromes and a bc1-like system characterized by a short ISP and a multiheme *c* cytochrome (Butler et al. 2010, Feist et al. 2014), which is not homologous to the membrane-bound cytochrome *c*1 subunit of the *petabc* operon. This protein has four or five cytochrome *c*-binding motifs and is probably related to similar proteins of metagenomic organisms of the Nitrospirae *phylum*. It is also similar to menaquinol-oxidizing pentaheme NrfH (Simon and Klotz 2013). So far, information on the *Geobacter* bc1-like system, which is absent in other deltaproteobacteria, remains limited to bioinformatic inference (Butler et al. 2010, Feist et al. 2014). The function of this system is crucial for transferring reducing equivalents from the TCA cycle and the cytoplasmic membrane to the external face of the outer membrane via a relay of *c* cytochromes (Butler et al. 2010). Of note is that some *Geobacter* species also have a b6f-like gene cluster, of which very little is known. In any case, an analysis of the rapidly expanding literature on the external electron transfer capacity of *Geobacter* is beyond the scope of the present chapter, also because the molecular mechanisms of this capacity are not well understood yet. The subject has great biotechnological applications and is well presented in recent articles (Ishii et al. 2013, Estevez-Canales et al. 2015, White et al. 2016) to which the reader's curiosity is redirected.

The respiratory chain of Myxococcales

The respiratory chain of predatory aerobic Myxococcales, which form the other major group of deltaproteobacteria (Kaiser et al. 2010), is much less known than that of SRB and is therefore presented in detail here (Fig. 3B). Thanks to early biochemical studies (Watson and Dworkin 1968), *Myxococcus xanthus* is the best known predatory organism in terms of energy metabolism (Goldman et al. 2007). The genome of the related, facultatively anaerobic *Anaeromyxobacter* has provided additional valuable information (Thomas et al. 2008), also in regard to the natural resistance of complex I (Degli Esposti 2015) to the powerful quinone-antagonist antibiotics produced by many Myxococcales (Reichenbach

1984). *Anaeromyxobacter*, in fact, does not produce the fruiting bodies in which these antibiotics are stored by other Myxococcales, a process that is linked to their aerobic metabolism (Goldman et al. 2007, Thomas et al. 2008, Kaiser et al. 2010). However, the terminal part of the respiratory chain of *Anaeromyxobacter* is essentially equivalent to that of *Myxococcus*, both having the three types of terminal oxidases that are generally present in alphaproteobacteria (Degli Esposti et al. 2014): the bd ubiquinol oxidase (Borisov et al. 2011), the cbb3 cytochrome *c* oxidase (HCO family C—Sousa et al. 2012) and aa3 cytochrome *c* oxidase (HCO family A—Sousa et al. 2012, Fig. 3B).

The dehydrogenases that reduce MQ are also similar in Myxococcales and alphaproteobacteria (Goldman et al. 2007, Thomas et al. 2008), with the distinction that complex II has a transmembrane di-heme cytochrome *b* which is equivalent to that of the fumarate reductase of epsilonproteobacteria (Kröger et al. 2002), rather than that of alphaproteobacteria (Goldman et al. 2007). However, the big difference with respect to the respiratory chain of alphaproteobacteria and mitochondria is the absence of a proper bc1 complex that re-oxidizes menaquinol and reduces soluble *c* cytochromes. Thus, the previous suggestion that a similar enzyme catalyzing the quinol-cytochrome *c* reaction is present in *Myxococcus* as in mitochondria (Goldman et al. 2007) is incorrect. Indeed, Myxococcales produce powerful universal inhibitors of this complex, for example myxothiazol (Reichenbach 1984), which would functionally block the eventual presence of such an enzyme in their respiratory chain.

Menaquinol is oxidized instead by the following electron transport systems in Myxococcales (Fig. 3B): a short bd ubiquinol oxidase of CIO type; a nitrite reductase of the Nrf type; a thiosulfate reductase (Tsr), member of the CISM superfamily of membrane oxidoreductases (Marreiros et al. 2016) and; an Act complex. The latter system is genetically linked to the gene cluster for a family A cytochrome *c* oxidase that is characteristic of Myxococcales, together forming an operon which is equivalent to COX operon subtype a-I of alphaproteobacteria (Degli Esposti et al. 2014). Homologs of *Desulfovibrio* soluble cytochrome *c*-553 are likely to function as the main substrates for the terminal oxidases of both family C and A in the respiratory chain of Myxococcales (Fig. 3B). The cbb3 oxidase, however, is characterized by the rare fusion of subunit I (*fixN*) with subunit II (*fixO*), which we previously encountered in CFB organisms (Chapter three, cf. Ducluzeau et al. 2008). In sum, Fig. 3B is currently the most detailed and accurate representation of the respiratory chain of aerobic Myxococcales, which looks surprisingly similar to the respiratory chain of aerobic Flavobacteriales (Fig. 1C in previous Chapter three). Comparison with the respiratory chain of deltaproteobacterial SRB (Fig. 3A) emphasizes the main differences with other deltaproteobacteria, which are mostly related to anaerobic sulfur metabolism. Another fundamental difference is that the overall redox balance of the respiratory chain of Myxococcales is shifted toward menaquinol oxidation, which can be accomplished under both normoxic and micro-oxic conditions as in facultatively anaerobic bacteria such as *E. coli*.

Phylogenetic features of deltaproteobacteria

Phylogenetically, the delta class has been generally positioned at the root of the whole *phylum* of proteobacteria (Woese 1987, Gupta 2000, Gupta and Sneath 2007). However, the class is likely to be polyphyletic, since the proteobacterial affiliation of its organisms has been increasingly questioned (Waite et al. 2017). Indeed, recent trees depict deltaproteobacteria as a separate *phylum*, mixed with Acidobacteria or other groups of Gram negative bacteria (Yarza et al. 2014, Hug et al. 2016). Very recently, predatory organisms of the *Bdellovibrio* and *Bacteriovorax* genera, previously classified among deltaproteobacteria, have been proposed to form another class, the Oligoflexia (Hahn et al. 2017), which was originally defined by environmental organisms (Nakai et al. 2014, 2016). Consequently, the phylogeny of deltaproteobacteria is at present unsettled and may further change in the near future, given the ongoing increase in its phylogenetic depth due to metagenomic data (Schulz et al. 2017).

Even the relative phylogenetic position of the major orders of deltaproteobacteria is far from established, with the SRB often forming a sister clade to Myxococcales (Thomas et al. 2008, Hahn et al. 2017), as shown in Fig. 4A—representing the tree of cytochrome *c* oxidase subunit 1, COX1. However, phylogenetic trees of the 16S rRNA (Singer et al. 2011, Müller et al. 2015) or of other bioenergetic proteins (Degli Esposti and Martinez-Romero 2017) show *Desulfovibrio* and related anaerobic SRB in deep branches, while Myxococcales appear to form late-diverging branches. Such a phylogeny would be consistent with the enlarged genome and presumably acquired aerobic metabolism of Myxococcales (Goldman et al. 2007, Kaiser et al. 2010), but it is clearly dependent upon the structure, function and history of the proteins used as molecular markers. For instance, the trees of the conserved DsrAB proteins (Müller et al. 2015, Imhoff 2016) show Desulfovibrionales in a late divergent branch with respect to Syntrophobacterales, which instead form a sister group of Desulfovibrionales in 16S rRNA trees (Singer et al. 2011, Müller et al. 2015). Nevertheless, *Thermodesulfovibrio* organisms of the Nitrospirae occupy the precursor position of deltaproteobacteria SRB in every phylogenetic tree (Müller et al. 2015, Imhoff 2016), thereby indicating that the sulfur metabolism of deltaproteobacteria has been vertically inherited from Nitrospirae (see also Chapter three). Of note is that the respiratory chain of *Thermodesulfovibrio* sp. N1 is very similar to that of *Desulfovibrio* (Fig. 3A), with the notable exception of the absence of the aa3 terminal oxidase (Frank et al. 2016)—besides the additional absence of the MQ reductases QRC and SQR, which are specific to deltaproteobacteria (Marreiros et al. 2016). This observation strongly suggests that the A family cytochrome oxidase present in *Desulfovbrio* and *Geobacter* has been added to the common ancestor of deltaproteobacteria via a possible LGT event from an ancestral aerobic prokaryote, producing in the process the unique re-arrangement of the COX2 subunit towards the end of the gene cluster (Lamrabet et al. 2011). Despite such a genetic re-shuffle, the structure of the COX operons of *Desulfovibrio* and *Geobacter* fundamentally

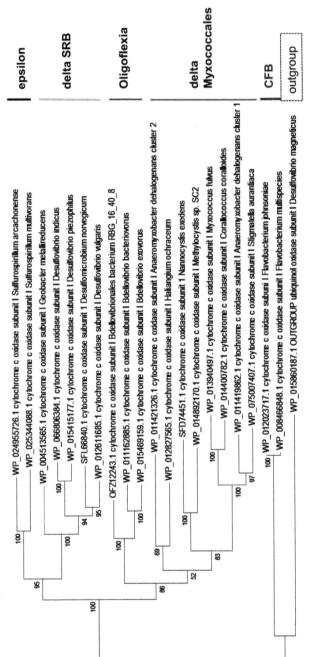

(A)

phylogenetic tree of COX1 of aa3-type cytochrome oxidase

Fig. 4 cont. ...

(B)

organism/strain	accession	up 3	up2	up1 - flank	COX2 (aa)	COX1	core CtaB-F COX3 & COD	COX4 (aa)	COX2 (aa) aI or down 1	down 2	down 3
Sulfurospirillum multivorans ε	WP_025344088	outer membrane porin, OprD		SCO - like	hypothetical protein 105 aa	COX1 - CoxA	COX3 Heme_Cu_Oxidase_III_like	CoxD 90 aa	COX2 fused with 2 cytochrome c, 444 aa	ctaB	HemK super family methylase
Desulfovibrio desulfuricans	DND132_3041	methyl-accepting chemotaxis	mono-heme cyt c	SCO		COX1 - CoxA	CoxC NorE-like	CoxD 105 aa	COX2 fused with 2 cytochrome c, 407 aa	ctaB	hypothetical protein 236 aa
Desulfovibrio vulgaris str. Hildenborough	WP_010939102		mono-heme cyt c	SCO		COX1 - CoxA	CoxC NorE-like	CoxD 96 aa	COX2 fused with 2 cytochrome c, 426 aa	ctaB	hypothetical protein 112 aa
Geobacter metallireducens GS-15	Gmet_0249	serine/threonine protein kinase	FdD	SCO		COX1 - CoxA	CoxC NorE-like	CoxD 98 aa	COX2 fused with cytochrome c, 302 aa	ctaB	UPF126-containing protein
Geobacter sulfurreducens KN400	KN400_0194	hypothetical protein 71 aa	Nitroreductase_3	SCO		COX1 - CoxA	CoxC NorE-like	CoxD 96 aa	COX2 fused with cytochrome c, 304 aa	ctaB	cupin protein
Bdellovibrio bacteriovorus HD100	Bd02877 / WP_011162885	alpha/beta hydrolase	mono-heme cyt c	SCO	COX2 fused with cytochrome c 318 aa	COX1 - CoxA	CoxC NorE-like	CoxD 111 aa		ctaB	MalA protein
Bdellovibrio exovorus JS	A1YQ_449 / WP_015469159		mono-heme cyt c	SCO	COX2 fused with cytochrome c 318 aa	COX1 - CoxA	CoxC NorE-like	CoxD 105 aa		ctaB	maleylacetoacetate isomerase
COX operon type a-I											
Anaeromyxobacter dehalogenans 2CP-C, cluster 1	WP_011419862 / A2CP1_RS04250	Act		SCO	COX2 fused with cytochrome c 350 aa	COX1 - CoxA	CoxC NorE-like	CoxD 131 aa	DUF1362 = ASRT pfam07100	hypothetical protein 204 aa	TPP_PYR_POX protein
Myxococcus xanthus DK 1622	MXAN_3868 / WP_011553878	Act		hypothetical protein 120 aa	COX2 fused with cytochrome c 346 aa	COX1 - CoxA	CoxC Heme_Cu_Oxidase III like	CoxD 121 aa	hypothetical protein 185 aa		
Myxococcales bacterium SG8_38_1	KPK51875	Act		mono-heme cyt c	COX2 - not fused 247 aa	COX1 - CoxA	COX3a Heme_Cu_Oxidase III like	CoxD 141 aa	COX3b Heme_Cu_Oxidase_II_like, 170 aa	COX3b hypothetical protein 89 aa	
Rhodothermus marinus SG0.5JP17-172, Bacteroidetes	Rhom172_0213 / WP_014065931	Act		SCO	COX2 fused with cytochrome c 316 aa	COX1 - CoxA	CoxC NorE-like	CoxD 122 aa	sema domain protein, hypothetical 103 a	methylenetetra hydrofolate reductase	cold-shock protein
Flavobacterium johnsoniae, Flavobacteriales, first part	BB050_RS04760 / WP_008466848	Act	deoxyribose-phosphate aldolase	hypothetical protein with TonB motif	COX2 - not fused 396 aa	COX1 - CoxA	COX3b Heme_Cu_Oxidase III like, 328 aa	CoxD 116 aa	RuvB Holliday junction branch recombinase	cytochrome P450	10 genes...
second part	FJOH_RS08605	10 genes...		hypothetical protein with	ctaB	COX3a Heme_Cu_Oxidase III like	COX3b Heme_Cu_Oxidase III like, 328 aa		hypothetical protein 238 aa	SCO	DUF420

Fig. 4. Phylogeny of A family cytochrome *c* oxidases in deltaproteobacteria and their COX operons. (A) Maximum Likelihood (ML) tree of COX1 proteins from CFB to deltaproteobacteria. The tree was constructed with the program MEGA5 and 500 bootstraps, the percentage values of which are reported by each node. *Methylocystis* is an alphaproteobacterial reference for COX operon subtype a-I (Degli Esposti 2014). (B) Gene clusters for COX operons in deltaproteobacteria and Bacteroidetes are organized as reported previously (Degli Esposti 2014).

resembles that of the oxidase enzyme in Myxococcales (Fig. 4B, cf. Lamrabet et al. 2011; see also Chapter seven).

Thomas et al. (2008) have argued that the common ancestor of deltaproteobacterial SRB and Myxococcales might have been microaerophilic because of the scattered presence of cytochrome *c* oxidases in different groups of the delta class (cf. Lamrabet et al. 2011). Considering that this intriguing hypothesis has not been followed much in the literature, it has been experimentally tested here by producing detailed phylogenetic trees of the catalytic subunit COX1 (Fig. 4), which is the fundamental marker for the evolution of HCO (Sousa et al. 2012). These trees show that COX1 proteins of extant deltaproteobacteria derive from a common ancestor (Fig. 4A and M.D.E. results not shown), as predicted by the hypothesis of Thomas et al. (2008). The common ancestor is related to current aerobic Flavobacteriia, which possess a previously unrecognized COX operon split in two gene clusters, separated by a minimum of 11 genes from each other. The upstream part of the cluster ends with COX1, followed by the Holliday junction recombinase protein (Yamada et al. 2002). The downstream part of the cluster starts instead with the assembly protein CtaB (protoheme farnesyl-transferase, see Chapter two), followed by two consecutive forms of COX3, a single COX4 protein and a hypothetical short protein, finally ending with the Cu-assembly protein SCO1. This looks like a hybrid split cluster between different operon subtypes of A family oxidases and appears to be characteristic of taxa belonging to the CFB *phylum* (Fig. 4B and M.D.E. unpublished results).

The SCO1 protein is required for the assembly of Cu in HCO oxidases and generally lies at the beginning of the COX operons that are present in deltaproteobacteria (Fig. 4B, cf. Lamrabet et al. 2011). Such operons characteristically contain the catalytic COX2 subunit, which often has two *c*-cytochromes fused at the C-terminus (Lobo et al. 2008), towards the end of the gene cluster as in *Desulfovibrio* and *Geobacter*. In contrast, the COX2 subunit is positioned at the beginning of most other forms of COX operons (Fig. 4B, cf. Lamrabet et al. 2011, Degli Esposti et al. 2014). In addition, the COX operon of Myxococcales is linked upstream to the complete gene cluster for the Act complex, as in subtype operon a-I of alphaproteobacteria (Fig. 4B, cf. Degli Esposti et al. 2014, see also Lamrabet et al. 2011). Further analysis is under way to define the ever increasing variants in the gene cluster for the A family of terminal oxidases, as partially reported in Chapter seven.

Important conclusions emerge from the novel analysis of COX operons presented here. First, the link between Act and subtype operon a-I for aa3 oxidase is much more ancient than previously considered (Degli Esposti 2014) and appears to have been vertically inherited from aerobic Flavobacteriales to proteobacteria via Myxococcales (Fig. 4A). Secondly, the genetic linkage with the Act complex has been lost in Bdellovibrionales (Fig. 4A), and consequently also in the newly proposed class of Oligoflexia, to which they have been recently assigned (Hahn et al. 2017). Thirdly, the loss of the Act linkage has been accompanied by a unique re-shuffling of the COX genes in sulfur-reacting epsilonproteobacteria of the

Sulfurospirillum genus and deltaproteobacteria of the *Desulfovibrio* and *Geobacter* groups (Fig. 4A). Fourth, this re-shuffling was entwined with the unusual fusion of the COX2 genes with a second *c*-type cytochrome (Lobo et al. 2008), except for the gene cluster of the *Geobacter* group (Fig. 4A, cf. Lamrabet et al. 2011). And finally, the reshuffled COX operon of epsilon- and deltaproteobacteria can be considered an evolutionary dead-end, since it has not been transmitted to other bacteria (so far).

Ultimately, the hypothesis of Thomas et al. (2008) regarding the ancestry of deltaproteobacteria now appears to be fundamentally right, even if it looks counter-intuitive in light of the predominantly anaerobic metabolism of these organisms (Goldman et al. 2007, Barton and Fauque 2009). All A family terminal oxidases of extant deltaproteobacteria do derive from a common ancestral COX operon, which is currently split in Flavobacteriales; these bacteria have also transmitted their unusual C family oxidase to the Myxococcales (see Chapter three). Hence, the respiratory chain of *Myxococcus* (Fig. 3B) has been vertically inherited rather than being shaped by cumulative LGT events as proposed earlier (Goldman et al. 2007). The same consideration may well apply to the respiratory chain of Bdellovibrionales, but not to the aa3 oxidases of *Sulfurospirillum, Desulfovibrio* and *Geobacter*, which appear to have been acquired from an ancestral organism related to contemporary *Flavobacterium*. Most likely, this acquisition derived from a LGT event because the underlying energy metabolism of deltaproteobacteria SRB is anaerobic and has been inherited from sulfate-reducing Nitrospirae, which are strictly anaerobes and generally lack terminal aa3 oxidases (Frank et al. 2016). Alternatively, the common progenitor of Nitrospirae and proteobacteria (Lin et al. 2017) might have evolved the subtype of A family oxidases that is found in contemporary deltaproteobacteria together with similar terminal oxidases that have been subsequently lost in most Nitrospirae, remaining scattered in metagenomic taxa affiliated to this *phylum*, as will be discussed in Chapter seven.

In conclusion, detailed genomic analysis has thrown new light on the aerobic metabolism of deltaproteobacteria and its evolutionary history.

Zetaproteobacteria—an Updated Overview

Zetaproteobacteria constitute the smallest and physiologically most homogeneous class of the *phylum* proteobacteria (Table 1). It was originally introduced following the description of a new lineage of deep ocean bacteria that live by oxidizing Fe^{II} in hydrothermal habitats rich in iron but poor in oxygen (Emerson et al. 2007, 2010, Singer et al. 2011, 2013). The term 'zeta' comes for the sixth letter of the Greek alphabet, for this is the sixth class described in the *phylum* proteobacteria (Emerson et al. 2007). The type genus is microaerophilic *Mariprofundus*, which has two different lifestyles: a motile stage that does not oxidize Fe^{II} and an attached stage that actively oxidizes Fe^{II}, producing encrusted helical stalks (Emerson et al. 2007, Singer et al. 2011). The motile stage has anaerobic metabolism and actively seeks marine zones with gradients of oxygen, so as to activate its chemolithotrophic metabolism that requires any form of reduced iron (Fe^{II}), from $FeCl_2$ to $FeSO_4$ (Singer et al. 2011). The oxidized iron products, Fe-oxyhydroxides, form helical precipitates outside the cells, encrusted in the stalk material that is produced by the bacteria and is more convoluted than that formed by freshwater Fe-oxidizing organisms such as betaproteobacterial *Gallionella* (Emerson et al. 2010). Bacteria similar to zetaproteobacteria were already present 1.9 billion years ago as deduced from geochemical evidence (Planavsky et al. 2009, Fullerton et al. 2017).

Contrary to other iron-oxidizing bacteria such as Acidithiobacillia, which oxidize Fe^{II} at low pH disfavouring its spontaneous reaction with oxygen, zetaproteobacteria are able to oxidize Fe^{II} under neutral or circumneutral conditions, thereby competing with the rapid spontaneous auto-oxidation of Fe^{II} at neutral pH (Emerson et al. 2010). This remarkable physiology depends upon functional properties that, albeit poorly known a few years ago (Emerson et al. 2010), are gradually emerging from genomic analysis of the organisms affiliated to zetaproteobacteria (Mori et al. 2017, Fullerton et al. 2017, Chiu et al. 2017), now also found in freshwater environments (McBeth et al. 2016, Emerson et al. 2016, Probst et al. 2017). Currently, there are over 30 full genomes of zetaproteobacteria (Table 1) and affiliated organisms such as Proteobacteria bacterium CG1_02_64_396 (Probst et al. 2017), provisionally called proteozeta1 (Degli Esposti 2017b). Although only two genomes are completed so far (Chiu et al. 2017), the estimated completeness of many other zetaproteobacterial genomes is over 95% (Singer et al. 2011, Degli Esposti 2017b—Table 2), thereby providing high quality genomic information that can be analysed in detail, as summarized below.

It has been discovered early that *Mariprofundus* strains fix carbon via the Benson-Calvin cycle (Emerson et al. 2007), using two different types of ribulose-1,5-biphosphate carboxylase (RuBisCo) enzymes with different affinity for CO_2 (Singer et al. 2011). This property is shared with photosynthetic purple bacteria and sulfur-oxidizing symbionts of marine invertebrates (see next Chapter five). Unlike such organisms, zetaproteobacteria are essentially specialized chemoautotrophs that utilize Fe^{II} for their energy source and survival, being unable to utilize sulfur or organic compounds as electron donors for their energy metabolism (Emerson et al. 2010, Chiu et al. 2017). However, metagenomic zetaproteobacteria found in underground aquifers very likely utilize H_2 as additional source of electrons under anaerobic conditions, because their genome contains the operons for different types of [NiFe]-hydrogenases (Probst et al. 2017), most of which transfer electrons to ferredoxin (Vignais and Billoud 2007, Greening et al. 2016). These hydrogenases, which were not considered in earlier studies, are also present in marine zetaproteobacteria such as *Mariprofudus* (Emerson et al. 2010, Singer et al. 2011, 2013, Makita et al. 2017, Mori et al. 2017, Chiu et al. 2017). Recently, the marine zetaproteobacterium *Ghiorsea bivora* (previously called zetaproteobacteria bacterium TAG-1) has been reported to utilize both H_2 and Fe^{II} as electron donor for its chemolithotrophic lifestyle (Mori et al. 2017). Genomic analysis by the author indicates that this physiology may well apply to most zetaproteobacteria. Indeed, *Mariprofundus* and other zetaproteobacteria contain various soluble ferredoxins of

Table 2. Bioenergy systems in zetaproteobacteria A. The table lists the bioenergetic enzymes used under anaerobic conditions: complex I, complex II (SDH), bc1, Rnf complex, Pyruvate-Ferredoxin Oxidoreductases (PFO), and [NiFe]-hydrogenases, as well as the cyc2 homologs for Fe^{II} oxidation. The organisms in bold are marine while the others are from underground aquifers (Emerson et al. 2016, Probst et al. 2017). B. Variations in the terminal oxidases of the organisms in A and of other marine zetaproteobacteria. While the cbb3 oxidase can be considered as part of the core genome of the class (Barco et al. 2015, Mori et al. 2017, Fullerton et al. 2017), other HCO oxidases appear to be present in marine zetaproteobacteria (Field et al. 2015, Fullerton et al. 2017, Mori et al. 2017, Chiu et al. 2017, Tully et al. 2018). Note that heme a synthase, CtaA, is unknown in metagenomic taxa (Zetabin) reported by Fullerton et al. (2017). Only high quality genomes (> 95% complete) are listed.

A. anaerobic traits

organism	Proteins	% Completeness*	Complex I	SDH	bc1	Rnf complex	PFO	other anaerobic traits
Mariprofundus ferrooxydans PV-1	2,694	94.6 #, 98.5^	yes	yes	yes	ABCDGE	yes	**Act, NirD, NiFe hydrogenases, cyc2**
Mariprofundus micogutta	2,417	100^	yes	yes	yes	ABCDGE		NirBD, NiFe hydrogenases, cyc2...
Zetaproteobacteria bacterium CG2_30_59_37	2,247	98.6	yes	yes	yes	ABCDGE		NiFe hydrogenases, cyc2...
Zetaproteobacteria bacterium CG1_02_49_23	2,550	97.1	yes	yes	yes	ABCDG, DGEH	yes	NiFe hydrogenases, cyc2...
Zetaproteobacteria bacterium CG1_02_55_237	2,209	98.6	yes	yes	yes	ABCDGE		NiFe hydrogenases, cyc2...
Zetaproteobacteria bacterium CG2_30_46_52	2,190	96.4	yes	yes	yes	ACDGE, B		NiFe hydrogenases, cyc2...
Zetaproteobacteria bacterium TAG-1 Ghiorsea	2,111	96.4	yes	yes	yes	ABD, C, E		**Act, NirB, NiFe hydrogenases, cyc2**
Proteobacterium bacterium CG1_02_64_396	2,532	94.2*, 94.6#, 98^	yes	yes	yes	ABCDGE	yes	cyc2 OM cytochrome c Fe-oxidizing

B. aerobic traits

organism	Proteins	% Completeness*	HCO family			CtaA	bd oxidase
			C cbb3	A aa3	B ba3		
Mariprofundus ferrooxydans PV-1	2,694	94.6 #, 98.5^	2				yes
Zetaproteobacteria bacterium CG2_30_59_37	2,247	98.6	2				yes
Zetaproteobacteria bacterium CG1_02_49_23	2,550	97.1	yes				
Zetaproteobacteria bacterium CG1_02_55_237	2,209	98.6	yes				
Zetaproteobacteria bacterium CG2_30_46_52	2,190	96.4	yes				
Zetaproteobacteria bacterium TAG-1 Ghiorsea	2,111	96.4	yes		yes		
Mariprofundus micogutta	2,417	100^	2	yes	yes	yes	
Mariprofundus aestuarium	2,427	100	partial	yes	yes	yes	
Mariprofundus ferrinatatus	2,237	100	partial	yes	yes	yes	
Zetabin040	2,812	98.74^	3			?	
Zetabin042	2,707	97.06^	4			?	
Zetabin066	4,265	99.58^	4			?	
Zetabin011	2,472	96.64^	3	2		?	
Zetabin050	4,118	95.66^	3	2		?	
Zetabin084	5,555	95.77^	6	yes		?	
Zetabin089	8,861	98.59^	5	2		?	
Zetaproteobacteria bacterium isolate NORP15	2,509	98.28^	yes			yes	yes

* Usually calculated by the approach of Rinke et al. (2013); #Calculated by the program BUSCO (Simão et al. 2015); ^Estimated with other methods.

both 2Fe2S and 4Fe4S type (Barco et al. 2015, Mori et al. 2017), an example for the latter being a NuoI-like protein of 84 aa of *Mariprofundus ferrooxydans* strain JV-1 (accession N48680). Soluble ferredoxins could function as electron acceptors for [NiFe]-hydrogenases, as well as pyruvate- or 2-oxoglutarate-ferredoxin oxidoreductases, the genes of which are commonly present in the genomes of zetaproteobacteria (e.g., EAU55406 of pyruvate:ferredoxin oxidoreductase, PFO, in *Mariprofundus ferrooxydans* strain PV-1, Barco et al. 2015–Table 2A). These enzymes form part of the rTCA cycle in anaerobic bacteria (see the section on epsilonproteobacteria earlier in this chapter, cf. Campbell et al. 2006) but have hardly been considered before in zetaproteobacteria (Emerson et al. 2010, Singer et al. 2011, 2013, Barco et al. 2015).

Interestingly, all genomes of zetaproteobacteria analysed thus far code for the Rnf complex, generally in its entire RnfABCDGE operon (Chiu et al. 2017); only in freshwater zetaproteobacteria bacterium CG1_02_49_23 and marine *Ghiorsea bivora,* the Rnf operon is split into different gene clusters (Table 2A). The Rnf complex is probably the most ancient protonmotive electron transport system (see Chapter two) and normally re-oxidizes reduced ferredoxin to produce NADH and membrane potential, as we have previously seen in the respiratory chain of Bacteroidetes (Fig. 1 in Chapter three of the book). Therefore, it constitutes the energetically favoured channel for discharging the electrons accumulated in soluble ferredoxins by the concerted action of rTCA enzymes and [NiFe]-hydrogenases. At the same time, the Rnf complex produces NADH, which can be used for energy storage in the form of polyhydroxyalkanoates (PHA, Singer et al. 2011) or for biosynthetic purposes, generally after exchange with $NADP^+$ via transhydrogenase. NADH is also the substrate for complex I, which is present with a nuo14 operon in all zetaproteobacteria (Table 2A) and can contribute substantial protonmotive force if its product ubiquinol (QH_2) is sufficiently re-oxidized downstream. Of note, zetaproteobacteria are the first group of bacteria we have encountered in this book that synthesize ubiquinone (Q) and not menaquinone as their membrane electron carrier (Degli Esposti 2017a,b, see also below).

The illustration shown in Fig. 5A represents the most detailed reconstruction of the anaerobic respiratory chain of typical zetaproteobacteria, as deduced from the bioenergy systems found in currently available genomes (Table 2A). Table 2B lists instead the variations in the distribution of terminal oxidases in recently described zetaproteobacteria. While the cbb3 oxidase can be considered as part of the core genome of the class (Barco et al. 2015, Mori et al. 2017, Fullerton et al. 2017), other HCO oxidases have been recently reported to be present in subgroups of marine zetaproteobacteria (Field et al. 2015, Fullerton et al. 2017, Mori et al. 2017, Chiu et al. 2017, Tully et al. 2018–Table 2B). These oxidases have been considered to be of the aa3 type, hence belonging to the A family of HCO (Field et al. 2015, Fullerton et al. 2017). However, some of the reported oxidases in marine zetaproteobacteria appear to have signatures typical of the B family of HCO (ba3 type). Table 2B also lists the presence of heme *a* synthase, CtaA, in zetaproteobacteria. This enzyme is necessary for both family A and B of HCO

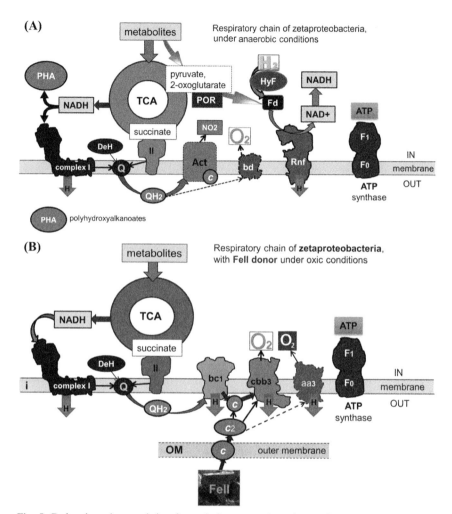

Fig. 5. Deduced respiratory chain of a typical zetaproteobacterium such as *Mariprofundus*. (A) Bioenergetics under anaerobic conditions. Dashed systems are not conserved in all zetaproteobacteria. POR, pyruvate or oxoglutarate oxidoreductase. HyF symbolizes ferredoxin-reducing hydrogenases. (B) Bioenergetic model of zetaproteobacteria under (micro)aerobic conditions with Fe^{II} as electron donor. The symbol *c2* indicates a homolog of cyt *c2* of *Rhodopseudomonas* that is involved in Fe^{II} oxidation. Family A aa3 oxidases are present in some zetaproteobacteria (Table 2B) and may function as terminal oxidases at levels of oxygen higher than micro-oxic (dashed arrow; Chiu et al. 2017, Fullerton et al. 2017).

oxidases (see the section on terminal oxidases in Chapter two), but its distribution is restricted to five zetaproteobacterial taxa which do have genes for family A oxidases (Table 2B). In principle, then, only these zetaproteobacteria would have *bona fide* heme *a*-containing cytochrome oxidases, leaving the rest of the class in uncertain terms regarding their A family oxidases (see further discussion on the topic in Chapter seven). Notably, the A family oxidases found in zetaproteobacteria do not conform to the molecular features characteristic of A1 type, low affinity cytochrome *c* oxidases (Pereira et al. 2001, Fullerton et al. 2017).

Figure 5A represents the essential energy metabolism of the anaerobic motile phase of *Mariprofundus* in a way which is significantly different from the previously reported cartoons for the respiratory chain of zetaproteobacteria (Singer et al. 2011, 2013, Barco et al. 2015, Chiu et al. 2017). Such representations ignored the ferredoxin relay of electrons via protonmotive Rnf complex, which is now sustained by a recent study (Mori et al. 2017), assuming that both the bc1 complex and complex I would work in reverse, to ultimately produce NADH for energy storage and biosynthetic needs (Singer et al. 2011, Barco et al. 2015, Chiu et al. 2017). This situation would resemble the *Nitrospira* scheme of electron transport (Fig. 3 in Chapter three) and is similar to the earlier scheme reported for the respiratory chain of another Fe-oxidizing organism, *Acidithiobacillus* (Elbehti et al. 2000). Of note is that *Acidithiobacillus* contains various [NiFe]-hydrogenases too (Greening et al. 2016). Conversely, *Mariprofundus ferrooxidans* contains a bd ubiquinol oxidase that could efficiently re-oxidize the ubiquinol produced by complex I and other dehydrogenases such as complex II, even at very low levels of ambient oxygen. Such a system would bypass the bc1 complex, as well as the alternative complex III (Act) which has been considered fundamental in the energy metabolism of *Mariprofundus* (Singer et al. 2011, Barco et al. 2015). However, both the bd oxidase and the Act complex are scarcely present in other zetaproteobacteria (Table 2). It follows that the respiratory chain of most zetaproteobacteria would rely on the bc1 complex and the cbb3 oxidase to re-oxidize ubiquinol produced by Complex I or other Q-reaching dehydrogenases at low ambient levels of oxygen (Fig. 5B). In the absence of oxygen, the accumulated NADH would be diverted instead to storage products or biosynthetic reactions (Fig. 5A), as in the case of nodulating symbionts experiencing a similar accumulation of ubiquinol (Degli Esposti and Martinez-Romero 2016).

Generally, the Act complex reacts with MQ (Marreiros et al. 2106), which is not present in zetaproteobacteria (Degli Esposti 2017b). Nevertheless, under the Fe^{II}-oxidizing conditions used to cultivate this organism, the Act complex has been postulated to transfer electrons from periplasmic *c* cytochromes, reduced upstream by the Fe^{II} electron donor, to the Q pool, from which it would stimulate reverse electron transfer within complex I (Barco et al. 2015). Thermodynamically, such a postulation looks unlikely, given the large midpoint potential span from soluble *c* cytochromes to the N2 cluster of complex I, which reacts directly with Q (Degli Esposti 2015, 2017a). Moreover, the combined reverse electron transfer via the bc1 complex, which is much more common than the Act complex in zetaproteobacteria (Table 2A), plus complex I would require a minimum of three protons pumped inward (against the normal membrane potential of the cytoplasmic membrane) per electron transported. Such an energy drain could be hardly compensated by the protonmotive activity associated with downstream oxidation of reduced periplasmic *c* cytochromes by either cbb3 or aa3 terminal oxidases, which pump one proton per electron (Sousa et al. 2012).

The considerations above would imply that previous schemes for the respiratory chain of *Mariprofundus* (Barco et al. 2015) may be bioenergetically impossible, as other models reported by Ilbert and Bonnefoy (2013). ATP might promote the reverse ATPase reaction of the F1Fo ATP synthase, thereby pumping protons inside and sustaining reverse electron transport, as postulated earlier for *Acidithiobacillus* (Elbehti et al. 2000). *Mariprofundus* would thus need to consume its energy stores (Singer et al. 2011) to sustain reverse electron transport via Q while physiologically respiring Fe^{II} in a bioenergetically useless process, since no net ATP would ultimately be generated. Finding this proposition improbable, the author elaborated an alternative view of Fe^{II} oxidation in zetaproteobacteria. Figure 5B specifically illustrates this view by depicting the respiratory chain in the microaerophilic phase of zetaproteobacteria using Fe^{II} as major electron donor; no reverse electron transfer is required, while all the core bioenergy systems of the zeta class (Table 2) could contribute to protonmotive reactions sustaining ATP synthesis. The normally functioning TCA cycle, which is present in zetaproteobacteria (Singer et al. 2011, Makita et al. 2017), would produce all the necessary NADH and metabolites to sustain life under the same conditions. The electrons deriving from the TCA cycle and those extracted from external Fe^{II} converging into *c* cytochromes would be funneled to oxygen by the cbb3 oxidase (Barco et al. 2015, Fig. 5B). When O_2 levels are above micro-oxic conditions, the activity of cbb3 oxidase can be supplemented by that of other terminal oxidases having less affinity for oxygen, in particular those of the A family of HCO (Chiu et al. 2017). This would explain the much larger distribution of aa3 oxidases than that of bd ubiquinol oxidases in zetaproteobacteria (Table 2B), given that ubiquinol oxidase has an oxygen affinity comparable with that of cbb3 oxidase (see Chapter two).

Phylogeny of zetaproteobacteria

Zetaproteobacteria constitute a monophyletic group which lies close to the class of gammaproteobacteria (Fig. 6A). Indeed, in early studies, *Mariprofundus* strains were tentatively classified within gammaproteobacteria (Emerson et al. 2007, Singer et al. 2011). Generally, the position of the zeta class within the *phylum* proteobacteria is intermediate between deep branching alphaproteobacteria such as *Magnetococcus* and deep branching gammaproteobacteria such as *Thiohalorhabdus* (see Chapter five). Proteins of *Thiohalorhabdus*, moreover, often cluster with those of Proteobacteria bacterium CG1_02_64_396 (Table 2A), now affiliated with the zeta class as Proteozeta 1 (Degli Esposti 2017b). The same intermediate position emerges from the trees of proteins belonging to Fe-oxidizing bacteria (Fig. 6A). Hence, the phylogeny of zetaproteobacteria appears to be the clearest and probably the most solid among the classes of proteobacteria discussed in the present chapter.

Having concluded this overview of zeta and other classes of proteobacteria, it is important to visualize their overall phylogeny. Figure 6A attempts to undertake this experimentally, using the beta subunit of the F1F0 ATP synthase complex, a conserved protein which is present in all the bacterial groups mentioned in this

(A) phylogenetic tree of the beta subunit of F1F0 ATP synthase

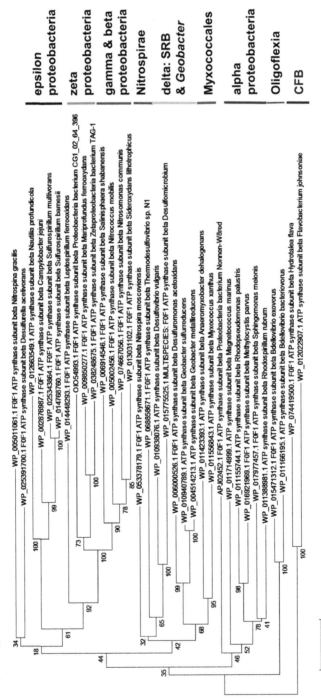

Fig. 6 contd. ...

...*Fig. 6 contd.*

(B)

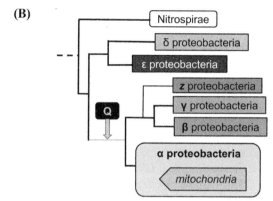

Fig. 6. Phylogenetic sequence of the bacterial taxa examined in this chapter and also previous chapter three. (A) The NJ tree of F1F0 ATP synthase subunit beta as universal protein marker was obtained with 1000 bootstraps, the percentage value of which is reported close to each node. (B) Sketch of the consensus phylogeny of proteobacteria. Q, ubiquinone—see Fig. 7 for the evolution of its biosynthesis.

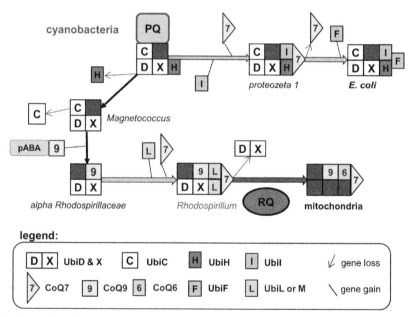

Fig. 7. Model for the evolution of the pathway of ubiquinone and rhodoquinone (RQ) biosynthesis in bacteria. pABA, *p*-aminobenzoate, the alternative precursor for the Q ring in alphaproteobacteria and mitochondria (Degli Esposti 2017b).

chapter, and also those discussed in previous Chapter three. The representative phylogenetic tree obtained with this protein (Fig. 6A) is then compared with the consensus view of the phylogeny of proteobacteria, sketched in Fig. 6B (cf. Degli

Esposti 2016), for there appears to be no single protein marker or concatenated series of proteins that can accurately visualize such a pattern.

Biosynthesis of Ubiquinone, a Summary

Zetaproteobacteria possess all the genes known to be involved in the biosynthetic pathway of ubiquinone in *E. coli* (Aussel et al. 2014, Degli Esposti 2017b). Proteozeta1, moreover, shows the same syntenic groups as *E. coli* but appears to have the deepest branching forms of many proteins encoded by the *Ubi* genes (Degli Esposti 2017b). However, this pathway of Q biosynthesis originally evolved for producing the Q analogue plastoquinone (PQ) in cyanobacteria (Pfaff et al. 2014, Degli Esposti 2017b), as illustrated in Fig. 7. Previous speculations suggesting that the pathway of Q biosynthesis evolved as an adaptation of the futalosine pathway for MQ biosynthesis (see Chapter two; cf. Ravcheev and Thiele 2016) can thus be discounted. Notably, the pathway of Q biosynthesis in mitochondria derives from a parallel system that originated in alphaproteobacteria of the Rhodospirillaceae family, which also invented the biosynthesis of the Q derivative, rhodoquinone (RQ), as sketched in Fig. 7 (Degli Esposti 2017b).

Acknowledgements

The work for this chapter has benefited from discussion with several colleagues at CCG (Mexico) and around the world along many years of research in bioenergetics and bacterial genomics.

References

Anantharaman, K., C.T. Brown, L.A. Hug, I. Sharon, C.J. Castelle, A.J. Probst et al. 2016. Thousands of microbial genomes shed light on interconnected biogeochemical processes in an aquifer system. Nat. Commun. 7: 13219. doi:10.1038/ncomms13219.

Aussel, L., F. Pierrel, L. Loiseau, M. Lombard, M. Fontecave and F. Barras. 2014. Biosynthesis and physiology of coenzyme Q in bacteria. Biochim. Biophys. Acta. 1837: 1004–1011. doi:10.1016/j.bbabio.2014.01.015.

Barco, R.A., D. Emerson, J.B. Sylvan, B.N. Orcutt, M.E. Jacobson Meyers, G.A. Ramírez et al. 2015. New Insight into Microbial Iron Oxidation as Revealed by the Proteomic Profile of an Obligate Iron-Oxidizing Chemolithoautotroph. Appl. Environ. Microbiol. 81: 5927–5937. doi:10.1128/AEM.01374-15.

Barton, L.L. and G.D. Fauque. 20009. Biochemistry, physiology and biotechnology of sulfate-reducing bacteria. Adv. Appl. Microbiol. 68: 41–98. doi:10.1016/S0065-2164(09)01202-7.

Borisov, V.B., R.B. Gennis, J. Hemp and M.I. Verkhovsky. 2011. The cytochrome *bd* respiratory oxygen reductases. Biochim. Biophys. Acta. 1807: 1398–413.

Butler, J.E., N.D. Young and D.R. Lovley. 2010. Evolution of electron transfer out of the cell: comparative genomics of six Geobacter genomes. BMC Genomics 11: 40. doi:10.1186/1471-2164-11-40.

Campbell, B.J., A.S. Engel, M.L. Porter and K. Takai. 2006. The versatile epsilon-proteobacteria: key players in sulphidic habitats. Nat. Rev. Microbiol. 4: 458–468.

Campbell, B.J., J.L. Smith, T.E. Hanson, M.G. Klotz, L.Y. Stein et al. 2009. Adaptations to submarine hydrothermal environments exemplified by the genome of Nautilia profundicola. PLoS Genet. 5: e1000362.

Chistoserdova, L. 2017. Application of omics approaches to studying methylotrophs and methylotroph comunities. Curr. Issues Mol. Biol. 24: 119–142. doi:10.21775/cimb.024.119.

Chiu, B.K., S. Kato, S.M. McAllister, E.K. Field and C.S. Chan. 2017. Novel pelagic iron-oxidizing zetaproteobacteria from the chesapeake bay oxic-anoxic transition zone. Front. Microbiol. 8: 1280. doi:10.3389/fmicb.2017.01280.

Ciccarelli, F.D., T. Doerks, C. von Mering, C.J. Creevey, B. Snel and P. Bork. 2006. Toward automatic reconstruction of a highly resolved tree of life. Science 311: 1283–1287.

Colston, T.J. and C.R. Jackson. 2016. Microbiome evolution along divergent branches of the vertebrate tree of life: What is known and unknown. Mol. Ecol. 25: 3776–3800. doi:10.1111/mec.13730.

Coppi, M.V. 2005. The hydrogenases of Geobacter sulfurreducens: a comparative genomic perspective. Microbiology 151: 1239–1254.

Dahl, C. 2015. Cytoplasmic sulfur trafficking in sulfur-oxidizing prokaryotes. IUBMB Life 67: 268–274. doi:10.1002/iub.1371.

da Silva, S.M., I. Pacheco and I.A. Pereira. 2012. Electron transfer between periplasmic formate dehydrogenase and cytochromes c in Desulfovibrio desulfuricans ATCC 27774. J. Biol. Inorg. Chem. 17: 831–838. doi:10.1007/s00775-012-0900-5.

Davidov, Y., D. Huchon, S.F. Koval and E. Jurkevitch. 2006. A new alpha-proteobacterial clade of Bdellovibrio-like predators: implications for the mitochondrial endosymbiotic theory. Environ. Microbiol. 8: 2179–2188.

Degli Esposti, M. 2014. Bioenergetic evolution in proteobacteria and mitochondria. Genome Biol. Evol. 6: 3238–3251. doi:10.1093/gbe/evu257.

Degli Esposti, M., B. Chouaia, F. Comandatore, E. Crotti, D. Sassera, P.M. Lievens et al. 2014. Evolution of mitochondria reconstructed from the energy metabolism of living bacteria. PLoS One. 9: e96566. doi: 10.1371/journal.pone.0096566.

Degli Esposti, M. 2015. Genome Analysis of Structure-Function Relationships in Respiratory Complex I, an Ancient Bioenergetic Enzyme. Genome Biol. Evol. 8: 126–147. doi:10.1093/gbe/evv239.

Degli Esposti, M., T. Rosas-Pérez, L.E. Servín-Garciadueñas, L.M. Bolaños, M. Rosenblueth and E. Martínez-Romero. 2015. Molecular evolution of cytochrome bd oxidases across proteobacterial genomes. Genome Biol. Evol. 7: 801–820. doi:10.1093/gbe/evv032.

Degli Esposti, M., D. Cortez, L. Lozano, S. Rasmussen, H.B. Nielsen and E. Martinez Romero. 2016. Alpha proteobacterial ancestry of the [Fe-Fe]-hydrogenases in anaerobic eukaryotes. Biol. Direct. 11: 34. doi:10.1186/s13062-016-0136-3.

Degli Esposti, M. and E. Martinez-Romero. 2016. A survey of the energy metabolism of nodulating symbionts reveals a new form of respiratory complex I. FEMS Microbiol. Ecol. 92: fiw084. doi:10.1093/femsec/fiw084.

Degli Esposti, M. 2016. Late mitochondrial acquisition, Really? Genome Biol. Evol. 8: 2031–2035. doi:10.1093/gbe/evw130.

Degli Esposti, M. and E. Martinez-Romero. 2017. The functional microbiome of arthropods. PLoS One 12: e0176573. doi:10.1371/journal.pone.0176573.

Degli Esposti, M. 2017a. The long story of mitochondrial DNA and respiratory complex I. Front. Biosci. (Landmark Ed). 22: 722–731.

Degli Esposti, M. 2017b. A Journey across Genomes Uncovers the Origin of Ubiquinone in Cyanobacteria. Genome Biol. Evol. 9: 3039–3053. doi:10.1093/gbe/evx225.

Devers-Lamrani, M., A. Spor, A. Mounier and F. Martin-Laurent. 2016. Draft genome sequence of *Pseudomonas* sp. strain ADP, a bacterial model for studying the degradation of the herbicide atrazine. Genome Announc. 4. pii: e01733-15. doi:10.1128/genomeA.01733-15.

Ducluzeau, A.L., S. Ouchane and W. Nitschke. 2008. The cbb3 oxidases are an ancient innovation of the domain bacteria. Mol. Biol. Evol. 25: 1158–1166. doi:10.1093/molbev/msn062.

Elbehti, A., G. Brasseur and D. Lemesle-Meunier. 2000. First evidence for existence of an uphill electron transfer through the bc(1) and NADH-Q oxidoreductase complexes of the acidophilic obligate chemolithotrophic ferrous ion-oxidizing bacterium Thiobacillus ferrooxidans. J. Bacteriol. 182: 3602–3606.

Emerson, D.M., J.A. Rentz, T.G. Lilburn, R.E. Davis, H. Aldrich, C. Chan et al. 2007. A novel lineage of proteobacteria involved in formation of marine Fe-oxidizing microbial mat communities. PLoS One 2: e667.

Emerson, D., E.J. Fleming and J.M. McBeth. 2010. Iron-oxidizing bacteria: an environmental and genomic perspective. Annu. Rev. Microbiol. 64:561–83. doi:10.1146/annurev.micro.112408.134208.

Emerson, J.B., B.C. Thomas, W. Alvarez and J.F. Banfield. 2016. Metagenomic analysis of a high carbon dioxide subsurface microbial community populated by chemolithoautotrophs and bacteria and archaea from candidate phyla. Environ. Microbiol. 18: 1686–1703. doi:10.1111/1462-2920.12817.

Estevez-Canales, M., A. Kuzume, Z. Borjas, M. Füeg, D. Lovley, T. Wandlowski et al. 2015. A severe reduction in the cytochrome C content of Geobacter sulfurreducens eliminates its capacity for extracellular electron transfer. Environ. Microbiol. Rep. 7: 219–226. doi:10.1111/1758-2229.12230.

Feist, A.M., H. Nagarajan, A.E. Rotaru, P.L. Tremblay, T. Zhang, K.P. Nevin et al. 2014. Constraint-based modeling of carbon fixation and the energetics of electron transfer in Geobacter metallireducens. PLoS Comput. Biol. 10:e1003575. doi:10.1371/journal.pcbi.1003575.

Field, E.K., A. Sczyrba, A.E. Lyman, C.C. Harris, T. Woyke, R. Stepanauskas et al. 2015. Genomic insights into the uncultivated marine Zetaproteobacteria at Loihi Seamount. ISME J. 9: 857–870. doi:10.1038/ismej.2014.183.

Florentino, A.P., A.J. Stams and I. Sánchez-Andrea. 2017. Genome sequence of Desulfurella amilsii Strain TR1 and Comparative Genomics of Desulfurellaceae Family. Front. Microbiol. 8: 222. doi:10.3389/fmicb.2017.00222.

Frank, Y.A., V.V. Kadnikov, A.P. Lukina, D. Banks, A.V. Beletsky, A.V. Mardanov et al. 2016. Characterization and genome analysis of the first facultatively alkaliphilic thermodesulfovibrio isolated from the deep terrestrial subsurface. Front. Microbiol. 7: 2000. doi:10.3389/fmicb.2016.02000.

Fullerton, H., K.W. Hager, S.M. McAllister and C.L. Moyer. 2017. Hidden diversity revealed by genome-resolved metagenomics of iron-oxidizing microbial mats from Lō'ihi Seamount, Hawai'i. ISME J.11: 1900–1914. doi:10.1038/ismej.2017.40.

Goldman, B., S. Bhat and LJ. Shimkets. 2007. Genome evolution and the emergence of fruiting body development in Myxococcus xanthus. PLoS One 2: e1329.

Greening, C., A. Biswas, C.R. Carere, C.J. Jackson, M.C. Taylor, M.B. Stott et al. 2016. Genomic and metagenomic surveys of hydrogenase distribution indicate H2 is a widely utilised energy source for microbial growth and survival. ISME J. 10: 761–777. doi:10.1038/ismej.2015.153.

Gupta, R.S. 2000. The phylogeny of proteobacteria: relationships to other eubacterial phyla and eukaryotes. FEMS Microbiol. Rev. 24: 367–402.

Gupta, R.S. and P.H. Sneath. 2007. Application of the character compatibility approach to generalized molecular sequence data: branching order of the proteobacterial subdivisions. J. Mol/ Evol. 64: 90–100.

Haase, D., B. Hermann, O. Einsle and J. Simon. 2017. Epsilonproteobacterial hydroxylamine oxidoreductase (εHao): characterization of a 'missing link' in the multihaem cytochrome c family. Mol. Microbiol. 105: 127–138. doi:10.1111/mmi.13690.

Hahn, M.W., J. Schmidt, U. Koll, M. Rohde, S. Verbarg, A. Pitt et al. 2017. *Silvanigrella aquatica* gen. nov., sp. nov., isolated from a freshwater lake, description of *Silvanigrellaceae* fam. nov. and *Silvanigrellales* ord. nov., reclassification of the order Bdellovibrionales in the class *Oligoflexia*, reclassification of the families *Bacteriovoracaceae* and *Halobacteriovoraceae* in the new order Bacteriovoracales ord. nov., and reclassification of the family *Pseudobacteriovoracaceae* in the order Oligoflexales. Int. J. Syst. Evol. Microbiol. 67: 2555–2568. doi:10.1099/ijsem.0.001965.

Herrmann, J., A. Abou Fayad and R. Müller. 2017. Natural products from myxobacteria: novel metabolites and bioactivities. Nat. Prod. Rep. 34: 135–160.

Hofreuter, D.M. Defining the metabolic requirements for the growth and colonization capacity of *Campylobacter jejuni*. Front. Cell. Infec. Microbiol. 4: 137. doi:10.3389/fcimb.2014.00137.

Hug, L.A., A. Stechmann and A.J. Roger. 2010. Phylogenetic distributions and histories of proteins involved in anaerobic pyruvate metabolism in eukaryotes. Mol. Biol. Evol. 27: 311–324.

Hug, L.A., B.J. Baker, K. Anantharaman, C.T. Brown, A.J. Probst, C.J. Castelle et al. 2016. A new view of the tree of life. Nat. Microbiol. 1: 16048. doi:10.1038/nmicrobiol.2016.48.

Ikeda-Ohtsubo, W., J.F. Strassert, T. Köhler, A. Mikaelyan, I. Gregor, A.C. McHardy et al. 2016. 'Candidatus Adiutrix intracellularis', an endosymbiont of termite gut flagellates, is the first representative of a deep-branching clade of *Deltaproteobacteria* and a putative homoacetogen. Environ. Microbiol. 18: 2548–2564. doi:10.1111/1462-2920.13234

Ilbert, M. and V. Bonnefoy. 2013. Insight into the evolution of the iron oxidation pathways. Biochim. Biophys. Acta. 1827: 161–175. doi:10.1016/j.bbabio.2012.10.001.

Imhoff, J.F. 2016. New Dimensions in Microbial Ecology-Functional Genes in Studies to Unravel the Biodiversity and Role of Functional Microbial Groups in the Environment. Microorganisms. 4. pii: E19. doi:10.3390/microorganisms4020019.

Ishii, S., S. Suzuki, T.M. Norden-Krichmar, A. Tenney, P.S. Chain, M.B. Scholz et al. 2013. A novel metatranscriptomic approach to identify gene expression dynamics during extracellular electron transfer. Nat. Commun. 4: 1601. doi:10.1038/ncomms2615.

Jormakka, M., K. Yokoyama, T. Yano, M. Tamakoshi, S. Akimoto, T. Shimamura et al. 2008. Molecular mechanism of energy conservation in polysulfide respiration. Nat. Struct. Mol. Biol. 15: 730–737. doi:10.1038/nsmb.1434.

Kaiser, D., M. Robinson and L. Kroos. 2010. Myxobacteria, polarity, and multicellular morphogenesis. Cold Spring Harb. Perspect. Biol. 2: a000380. doi:10.1101/cshperspect.a000380.

Karlsson, F.H., D.W. Ussery, J. Nielsen and I. Nookaew. 2011. A closer look at bacteroides: phylogenetic relationship and genomic implications of a life in the human gut. Microb. Ecol. 61: 473–485. doi:10.1007/s00248-010-9796-1.

Kern, M. and J. Simon. 2009. Electron transport chains and bioenergetics of respiratory nitrogen metabolism in *Wolinella succinogenes* and other *Epsilonproteobacteria*. Biochim. Biophys. Acta. 1787: 646–656. doi:10.1016/j.bbabio.2008.12.010.

Kröger, A., S. Biel, J. Simon, R. Gross, G. Unden and C.R. Lancaster. 2002. Fumarate respiration of Wolinella succinogenes: Enzymology, energetics and coupling mechanism. Biochim. Biophys. Acta. 1553: 23–38.

Kuever, J., F.A. Rainey and F. Widdel. 2005. Class IV. *Deltaproteobacteria* class nov. Bergey's Manual® of Systematic Bacteriology. Springer US, pp. 922–1144.

Lamrabet, O., L. Pieulle, C. Aubert, F. Mouhamar, P. Stocker, A. Dolla et al. 2011. Oxygen reduction in the strict anaerobe *Desulfovibrio vulgaris* Hildenborough: Characterization of two membrane-bound oxygen reductases. Microbiology 157: 2720–2732. doi:10.1099/mic.0.049171-0.

Land, M., L. Hauser, S.R. Jun, I. Nookaew, M.R. Leuze, T.H. Ahn et al. 2015. Insights from 20 years of bacterial genome sequencing. Funct. Integr. Genomics 15: 141–161. doi:10.1007/s10142-015-0433-4.

Lau, E., M.C. Fisher, P.A. Steudler and C.M. Cavanaugh. 2013. The methanol dehydrogenase gene, mxaF, as a functional and phylogenetic marker for proteobacterial methanotrophs in natural environments. PLoS One 8: e56993. doi:10.1371/journal.pone.0056993.

Lin, W., G.A. Paterson, Q. Zhu, Y. Wang, E. Kopylova, Y. Li et al. 2017. Origin of microbial biomineralization and magnetotaxis during the Archean. Proc. Natl. Acad. Sci. USA 114: 2171–2176. doi:10.1073/pnas.1614654114.

Lobo, S.A., C.C. Almeida, J.N. Carita, M. Teixeira and L.M. Saraiva. 2008. The haem-copper oxygen reductase of Desulfovibrio vulgaris contains a dihaem cytochrome c in subunit II. Biochim. Biophys. Acta. 1777: 1528–1534. doi:10.1016/j.bbabio.2008.09.007.

Lovley, D.R., D.E. Holmes and K.P. Nevin. 2004. Dissimilatory Fe(III) and Mn(IV) reduction. Adv. Microb. Physiol. 49: 219–86.

Makita. H., E. Tanaka, S. Mitsunobu, M. Miyazaki, T. Nunoura, K. Uematsu et al. 2017. Mariprofundus *Micogutta* sp. nov., a novel iron-oxidizing *Zetaproteobacterium* isolated from a deep-sea hydrothermal field at the Bayonnaise knoll of the Izu-Ogasawara arc, and a description of *Mariprofundales* ord. nov. and *Zetaproteobacteria classis* nov. Arch. Microbiol. 199: 335–346. doi:10.1007/s00203-016-1307-4.

Marreiros, B.C., A.P. Batista, A.M. Duarte and M.M. Pereira. 2013. A missing link between complex I and group 4 membrane-bound [NiFe] hydrogenases. Biochim. Biophys. Acta. 1827: 198–209.

Marreiros, B.C., F. Calisto, P.J. Castro, A.M. Duarte, F.V. Sena, A.F. Silva et al. 2016. Exploring membrane respiratory chains. Biochim. Biophys. Acta. 1857: 1039–1067. doi:10.1016/j.bbabio.2016.

McBeth, J.M., E.J. Fleming and D. Emerson. 2013. The transition from freshwater to marine iron-oxidizing bacterial lineages along a salinity gradient on the Sheepscot River, Maine, USA. Environ. Microbiol. Rep. 5: 453–463. doi:10.1111/1758-2229.12033.

McInerney, M.J., L. Rohlin, H. Mouttaki, U. Kim, R.S. Krupp, L. Rios-Hernandez et al. 2007. The genome of Syntrophus aciditrophicus: Life at the thermodynamic limit of microbial growth. Proc. Natl. Acad. Sci. USA 104: 7600–7605.

Meyer, J.L. and J.A. Huber. 2014. Strain-level genomic variation in natural populations of *Lebetimonas* from an erupting deep-sea volcano. ISME J. 8: 867–880. doi:10.1038/ismej.2013.206.

Miroshnichenko, M.L., F.A. Rainey, M. Rhode and E.A. Bonch-Osmolovskaya. 1999. *Hippea maritima* gen. nov., sp. nov., a new genus of thermophilic, sulfur-reducing bacterium from submarine hot vents. Int. J. Syst. Bacteriol. 49 Pt 3: 1033-8.

Mori, J.F., J.J. Scott, K.W. Hager, C.L. Moyer, K. Küsel and D. Emerson. 2017. Physiological and ecological implications of an iron- or hydrogen-oxidizing member of the Zetaproteobacteria, *Ghiorsea bivora*, gen. nov., sp. nov. ISME J. 11: 2624–2636. doi:10.1038/ismej.2017.132.

Müller, A.L., K.U. Kjeldsen, T. Rattei, M. Pester and A. Loy. 2015. Phylogenetic and environmental diversity of DsrAB-type dissimilatory (bi)sulfite reductases. ISME J. 9: 1152–1165. doi:10.1038/ismej.2014.208.

Muyzer, G. and A.J. Stams. 2008. The ecology and biotechnology of sulfate-reducing bacteria. Nat. Rev. Microbiol. 6: 441–454. doi:10.1038/nrmicro1892.

Nakai, R., M. Nishijima, N. Tazato, Y. Handa, F. Karray, S. Sayadi et al. 2014. *Oligoflexus tunisiensis* gen. nov., sp. nov., a Gram-negative, aerobic, filamentous bacterium of a novel proteobacterial lineage, and description of *Oligoflexaceae* fam. nov., *Oligoflexales* ord. nov. and *Oligoflexia classis* nov. Int. J. Syst. Evol. Microbiol. 64: 3353–3359. doi:10.1099/ijs.0.060798-0.

Nakai, R., T. Fujisawa, Y. Nakamura, T. Baba, M. Nishijima, F. Karray et al. 2016. Genome sequence and overview of Oligoflexus tunisiensis Shr3(T) in the eighth class Oligoflexia of the phylum Proteobacteria. Stand Genomic Sci. 11: 90. doi: 10.1186/s40793-016-0210-6.

Ormerod, K.L., D.L. Wood, N. Lachner, S.L. Gellatly, J.N. Daly, J.D. Parsons et al. 2016. Genomic characterization of the uncultured Bacteroidales family S24-7 inhabiting the guts of homeothermic animals. Microbiome 4: 36. doi:10.1186/s40168-016-0181-2.

Oyaizu, H. and C.R. Woese. 1985. Phylogenetic relationships among the sulfate respiring bacteria, myxobacteria and purple bacteria. Syst. Appl. Microbiol. 6: 257–263.

Pagani, I., A. Lapidus, M. Nolan, S. Lucas, N. Hammon, S. Deshpande et al. 2011. Complete genome sequence of *Desulfobulbus propionicus* type strain (1pr3). Stand Genomic Sci. 4: 100–110. doi:10.4056/sigs.1613929.

Parks, D.H., C. Rinke, M. Chuvochina, P.A. Chaumeil, B.J. Woodcroft, P.N. Evans et al. 2017. Recovery of nearly 8,000 metagenome-assembled genomes substantially expands the tree of life. Nat. Microbiol. doi:10.1038/s41564-017-0012-7.

Pfaff, C., N. Glindemann, J. Gruber, M. Frentzen and R. Sadre. 2014. Chorismate pyruvate-lyase and 4-hydroxy-3-solanesylbenzoate decarboxylase are required for plastoquinone biosynthesis in the cyanobacterium *Synechocystis* sp. PCC6803. J. Biol. Chem. 289: 2675–2686. doi:10.1074/jbc.M113.511709.

Planavsky, N., O. Rouxel, A. Bekker, R. Shapiro, P. Fralick and A. Knudsen. 2009. Iron-oxidizing microbial ecosystems thrived in late *Paleoproterozoic* redox-stratified oceans. Earth and Planetary Science Letters 286: 230–242.

Probst, A.J., C.J. Castelle, A. Singh, C.T. Brown, K. Anantharaman, I. Sharon et al. 2017. Genomic resolution of a cold subsurface aquifer community provides metabolic insights for novel microbes adapted to high CO(2) concentrations. Environ. Microbiol. 19: 459–474. doi:10.1111/1462-2920.13362.

Ramel, F., G. Brasseur, L. Pieulle, O. Valette, A. Hirschler-Réa, M.L. Fardeau et al. 2015. Growth of the obligate anaerobe Desulfovibrio vulgaris Hildenborough under continuous low oxygen concentration sparging: impact of the membrane-bound oxygen reductases. PLoS One 10: e0123455. doi:10.1371/journal.pone.0123455.

Rainey, F.A., R.T. Toalster and E. Stackebrandt. 1993. Desulfurella acetivorans, a thermophilic, acetate-oxidizing and sulfur-reducing organism, represents a distinct lineage within the *Proteobacteria*. Syst. Appl. Microbiol. 16: 373–379.

Ravcheev, D.A. and I. Thiele. 2016. Genomic analysis of the human gut microbiome suggests novel enzymes involved in quinone biosynthesis. Front. Microbiol. 7: 128. doi:10.3389/fmicb.2016.00128.

Reichenbach, H. 1984. Myxobacteria: a most peculiar group of social prokaryotes. pp. 1–50. *In*: Rosenberg E.(ed.). Myxobacteria, Springer-Verlag, New York.

Rinke, C., P. Schwientek, A. Sczyrba, N.N. Ivanova, I.J. Anderson, J.F. Cheng et al. 2013. Insights into the phylogeny and coding potential of microbial dark matter. Nature 499: 431–437. doi:10.1038/nature12352.

Schulz, F., E.A. Eloe-Fadrosh, R.M. Bowers, J. Jarett, T. Nielsen, N.N. Ivanova et al. 2017. Towards a balanced view of the bacterial tree of life. Microbiome 5: 140. doi:10.1186/s40168-017-0360-9.

Simão, F.A., R.M. Waterhouse, P. Ioannidis, E.V. Kriventseva and E.M. Zdobnov. 2015. BUSCO: Assessing genome assembly and annotation completeness with single-copy orthologs. Bioinformatics 31: 3210–3212.

Singer, E., D. Emerson, E.A. Webb, R.A. Barco, J.G. Kuenen, W.C. Nelson et al. 2011. Mariprofundus ferrooxydans PV-1 the first genome of a marine Fe(II) oxidizing *Zetaproteobacterium*. PLoS One 6: e25386. doi:10.1371/journal.pone.0025386.

Singer, E., J.F. Heidelberg, A. Dhillon and K.J. Edwards. 2013. Metagenomic insights into the dominant Fe(II) oxidizing Zetaproteobacteria from an iron mat at Lō´ihi, Hawai´i. Front. Microbiol. 4: 52. doi:10.3389/fmicb.2013.00052.

Slaby, B.M., T. Hackl, H. Horn, K. Bayer and U. Hentschel. 2017. Metagenomic binning of a marine sponge microbiome reveals unity in defense but metabolic specialization. ISME J. 11: 2465–2478. doi:10.1038/ismej.2017.101.

Sockett, R.E. 2009. Predatory lifestyle of Bdellovibrio bacteriovorus. Annu. Rev. Microbiol. 63: 523–539. doi:10.1146/annurev.micro.091208.073346.

Sousa, F.L., R.J. Alves, M.A. Ribeiro, J.B. Pereira-Leal, M. Teixeira and M.M. Pereira. 2012.The superfamily of heme-copper oxygen reductases: types and evolutionary considerations. Biochim. Biophys. Acta. 1817: 629–637. doi:10.1016/j.bbabio.2011.09.020.

Stackebrandt, E., R.G.E. Murray and H.G. Trüper. 1988. *Proteobacteria* classis nov., a name for the phylogenetic taxon that includes the "purple bacteria and their relatives". Int. J. Syst. Bacteriol. 38: 321–325.

Stolz, J.F., R.S. Oremland, B.J. Paster', F.E. Dewhirst and P. Vandamme. 2015. Sulfurospirillum. Bergey's Manual of Systematics of Archaea and Bacteria. 1–7. John Wiley & Sons, Inc., in association with Bergey's Manual Trust.

Tenover, F.C., C.L. Fennel, A. Tanner and B.J. Paster. 1992. Subdivision C5: Epsilon subclass. pp. 3499–3512. *In*: Balows, A., H.G. Troper, M. Dworkin and W. Harder (eds.). The Prokaryotes, 2nd Edn. Springer-Verlag, Berlin.

Thiergart, T., G. Landan and W.F. Martin. 2014. Concatenated alignments and the case of the disappearing tree. BMC Evol. Biol. 14: 266. doi:10.1186/s12862-014-0266-0.

Thomas, S.H., R.D. Wagner, A.K. Arakaki, J. Skolnick, J.R. Kirby, L.J. Shimkets et al. 2008. The mosaic genome of Anaeromyxobacter dehalogenans strain 2CP-C suggests an aerobic common ancestor to the delta-proteobacteria. PLoS One 3: e2103. doi:10.1371/journal.pone.0002103.

Tully, B.J. and J.F. Heidelberg. 2016. Potential mechanisms for microbial energy acquisition in oxic deep-sea sediments. Appl. Environ. Microbiol. 82: 4232–4243. doi:10.1128/AEM.01023-16.

Tully, B.J., R. Sachdeva, E.D. Graham and J.F. Heidelberg. 2017. 290 metagenome-assembled genomes from the Mediterranean Sea: a resource for marine microbiology. Peer J. 5: e3558. doi:10.7717/peerj.3558.

Tully, B.J., C.G. Wheat, B.T. Glazer and J.A. Huber. 2018. A dynamic microbial community with high functional redundancy inhabits the cold, oxic subseafloor aquifer. ISME J. 12: 1–16. doi:10.1038/ismej.2017.187.

Vignais, P.M. and B. Billoud. 2007. Occurrence, classification, and biological function of hydrogenases: an overview. Chem. Rev. 107: 4206–4272.

Waite, D.W., I. Vanwonterghem, C. Rinke, D.H. Parks, Y. Zhang, K. Takai et al. Comparative genomic analysis of the class epsilonproteobacteria and proposed reclassification to epsilonbacteraeota (phyl. nov.). Front. Microbiol. 8: 682. doi:10.3389/fmicb.2017.00682.

Weerakoon, D.R. and J.W. Olson. 2008. The Campylobacter jejuni NADH:ubiquinone oxidoreductase (complex I) utilizes flavodoxin rather than NADH. J. Bacteriol. 190: 915–925.

White, G.F., M.J. Edwards, L. Gomez-Perez, D.J. Richardson, J.N. Butt and T.A. Clarke. 2016. Mechanisms of bacterial extracellular electron exchange. Adv. Microb. Physiol. 68: 87–138. doi:10.1016/bs.ampbs.2016.02.002.

Williams, K.P. and D.P. Kelly. 2013. Proposal for a new class within the phylum Proteobacteria, *Acidithiobacillia classis* nov., with the type order *Acidithiobacillales*, and emended description of the class Gammaproteobacteria. Int. J. Syst. Evol. Microbiol. 63: 2901–2906. doi:10.1099/ijs.0.049270-0.

Woese, C.R. 1987. Bacterial evolution. Microbiol. Rev. 51: 221–271.

Yamada, K., T. Miyata, D. Tsuchiya, T. Oyama, Y. Fujiwara and T. Ohnishi. 2002. Crystal structure of the RuvA-RuvB complex: a structural basis for the Holliday junction migrating motor machinery. Mol. Cell. 10: 671–681.

Yarza, P., P. Yilmaz, E. Pruesse, F.O. Glöckner, W. Ludwig, K.H. Schleifer et al. 2014. Uniting the classification of cultured and uncultured bacteria and archaea using 16S rRNA gene sequences. Nat. Rev. Microbiol. 12: 635–645. doi:10.1038/nrmicro3330.

Zhang, Y. and S.M. Sievert. 2014. Pan-genome analyses identify lineage- and niche-specific markers of evolution and adaptation in *Epsilonproteobacteria*. Front. Microbiol. 5: 110. doi:10.3389/fmicb.2014.00110.

Zhou, J., Q. He, C.L. Hemme, A. Mukhopadhyay, K. Hillesland, A. Zhou et al. 2011. How sulfate-reducing microorganisms cope with stress: lessons from systems biology. Nat. Rev. Microbiol. 9: 452–466. doi:10.1038/nrmicro2575.

5

Gammaproteobacteria

*Mauro Degli Esposti** and *Esperanza Martínez-Romero*

Introduction: Outlook and Frame of the Chapter

The class of gammaproteobacteria has the largest number of genomes and protein sequences in NCBI databases (Garrity et al. 2005, Land et al. 2015), as well as in terms of taxonomic richness (Schulz et al. 2017). This class contains important human pathogens and two of the most studied bacterial species: *Pseudomonas aeruginosa* and *Escherchia coli*, hereafter abbreviated as *E. coli*. The name *Pseudomonas* has been frequently used in microbiology to describe diverse bacterial species, which were later classified in different families, orders or even classes from the current genus that belongs to the vast order of Pseudomonadales (Stackebrandt 1985, Woese 1987, Peix et al. 2009, 2018, Williams et al. 2010). *E. coli* has been the model system for biochemical and genetic studies, then becoming the workhorse for molecular biology and biotechnological applications (Alberts et al. 1994). It is a facultatively anaerobe with an impressive array of metabolic traits for adapting to microaerobic conditions and using diverse electron donors or acceptors (Simon and Klotz 2013). The respiratory chain of *E. coli* thus contains most of electron transport systems we have seen in other bacteria, except for the bc1 complex and heme *a*-containing terminal oxidases (Gennis 1991, Simon and Klotz 2013, Borisov and Verkhovski 2015).

There are so many articles on *E. coli* and *Pseudomonas* that it would be of little interest to add here another essay on these species. Instead, this chapter presents a view of the class of gammaproteobacteria that focuses on taxa that are not pathogenic to animals or humans. Indeed, gammaproteobacteria include several different insect endosymbionts with reduced genomes, as well as symbionts of

Center for Genomic Sciences, UNAM Campus de Cuernavaca, Cuernavaca, 62130 Morelos, Mexico.
* Corresponding author: mauro1italia@gmail.com

marine animals that expand the leitmotif of symbiosis which underlies this and all subsequent chapters of the book. Human pathogens, which are so common among gammaproteobacteria, will not be described for multiple reasons. One major reason is that pathogenic taxa are generally late branching in phylogenetic trees, especially in the case of Enterobacterales such as *E. coli, Salmonella, Klebsiella, Yersinia,* etc. (Williams et al. 2010). Secondly, their analysis would be much more appropriate for a text of medical microbiology, an area in which the authors of this book have very little experience. Our direct experience on gammaproteobacteria initially derived from phylogenetic studies on the nearly ubiquitous bd oxidase that led to a detailed phylogenomic analysis of all gammaproteobacteria (Degli Esposti et al. 2015). Subsequently, our knowledge of the gamma class has expanded by genome-wide searches of metabolic traits for all proteobacteria present in the microbiome of arthropods (Degli Esposti and Martinez-Romero 2017). Such studies enhanced our awareness of metabolic traits underlying phylogenetic relationships among proteobacterial taxa which, more frequently than not, are not evident in traditional phylogenetic trees based upon rRNA markers (see below). Consequently, the phylogenetic inferences and evolutionary considerations presented here fundamentally derive from the analysis of proteins, especially those that are crucial in the bioenergetics of both bacteria and mitochondria.

Our outlook on gammaproteobacteria will be probably unfamiliar to scientists working on diverse species of this class or have matured a comprehensive vision of gammaproteobacteria based upon established phylogenetic analysis of rRNA markers (Land et al. 2015, Schulz et al. 2017). This theme will be discussed in detail later on in this and other chapters, the frame of which is generally organized as follows. An initial introduction of the proteobacterial class is followed by a succinct presentation of the phylogenetic profile of the class, fundamentally based upon the concepts discussed above. Then representative members of the class are discussed in specific sections; such members have been chosen often because they are symbionts of animals or plants, or have peculiar features that illuminate aspects of the bioenergetics of bacteria that are of interest to the evolution of mitochondria, adhering to the leitmotif of the book.

Phylogenesis of Gammaproteobacteria

A brief premise on the tools of molecular phylogeny is necessary before discussing aspects of gammaproteobacterial phylogenesis. Since all bacteria contain the genes encoding for ribosomal RNA (23S, 16S and 5S), these genes are considered particularly useful to determine evolutionary relationships between different species (Woese 1987, Land et al. 2015, Schulz et al. 2017). The gene for 16S rRNA contains both conserved and variable regions, which has made it a key molecule to establish taxonomic relationships (Woese 1987, Schulz et al. 2017). Dave Ussery and colleagues have recently attempted to visualize phylogenetic relationships of gammaproteobacteria by building trees of 52 family type strains based on complete 16S rRNA genes and the results indicate that Xanthomonadales, *Salinisphaera*

(Crespo-Medina et al. 2009) and some Thiotrichales form the basal branches of the class (D.W. Ussery, personal communication). However, the branching order and phylogenetic relationships among several other families is unclear and in places appears to contradict established relationships among certain taxonomic groups of gammaproteobacteria. This indicates that the taxonomic divisions of the class only partially coincide with phylogenetic features inherent to the 16S rRNA gene (D.W. Ussery, personal communication), consistent with previous findings (Williams et al. 2010). Using the phylogenomic approach of Segata et al. (2013), which is based upon the combined tree of *ca.* 400 proteins, we found essentially similar results (Degli Esposti et al. 2015) leading to the following presentation.

Phylogenetic trees of gammaproteobacteria usually show Pseudomonades at the center and Enterobacterales including *E. coli* at the tip (Williams et al. 2010, Peix et al. 2018). As mentioned above, the base of these trees contains Thiotrichales (filamentous bacteria characterized by sulfur deposits, typified by *Beggiatoa*) and other sulfur-oxidizers, as well as Xanthomonadales, a vast order of terrestrial bacteria which are widespread in various ecosystems including animal microbiota (Degli Esposti and Martinez-Romero 2017). Later in the chapter, we shall examine a member of the latter order, the olive tree killer *Xylella*.

Gammaproteobacteria form a sister group to betaproteobacteria, both stemming from the ancestors of current alphaproteobacteria after the separation of zetaproteobacteria (see last part of Chapter four). They share the metabolic trait of magnetotaxis, methanotrophy and nitrite oxidation with alphaproteobacteria. We thus surmise that the common ancestor of gamma and alphaproteobacteria was an organism resembling more a magnetotactic Nitrospirae or *Magnetococcus*, a basal alphaproteobacterium (Bazylinski et al. 2013), than other contemporary bacteria. This ancestral prokaryote must have possessed the metabolic trait of sulfur oxidation, which is present in *Magnetococcus* and common among deep branching taxa of gammaproteobacteria, as discussed later. Sulfur-oxidizing gammaproteobacteria are taxonomically grouped in the recognized orders of Chromatiales and Thiotrichales, which contain the bacteria in which chemosynthesis was first described (see Chapter two), and various unclassified taxa including a growing group of symbionts of marine invertebrates, originally found in hydrothermal vents (Cavanaugh et al. 1981, Dubilier et al. 2008).

Marine species of the *Salinisphaera* genus have also been considered to represent a deep branching lineage among gammaproteobacteria (Crespo-Medina et al. 2009), a notion based upon 16S rRNA trees (see above). Intriguingly, they contain an ancestral form of the gene cluster for cytochrome bd oxidase that is present only in a few alphaproteobacteria, in particular *Acidocella* (Degli Esposti et al. 2015). However, recent phylogenetic analysis integrated with functional genomics indicates that Salinisphaerales may not lie in the deepest branches of gammaproteobacteria (Degli Esposti and Martinez-Romero 2017). The same may apply to Xanthomonadales, as discussed below (Figs. 1–3).

Finally, *Thiohalorhabdus* appears to constitute another deep branching lineage of the gammaproteobacterial class (Sorokin et al. 2008), since its proteins

(A)

taxon	proteins	aerobic traits		anaerobic traits						Group 1 hydrogen.		MQ
		O2 reductase	NOR	NarG	NapA	NapH	NirB/D	PFO	Rnf			
Shewanella benthica KT99	4,235	cbb3, COX		yes	yes		NirB	yes	ABCDGE	Hyb	b	yes
Dyella-like sp. DHo	3,392	bo3, COX			yes, 2		NirB		ABDE	Hyb	6c	?
Xanthomonadaceae bacterium SCN 69-123	6,825	cbb3, COX	qNOR					OAFO	B only			
Xylella fastidiosa 9a5c	2,379	bo3 only										
Photobacterium damselae subsp. piscicida	3,815	cbb3, COX			yes		NirB	yes	ABCDGE	Hyb	b	likely
Vibrio breoganii ZF-55	3,918	cbb3					NirB	yes				likely
Thioalkalivibrio sp. K90mix	5,631	cbb3, COX			NasA?		NirB		ABCDGE			likely
Thioalkalivibrio sp. HK1	2,892	COX			NasA?			OAFO	ABCDGE			likely
uncultured Thiohalocapsa sp. PB-PSB1	6,202	cbb3, partial	yes		?	yes		yes	BCDGE	Isp	6C	likely
Hydrogenovibrio crunogenus XCL-2 (Thiomicrospira crunogena)	2,244	cbb3		yes	yes	yes	NirB	OAFO	ABCDGE			
Candidatus Thioglobus singularis PS1	1,686	cbb3, COX			NasA?	yes			B only			
Candidatus Thioglobus autotrophicus	1,600	cbb3, COX	yes	yes	yes	yes			ABCDGE			
Bathymodiolus thermophilus thioautotrophic gill symbiont	3,045	cbb3, COX	yes	yes	NasA?		NirB	yes	ACDG, E		6C	
Bathymodiolus azoricus thioautotrophic gill symbiont	1,785	cbb3, COX	yes	yes	NasA?		NirB	yes	A only		6C	
endosymbiont of Bathymodiolus septemdierum str. Myojin knoll	1,508	cbb3, COX	yes	yes	?		NirB	yes	ABCDGE			
Calyptogena okutanii thioautotrophic gill symbiont	2,119	cbb3, COX	yes	yes			NirB	yes	ABCDGE			
Candidatus Ruthia magnifica str. Cm	1,034	cbb3, COX	yes	yes	NasA?		NirB		ABCDGE			
Solemya velum gill symbiont strain WH	2,716	cbb3, COX, ba3	yes	yes	yes	yes	NirB	yes	2 ABCDGE	Isp		likely
endosymbiont of Tevnia jerichonana	3,230	cbb3	yes	yes	yes	yes	NirB	yes	ABCDGE	Isp		likely
endosymbiont of Riftia pachyptila	3,182	cbb3	yes	yes	yes	yes	NirB	yes	ABCDGE	Isp		likely
endosymbiont of Ridgeia piscesae	2,768	bo3, COX	yes	yes	yes	yes, 2		yes	ABCDGE	Isp		likely
Thiolapillus brandeum (Hiromi 1)	2,929	COX	yes	yes	yes	yes	NirD		ABCDGE	Isp		likely
endosymbiont of unidentified scaly snail isolate Monju	2,313	cbb3, 2	yes	yes	yes	yes	NirB		ABCDGE	Isp	6C, b	likely
Candidatus Thiodiazotropha endolucinida	4,135	cbb3, COX	yes	yes	NasA?	yes		yes	ABCDGE	Hyb	6C	
Sedimenticola thiotaurini Strain SIP-G1	3,597	cbb3, COX	yes	yes	yes		NirB	yes	ABCDGE	Hyb		yes
Sedimenticola selenatireducens	3,760	cbb3, COX	yes	yes	yes	yes		yes	ABCDGE	Hyb	6C	likely
Thiohalobacter thiocyanaticus	2,927	cbb3, COX	yes	yes			NirB		ABCDGE	Isp		
Thiohalorhabdus denitrificans	2,418	COX	yes	yes	yes				ABCDGE	Isp		

(B)

subgroup*	common name	clade#	E.coli	functional properties	gene sequence	taxonomic distribution
1a	ancestral	a		oxygen-sensitive, anaerobic metabolism	small-cyt b-large	Archaea and anaerobic bacteria
1b	prototypical	c, d		oxygen-sensitive, anaerobic metabolism low oxygen-tolerance, **NrfD** membrane	small-cyt b-large	predominantly delta- and epsilon-proteobacteria
1c	Hyb type	**HybA**	hyd-2	subunit, generally MQ reacting, probably bidirectional	small-HybA Fe-S, NrfD-large	*Magnetococcus* and some alphaproteobacteria, betaproteobacteria and gammaproteobacteria
1d	oxygen-tolerant	**6C**	hyd-1	oxygen tolerant, proximal **4S3S** cluster in small subunit, aerotolerant anaerobic respiration and 'Knallgas' reaction^	small-cyt b-large	CFB, Aquificae, alphaproteobacteria, betaproteobacteria and gammaproteobacteria
1e	Isp type	**Isp**		low oxygen-tolerance, associated with sulfur respiration	small-cyt b5TM, hdr-large	predominantly gammaproteobacteria

Fig. 1. (A) Metabolic traits of deep branching gammaproteobacteria plus other selected taxa. The selected taxa discussed in the text are highlighted in pale blue. Taxa highlighted in pale brownish correspond to sulfur oxidizing symbionts and close thiotrophs of clade 1. Taxa highlighted in pale yellow are sulfur oxidizing symbionts and close thiotrophs of clade 2. At the top, the list includes three Chromathiales and Xanthomonadales (with *Xylella* highlighted in pale blue) and two Vibrionales. The column of proteins is highlighted in yellow when the complete genome is available. MQ, menaquinone–deduced from the presence of genes for its biosynthesis (Degli Esposti 2017); **yes** indicates that MQ presence has also been demonstrated biochemically. Basically, all taxa have complex I, II (Sdh) and III (bc1), except for *Bathymodiolus* symbionts of clade 1, which do not have Sdh (Ponmudurai et al. 2017). Only *Sedimenticola* and *Thiohalorhabdus* additionally have bd type ubiquinol oxidases. Group 1 hydrogen. indicates various Group 1 hydrogenases as listed in B. (B) Classification of Group 1 hydrogenases. The recognized subgroups of Group 1 hydrogenases, which are also called Hup hydrogenases (Greening et al. 2016, Degli Esposti and Martinez-Romero 2016), are indicated according to the following references:

*Greening et al. (2016), #Pandelia et al. (2012). ^full oxidation of H_2 to O_2, cf. Pandelia et al. (2012).

frequently cluster with homologs from zetaproteobacteria (Degli Esposti 2017). Little information is currently available on this organism, besides phenotypic features regarding its sulfur oxidizing metabolism and halophilic character (Sorokin et al. 2008). Interestingly, *Thiohalorhabdus* genomes contain the Rnf complex, which is present in *Magnetococcus* (Chapter seven) and is nearly universal among zetaproteobacteria (Chapter four). The same complex is very common also among sulfur-oxidizing endosymbionts (Dmytrenko et al. 2014) and free-living thiotrophs (Flood et al. 2005, Oshiki et al. 2017, Boden et al. 2017), as well as in Thiotrichales, but not in Xanthomonadales (Fig. 1). The Rnf complex can thus be considered a basal trait for gammaproteobacteria (Fig. 1A).

Acidithiobacilli have been regularly found to be either the deepest branching taxa of gammaproteobacteria or to form their outgroup (Williams et al. 2010), similar to the situation of *Magnetococcus* for alphaproteobacteria (Bazylinski et al. 2013; see also Chapter seven). Acidithiobacilli have been subsequently proposed to constitute a separate class that would be intermediate between zetaproteobacteria and gammaproteobacteria, the Acidithiobacillia (Williams and Kelly 2013, González et al. 2016, Tran et al. 2017). Acidithiobacillia are chemolithoautotrophic acidophilic bacteria (often obligate–Boden et al. 2016) capable of extracting metals such as copper and gold from mineral deposits (Johnson 2014, Tran et al. 2017). They were previously classified among gammaproteobacteria on the basis of 16S rRNA (Garrity et al. 2005), but found later to segregate into a separate group by multiprotein markers analysis (Williams and Kelly 2013, González et al. 2016). This appears to be a recurrent theme that is connected to what was discussed in the Introduction.

All Acidithiobacillia extract energy from the oxidation of elemental sulfur and reduced inorganic sulfur compounds; however, they do not use the Sox pathway of thiosulfate oxidation which is described below (Friedrich et al. 2005). They instead oxidize thiosulfate to tetrathionate via a yet undefined dehydrogenase (Williams et al. 2010, Boden et al. 2016), while retaining diverse components of the Sox pathway intercalated by a characteristic conserved protein (DUF302, Boden et al. 2016). The genome of about one half of Acidithiobacillia contains genes for enzymes that enable them to oxidize Fe^{II} under acidic conditions, which reduce the spontaneous auto-oxidation of iron with oxygen (see Chapters two and four). Acidithiobacillia share the trait of sulfur oxidation and tetrathionate accumulation with *Thiohalorhabdus* (Sorokin et al. 2008), whereas symbionts of clams and other metazoans, as well as *Ca.* Thioglobus of the SUP05 clade contain a combination of the incomplete periplasmic Sox pathway with the cytoplasmic pathway of sulfide oxidation based upon the reverse reaction of the Dissimilatory sulfite reductase (rDsr), which drives sulfate reduction in other bacteria (see Chapters two and four). All these sulfur-oxidizing organisms are consistently placed in basal branches of trees obtained with conserved bioenergetic proteins such as NuoD of complex I, NADH-ubiquinone reductase (Fig. 2), or subunits of the bc1 complex

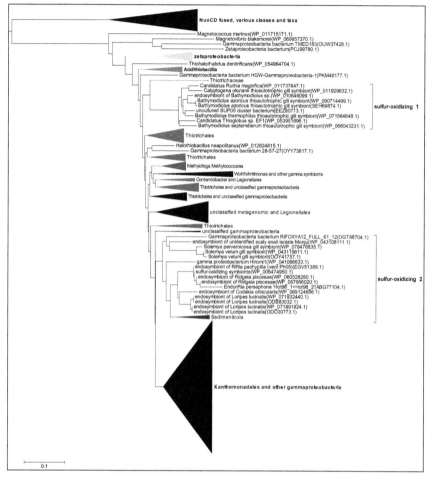

Fig. 2. Phylogenetic tree of deep branching gammaproteobacteria with NuoD subunit of complex I. The Neighbor Joining (NJ) tree was obtained from a DeltaBLAST1000 search over a phylogenetically broad set of taxa. Phylogenetic trees obtained with the program MEGA5 with about 30 sequences covering the span of this tree gave bootstrap values of at least 65% for the separation node between the two clades of sulfur-oxidizing symbionts.

(Fig. 3A,B). Sulfur oxidation and thiotrophy, therefore, may be considered characteristic ancestral traits of gammaproteobacteria, which have been subsequently lost or modified along the evolution of the taxa affiliated to the class (Dahl 2015). So far, the pathways of sulfur oxidation have been mentioned only cursorily in the book (Chapter two, section on the sulfur cycle); they are described in detail in the following section.

(A)

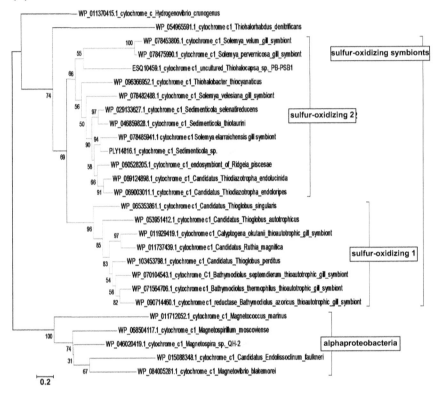

(B)

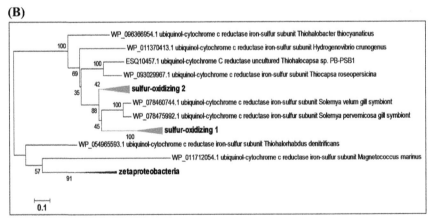

Fig. 3 contd. ...

...*Fig. 3 contd.*

(C)

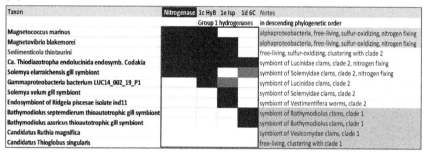

Taxon	Nitrogenase	1c HyB	1e Isp	1d 6C	Notes
		Group 1 hydrogenases			in descending phylogenetic order
Magnetococcus marinus					alphaproteobacteria, free-living, sulfur-oxidizing, nitrogen fixing
Magnetovibrio blakemorei					alphaproteobacteria, free-living, sulfur-oxidizing, nitrogen fixing
Sedimenticola thiotaurini					free-living, sulfur-oxidizing, clustering with clade 2
Ca. Thiodiazotropha endolucinida endosymb. Codakia					symbiont of Lucinidae clams, clade 2, nitrogen fixing
Solemya elarraichensis gill symbiont					symbiont of Solemyidae clams, clade 2, nitrogen fixing
Gammaproteobacteria bacterium LUC14_002_19_P1					symbiont of Lucinidae clams, clade 2
Solemya velum gill symbiont					symbiont of Solemyidae clams, clade 2
Endosymbiont of Ridgeia piscesae isolate ind11					symbiont of Vestimentifera worms, clade 2
Bathymodiolus septemdierum thioautotrophic gill symbiont					symbiont of Bathymodiolus clams, clade 1
Bathymodiolus azoricus thioautotrophic gill symbiont					symbiont of Bathymodiolus clams, clade 1
Candidatus Ruthia magnifica					symbiont of Vesicomyidae clams, clade 1
Candidatus Thioglobus singularis					free-living, clustering with clade 1

Fig. 3. Phylogenetic history of sulfur-oxidizing symbionts. (A) NJ tree of an alignment of cytochrome *c*1 sequences of the bc1 complex, obtained with1000 bootstraps. Of note is that the bootstrap values are better (75 between two clades) but the topology is the same using the Maximum Likelihood (ML) method with 500 bootstraps. (B) Compressed view of the NJ tree of the iron-sulfur protein (ISP) subunit of the bc1 complex. The tree was computed with 500 bootstraps of an alignment of 25 ISP sequences. (C) Distribution of Group 1 hydrogenases vs. nitrogenase in sulfur-oxidizing proteobacteria.

Sulfur Oxidation—Pathways of Thiotrophy and Bioenergy Production

Sulfur-oxidizers are common among proteobacteria and utilize two main metabolic pathways: (1) the Sox pathway, which uses thiosulfate as primary electron donor and is localized in the periplasmic space of Gram negative bacteria; (2) the rDsr pathway, which predominantly occurs in the cytoplasm but requires interaction with the cytoplasmic membrane (Friedrich et al. 2005, Dmytrenko et al. 2014, Nunoura et al. 2014, Dahl 2015). The complete Sox pathway is present in alphaproteobacteria such as *Paracoccus denitrificans* and *Rhodopseudomonas palustris* (Friedrich et al. 2005, Dmytrenko et al. 2014), as well as in the gammaproteobacterium *Thiomicrospira crunogena* (Scott et al. 2006), which has recently been reclassified in the genus *Hydrogenovibrio* (Boden et al. 2017). However, the complement of Sox genes is most commonly limited to the SoxAYB cluster that catalyzes the first oxidation step of thiosulfate to elemental sulfur (Friedrich et al. 2005, Dmytrenko et al. 2014, Nunoura et al. 2014, Dahl 2015). In the absence of downstream electron acceptors or oxygen, this reaction normally leads to the accumulation of sulfur deposits in the form of globules within the cytoplasm or at the cellular surface. When oxygen is relatively abundant, the sulfur deposits are mobilized by forming organic persulfites that are then transported into the cytoplasm for full oxidation to sulfite or sulfate (Dahl 2015). Persulfite compounds are first transferred to sulfur carrier proteins and then oxidized via the rDsr pathway, starting with the reduced form (persulfurated) of the carrier protein DsrC (Dahl 2015). Persulfurated DsrC is re-oxidized by the cytoplasmic DsrAB complex, which ultimately oxidizes sulfur to sulfite by donating electrons to the transmembrane DsrKMJOP complex. This transmembrane complex contains *b* and *c* cytochromes (see Chapter two) and in its reverse reaction reduces membrane quinones, either menaquinone (MQ) as in deltaproteobacteria (Chapter four) or ubiquinone (Q) as in the model sulfur-oxidizer, *Allochromatium vinosum* (another gammaproteobacterium, which belongs to the order Chromatiales–Dahl 2015). The DsrAB and DsrKMJOP complexes are shown in Fig. 4A, which illustrates the basic respiratory chain of a sulfur-oxidizing symbiont of the bivalve *Solemya* (Dmytrenko et al. 2014).

The sulfite produced by rDsr may be further oxidized to sulfate by the sequential activity of APS (3'-Phosphoadenosine-5'-phosphosulfate) reductase and ATP sulfurylase (Sat), generating substrate level ATP. Therefore, the cytoplasmic rDsr pathway generates bioenergy by sustaining both protonmotive reactions following quinone reduction in the membrane and substrate level ATP from the high energy compound APS (Dmytrenko et al. 2014). Combined with the periplasmic oxidation of thiosulfate by the SoxAYB system (Friedrich et al. 2005), which feeds electrons to the pool of periplasmic *c* cytochromes, either directly or via SoxEF (Fig. 4, cf. Nunoura et al. 2014), the rDsr pathway of sulfur oxidation is a metabolism with no rivals in terms of bioenergy yield among chemosynthetic organisms (Dmytrenko et al. 2014, Nunoura et al. 2014). As mentioned before, variations of these fundamental pathways seem to be present in Acidithiobacillia and other basal gammaproteobacteria such as *Thiohalorhabdus*. However, the combination of periplasmic SoxAYB with the cytoplasmic rDsr pathway is typically present in gammaproteobacterial sulfur-oxidizing symbionts, which are now described in relation to functional traits that may illuminate their basal position in the whole class (Figs. 1–3).

Sulfur-Oxidizing Symbionts of Mollusks and Other Marine Invertebrates

The sulfur-oxidizing symbionts of marine invertebrates constitute the first example of a chemoautotrophic mutualistic symbiosis (Dubilier et al. 2008). A large number of small subunit rRNA sequences has been collected from diverse invertebrates including tubeless Vestimentifera worms and giant clams inhabiting hydrothermal vents, as well as from other marine invertebrates inhabiting environments rich in sulfide, which can sustain the sulfur-oxidizing autotrophic lifestyle of their symbionts (Dubilier et al. 2008, Petersen et al. 2012, Kleiner et al. 2012, Dmytrenko et al. 2014). The symbionts are generally vertically transmitted and housed within cells of the enlarged gills called bacteriocytes (König et al. 2016, Ikuta et al. 2016), or in a specialized organ called trophosome (Dubilier et al. 2008). The diversity in the host-symbiont interactions has been interpreted to derive from multiple independent evolutionary events of interspecies association, which appear to be sustained by rRNA phylogenetic analysis (Dubilier et al. 2008, Kleiner et al. 2012).

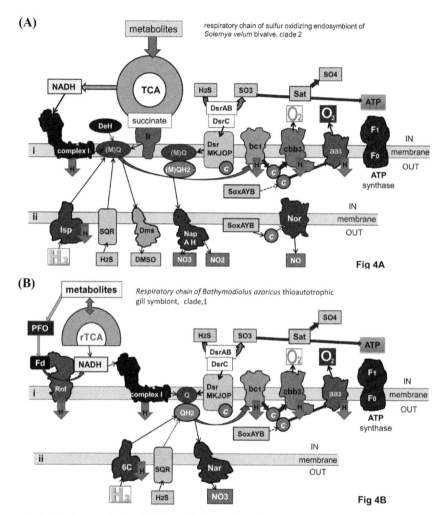

Fig. 4. Respiratory chains of sulfur-oxidizing endosymbionts. (A) Respiratory chain of *Solemya velum* gill as representative of clade 2 (Figs. 1 and 2). A very similar respiratory chain is present in related free-living strains of *Sedimenticola* (Fig. 1). The relay of ferredoxin via the Rnf complex is shown in part B. (B) Respiratory chain of *Bathymodiolus azoricus* gill symbiont (Petersen et al. 2011) as representative of clade 1 (Figs. 1 and 2). Note that other taxa of the same clade do not have group 1 hydrogenases (Fig. 1). The TCA is incomplete (Ponnudurai et al. 2017).

Functional insights into heterotrophic capacity have gradually emerged from the accumulation of genomic information on sulfur-oxidizing symbionts (Newton et al. 2007, Dubilier et al. 2008, Kuwahara et al. 2011, Dmytrenko et al. 2014, Nunoura et al. 2014, Nakagawa et al. 2014, Ikuta et al. 2016, König et al. 2016, Ponnudurai et al. 2017). A comprehensive overview of the metabolic traits of sulfur-oxidizing symbionts is presented here (Fig. 1), for an analysis of this kind is lacking in the literature. Following earlier genomic analysis (Degli Esposti and Martinez-Romero 2017), the deduced panorama of functional properties (Fig. 1) suggest a new evolutionary perspective for sulfur-oxidizing symbionts of gammaproteobacteria.

Currently, there are 27 genomes for taxa of sulfur-oxidizing symbionts of marine invertebrates that belong to the class of gammaproteobacteria. A selection of these genomes, including all those that have been completed to date, is shown in Fig. 1, which lists fundamental traits of anaerobic energy metabolism that complement autotrophic sulfur oxidation (thiotrophy). So far, the phylogenetic position of sulfur-oxidizing endosymbionts has been evaluated predominantly by trees based upon the 16S rRNA marker, which have led to the conclusion that these endosymbionts are dispersed in evolutionary diverse clades along the taxonomic span of gammaproteobacteria (Duvilier et al. 2008, Petersen et al. 2012, Kleiner et al. 2012). Such a conclusion implies that the strong similarity in many metabolic traits of these organisms may result from phenomena of convergent evolution (Petersen et al. 2012, Kleiner et al. 2012).

However, this implication is questionable and probably flawed due to the dichotomy between RNA genes and the physiology of bacteria (see Chapters one, two and four). Indeed, phylogenetic trees of conserved proteins of universal bioenergetic systems such as NuoD of complex I (Fig. 2), subunits of the cytochrome bc1 complex (Fig. 3) and the largest subunit of Heme Copper Oxygen reductases (HCO) show tight clustering of most homologs of sulfur-oxidizing symbionts in two clades. Clade 1 contains endosymbionts from the bivalves *Bathymodiolus* and *Calyptogena* of hydrothermal vents, together with free-living Candidatus *Thioglobus* and sometimes other members of the SUP05 clade living in oxygen-limiting zones of marine environments (Walsh et al. 2009, Anantharaman et al. 2013, Marshall and Morris 2013, Shah and Morris 2015). All these organisms have only Q as membrane electron carrier and HCO of family C and A, but no NOR - Nitrous Oxide Reductase (Fig. 1). Of note is that NOR enzymes are ancestral vs. other HCO (see Chapter two). They are characteristically present in clade 2 of sulfur-oxidizing symbionts, which includes the endosymbionts of Vestimentifera worms of hydrothermal vents (*Riftia*, *Tevnia* and *Ridgeia*), of a scaly snail (Nakagawa et al. 2014), and of Lucinidae or Solemyidae bivalves thriving in both deep ocean and shallow water environments (Dubilier et al. 2008, Dmytrenko et al. 2014, König et al. 2016, Russell et al. 2017), as well as the *Thiolapillus* relative to a gastropod endosymbiont (Nunoura et al. 2014). Free-living strains of the genus *Sedimenticola* (Flood et al. 2005, 2015) also cluster within clade 2 (Figs. 2 and 3, cf. Kleiner et al. 2012), the members of which show a more pronounced anaerobic physiology than those of clade 1 because their genomes contain: (i) other elements of dissimilatory N metabolism (Simon and Klotz 2013) besides NOR, including periplasmic nitrate reductase NapA, membrane-linked menaquinol-nitrite reductase NapH (Figs. 1 and 4) and, at least in one case, the full complement of denitrification enzymes (Nunoura et al. 2014);

(ii) the genes of the enzymes for the classical pathway of MQ biosynthesis (see Chapter two); (iii) the operon for one or more Group 1 Ni-Fe hydrogenases (Fig. 1; Pandelia et al. 2012, Nakagawa et al. 2014, Sargent 2016). Endosymbionts of *Bathymolinus* clams (Petersen et al. 2011) and free-living members of the SUP05 clade (Anantharaman et al. 2013) also contain hydrogenases (Fig. 1A). However, the particular type of Group 1 hydrogenases of these clade 1 bacteria has different characteristics from the hydrogenases that are most frequently present in clade 2 bacteria, as discussed in depth below.

Hydrogenases Define the Evolution of Sulfur-Oxidizing Taxa at the Root of Gammaproteobacteria

Membrane-bound hydrogenases have been reported to be present in sulfur-oxidizing symbionts and previously discussed as means for extracting energy from abundant levels of H_2 in hydrothermal vent environments (Petersen et al. 2011, Anantharaman et al. 2013, Ikuta et al. 2016). What was not noted before is that these hydrogenases belong to the most aerotolerant of the several subgroups that are now recognized within Group 1 hydrogenases (summarized in Fig. 1B), namely subgroup **1d** or **6C**, which includes *hyd-1* of *E. coli* (Pandelia et al. 2012, Sargent 2016, Greening et al. 2016). Published evidence indicates that the enzyme is used by sulfur-oxidizing symbionts for H_2 oxidation all the way to oxygen (Petersen et al. 2011), a pathway also called 'Knallgas' reaction (Pandelia et al. 2012). Subgroup **1d** hydrogenases are present also in clade 2 bacteria, for example in Ca. *Thiodiazotropha* and *Sedimenticola* (Fig. 1A and Fig. 3C), thereby suggesting that they might have been present in a common ancestor of sulfur-oxidizing gammaproteobacteria. However, thiotrophic gammaproteobacteria predominantly have membrane hydrogenases of subgroup **1e** or **Isp** (Fig. 1, cf. Greening et al. 2016).

Subgroup **1e** or **Isp** hydrogenases are fundamentally associated with sulfur metabolism (Pandelia et al. 2012, Greening et al. 2016), also because they contain an Fe-S subunit related to heterosulfide reductase (Pandelia et al. 2012). They can tolerate only low levels of oxygen compared with group **1d** enzymes and, therefore, are active under micro-oxic and anaerobic conditions, similar to group **1c** hydrogenases (Pandelia et al. 2012). Candidatus *Thiodiazotropha* and *Sedimenticola* additionally possess hydrogenases of the latter subgroup **1c** (Fig. 1), which is also called **Hyb** and includes *hyd-2* of *E. coli* (Pandelia et al. 2012, Sargent 2016, Greening et al. 2016). Characteristically, these enzymes have a membrane subunit related to NrfD nitrite reductases instead of a transmembrane *b* cytochrome, which preferentially reacts with MQ (Pandelia et al. 2012). The distribution of subgroup **1c** hydrogenases indeed matches that of MQ among the gammaproteobacteria examined in Fig. 1. In particular, *Sedimenticola thiotaurini* has both Q and MQ in its membranes as determined biochemically (Flood et al. 2005), consistent with the presence of the enzymes for MQ biosynthesis that are found in its genome (Flood et al. 2015, cf. Degli Esposti 2017).

Intriguingly, *Sedimenticola thiotaurini* also contains the operon for nitrogenase (Flood et al. 2015), thus exhibiting the relatively rare trait of N_2 fixation among thiotrophs (Fig. 3C). The same trait has subsequently been found in the endosymbiont of Lucinae clams, *Ca.* Thiodiazotropha endolucinica (König et al. 2016)–previously known as endosymbiont of *Codakia orbicularis* (Fig. 2, cf. Dubilier et al. 2008)—as well as in *Solemya elarraichensis* gill symbiont (Fig. 3C, cf. Russel et al. 2017). The antiquity of N_2 fixation (Chapter one) is congruent with the ancestral position of the **Hyb** clade in the evolutionary trees of Groups 1 hydrogenases (Pandelia et al. 2012). These hydrogenases are fundamental for removing the H_2 byproduct of the nitrogenase reaction (Degli Esposti and Martinez-Romero 2016), which consumes reduced ferredoxin produced by the Rnf complex or pyruvate-ferredoxin oxidoreductase (PFO), enzymes that are also present in the genome of both *Sedimenticola thiotaurini* and *Ca.* Thiodiazotropha (Fig. 1A). The distribution of the **Hyb** subgroup of membrane hydrogenases thus correlates with that of nitrogenase and the enzymes that are functionally associated with it, together forming a metabolic core that may be ancestral to sulfur-oxidizing taxa from either gammaproteobacteria or deep-branching alphaproteobacteria such as *Magnetococcus* (Fig. 3B,C). Indeed, the closest homologs of *Magnetococcus* **Hyb** hydrogenase are from *Sedimenticola* and *Ca.* Thiodiazotropha (not shown), forming a monophyletic clade among other hydrogenases (Greening et al. 2016).

The distribution pattern of nitrogenase and Groups 1 hydrogenases (Fig. 3C) now suggests that the two clades in which sulfur-oxidizing symbionts and related taxa segregate (Figs. 2 and 3A) may derive from a progressive loss of anaerobic functional traits originally associated with N_2 fixation. The combination of such traits was presumably present in common ancestors that might have been shared with deep-branching alphaproteobacteria such as *Magnetococcus*. Together with *Magnetovibrio*, *Magnetococcus* combines nitrogenase with membrane hydrogenases (of **Hyb** or **Isp** subgroup) that have an oxygen tolerance (Pandelia et al. 2012) equivalent to the optimal O_2 concentrations towards which these bacteria are attracted (Dufour et al. 2014; see later Chapter seven). These oxygen concentrations appear to sustain the growth of sulfur-oxidizing gammaproteobacteria as well (Dufour et al. 2014, König et al. 2016), whether nitrogenase activity is required or not for their normal physiology. It can be speculated that the abundance of ammonia and nitrate in deep ocean environments where the invertebrate hosts live (Dubilier et al. 2008, Dmytrenko et al. 2014) may not require a nutritional supplement of N compounds from the endosymbionts, contrary to the situation that occurs with some endosymbionts of land insects (see next chapter of the book). Hence, there has been little evolutionary pressure on sulfur-oxidizing symbionts to retain the trait of N_2 fixation, and its associated hydrogenases, once mutualistic associations with deep sea animals have been established. This would explain the progressive but discrete loss of anaerobic traits that defines the two clades of contemporary sulfur-oxidizing symbionts (Figs. 1–3) and their respiratory chain, which is illustrated in Fig. 4.

The above considerations throw new light on the evolution of sulfur-oxidizing symbionts, contrasting strongly with the popular concept that these bacteria might have derived from multiple, phylogenetically distant species (Dubilier et al. 2008, Petersen et al. 2012, Kleiner et al. 2012, Reveillaud et al. 2018, Bergin et al. 2018). One of such species is the previously mentioned *Thiomicrospira crunogena*, now reclassified as *Hydrogenovibrio crunogenus* (Boden et al. 2017), which has been associated with the endosymbionts of *Calyptogena* clams since the work of Williams et al. (2010), later confirmed by other studies (Kleiner et al. 2012, McGill and Barker 2017). The genome of *Thiomicrospira* has been the first to be completed for sulfur-oxidizing marine chemoautotrophs (Scott et al. 2006), but does not harbor genes for either MQ biosynthesis nor group1 hydrogenases (Fig. 1). These features would exclude *Thiomicrospira* from the ancestral lineage of thiotrophic gammaproteobacteria (cf. Fig. 3C), even if some of its bioenergetic proteins appear in deep-branching position vs. those of sulfur-oxidizing symbionts (Figs. 2 and 3). In any case, the physiological profile of *Thiomicrospira* seems to be superimposable to that of sulfur-oxidizing symbionts of clade 1 (Fig. 1), which are likely to be late-branching vs. those of clade 2 (Figs. 2 and 3).

To conclude, the analysis of anaerobic functional traits such as hydrogenases has provided a valuable functional pattern for the evolutionary origin not only of sulfur-oxidizing symbionts, but also of the whole class of gammaproteobacteria. The emerging taxonomic picture of sulfur-oxidizing symbionts is much simpler than that contemplating nine diverse groups previously deduced from the analysis of rRNA sequences (Dubilier et al. 2008). RNA-based phylogeny appears to provide a distorted, if not misleading picture of the underlying functional diversification of symbiotic taxa, possibly due to phenomena of long-branch attraction (Belda et al. 2005, Bodilis et al. 2011) derived from the vertical transmission of the endosymbionts. Such a transmission has insulated their genomes from the diversification occurring among free-living relatives (Philippe and Forterre 1999, Ku et al. 2015), a situation that resembles that described for the ancestors of mitochondria (see Chapters seven and eight of the book). Based upon the analysis presented here (Figs. 1 and 2), a conservative outlook would put the majority of sulfur-oxidizing symbionts within lineages of the Thiotrichales order of gammaproteobacteria, which appears to have a much wider phylogenetic diversity than previously thought (Figs. 2 and 3). Together with the as yet unclassified *Thiohalorhabdus* and some Chromatiales, Thiotrichales could therefore form the oldest taxonomic group among gammaproteobacteria. The clear similarity with the metabolic traits of deep-branching alphaproteobacteria (Fig. 3) sustains this possibility, since it suggests that the progenitors of sulfur-oxidizing bacteria were close to the common ancestor of alphaproteobacteria and gammaproteobacteria. Conversely, Xanthomonadales and the majority of Chromatiales, represented by various taxa at the top of the list in Fig. 1, display a more limited metabolic repertoire than Thiotrichales and sulfur-oxidizing symbionts, thereby suggesting that their members may have evolved later, as confirmed by phylogenetic trees of bioenergetic proteins (Fig. 2).

Gammaproteobacteria Endosymbionts of Insects

The class of gammaproteobacteria includes many species that live as endosymbionts of insects. Several aspects of these endosymbionts are of interest to the leitmotif of bacterial evolution into mitochondria, in particular genome reduction and biased composition of the DNA. Moreover, the recent description of many new genera of such insect symbionts reveals their large genetic diversity and shows the emergence of a fascinating research area on invertebrate microbiology. Endosymbionts may compensate diet deficiencies of insects that feed on sap or blood (Gosalbes et al. 2010). Others, considered secondary symbionts, may confer stress tolerance or defense as in the case of *Hamiltonella defensa* against wasps (Oliver et al. 2003). We are now aware that most gammaproteobacterial endosymbionts impact insect adaptability to new environmental conditions, or to colonize and feed on plants (Hansen and Moran 2014). Insects can harbor multiple species of endosymbionts and it is thus necessary to distinguish between primary and secondary symbionts among the increasing number of gammaproteobacteria intimately associated with insects (Baumann 2005, Husník et al. 2011). Remarkably, some primary endosymbionts are maternally transmitted (Moran and Baumann 2000) and may be found inside special insect cells that are called bacteriocytes, which are, in some cases, grouped in larger abdominal structures called bacteriomes (Buchner 1965). Although anatomically different, insect bacteriocytes are functionally analogous to the specialized bacteria-containing cells of mollusks and other marine invertebrates mentioned earlier in this chapter (Dubilier et al. 2008).

Within gammaproteobacteria, insect endosymbionts fundamentally segregate into two different phylogenetic groups: Enterobacterales and Oceanospirillales (Wernegreen 2002, Husník et al. 2011, Neave et al. 2016). The most studied gammaproteobacterial endosymbionts belong to the genus *Buchnera*, which is classified within the family of Erwiniaceae of the order of Enterobacterales (Husník et al. 2011), previously called gamma3-subdivision (Wernegreen 2002). *Buchnera* strains are found in the bacteriomes of many aphid species and are transmitted from mothers to offspring (Moran and Baumann 2000). Aphids feed on plant sap, which is nutritionally imbalanced because it lacks essential amino acids (Hansen and Moran 2011). So, it is not surprising that many insects feeding on sap have dependences on bacterial endosymbionts such as *Buchnera*, which provide amino acids and other nutrients to the host (Wernegreen 2002, Hansen and Moran 2011). The first genome reported from an endosymbiont was in 2000, from an aphid-colonizing *Buchnera* (Shigenobu et al. 2000). This report marked the beginning of a new era for the study of insect endosymbionts, showing genes for the synthesis of riboflavin and essential amino acids within a largely reduced genome (approx. one tenth of that of *E. coli*) with limited capacity of central metabolism but the presence of plasmids coding extra gene copies for the biosynthesis of essential amino acids (Shigenobu et al. 2000). Notably, *Buchnera* strains and their insect hosts cospeciate (Clark et al. 2000, Wernegreen 2002, Lo et al. 2003).

Mutualistic associations of gammaproteobacteria with insects may be as old as 160–280 million years (Moran et al. 1993), while other endosymbionts seem to have been acquired recently (Rosas-Pérez et al. 2017). Inside distinct insect bacteriomes, there are intracellular diverse primary endosymbionts and even secondary endosymbionts; metabolic complementation may occur among some of the different symbionts inhabiting a single host (Rao et al. 2015). The long and persistent symbioses have led to the evolution of well adapted gammaproteobacterial symbionts with highly reduced genomes, even smaller than that of *Buchnera*, for example *Ca.* Portiera (Sloan and Moran 2012, Santos-Garcia et al. 2012, Jiang et al. 2013) and *Ca.* Carsonella (Nakabachi et al. 2006). *Ca.* Portiera is an obligate endosymbiont from the whitefly *Bemisia tabaci* with a genome that is 358 kb long and enriched in genes encoding essential amino acid biosynthesis (Sloan and Moran 2012, Santos-Garcia et al. 2012). Conversely, *Ca.* Carsonella is an endosymbiont of psyllids with a 160 Kb genome, which for a while held the prize of the smallest bacterial genome (Nakabachi et al. 2006). Phylogenetically, *Ca.* Carsonella and *Portiera* belong to a distinct group from that including *Buchnera* and the majority of gammaproteobacterial endosymbionts of insects (Husník et al. 2011). This group is part of the order of Oceanospirillales, which includes free-living marine species such as the type genus *Oceanospirillum*, as well as widespread endosymbionts of marine organisms that belong to the family Endozoicomonadaceae (Neave et al. 2016).

Phylogenetic trees of the catalytic subunit of bo3 ubiquinol oxidase, the most frequent terminal oxidase in the respiratory chain of gammaproteobacterial symbionts of insects (Wernegreen 2002), show that the proteins from Oceanospirillales are in precursor branches vs. those found in *Buchnera* and other taxa affiliated to the Enterobacterales (Fig. 5A). This is consistent with the deeper branching order of Oceanospirillales vs. Enterobacterales in the phylogenesis of the whole gamma class (Williams et al. 2010). Currently, the majority of gammaproteobacterial symbionts of insects are classified within the family of Enterobacteriaceae under the grouping of ant, tsetse, mealybug, aphid, etc., endosymbionts—https://www.ncbi.nlm.nih.gov/Taxonomy/Browser/wwwtax.cgi?id=84563—which is distinct from the taxonomic group containing *Buchnera* strains (Husník et al. 2011). Single protein trees such as that shown in Fig. 5A can barely resolve the *Buchnera* clade from other clades within the Enterobacterales, which may subtend multiple origins of gammaproteobacterial endosymbiosis with insects (Husník et al. 2011, Husnik and McCutcheon 2016). Of note is that *Ca.* Carsonella does not have the genes for bo3 oxidase due to extreme reduction of its genome (Nakabachi et al. 2006). The same loss is seen in the genome of *Sodalis* SOPE (Oakerson et al. 2014), which has instead retained the bd ubiquinol oxidase that is present in other *Sodalis* species. Notably, the respiratory chain of *E. coli*, which belongs to the same family of Enterobacteriaceae as *Sodalis*, contains both bo3 and bd ubiquinol oxidase under aerobic or microaerobic conditions (Borisov and Verkhovski 2015).

Sodalis (SOPE) from grain seeds (Gil et al. 2008) provides the weevil with vitamins such as riboflavin, biotin and pantothenic acid, and apparently enhances

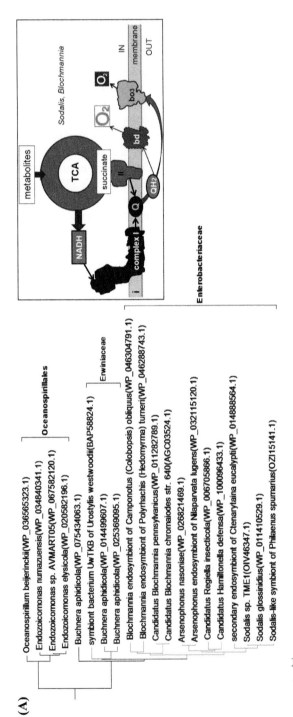

Fig. 5 contd. ...

...Fig. 5 contd.

(B)

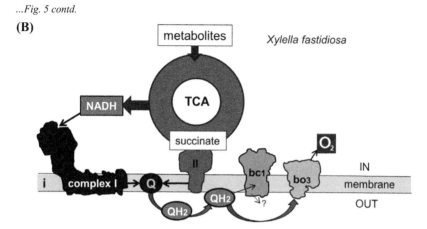

Fig. 5. Respiratory chain of endosymbionts and *Xylella*. (A) NJ tree of the largest subunit, homologous to COX1, of the bo3 ubiquinol oxidase of *Candidatus* Blochmannia chromaiodes str. 640 from a blast search against Oceanospirilalles and other endosymbionts of the Enterobacterales order of gammaproteobacteria. The basic respiratory chain of these endosymbionts is shown on the right. Note that in some strains of *Sodalis,* the bo3 oxidase has been lost (Oakeson et al. 2014), while in *Blochmannia* species it is the bd ubiquinol oxidases that has been lost upon genome reduction. (B) Respiratory chain of *Xylella fastidiosa*. The question mark indicates the absence of a soluble cytochrome *c* functioning as the electron acceptor for the bc1 complex (Bhattacharyya et al. 2002). Note the similarity with the respiratory chain of *Sodalis* in A.

mitochondrial oxidative phosphorylation of the host (Heddi et al. 1999); its genome is considered to be on the process of reduction vs. that of free-living relatives (Oakerson et al. 2014). Some *Sodalis* species and those of the unrelated *Arsenophonus* genus (see below) have genomes almost the size of those of free living bacteria; indeed, free-living *Sodalis* strains have been found (reviewed in Rosas-Perez et al. 2017). This situation is relatively unusual among insect endosymbionts and allows the cultivation of bacterial strains in laboratories, thereby indicating that the genus *Sodalis* has flexible lifestyle characters (Rosas-Perez et al. 2017). In scale insects, Enterobacteria related to *Sodalis* are common co-symbionts of Flavobacteria (Rosenblueth et al. 2018), as will be presented in Chapter six. Some of these enterobacterial symbionts could be clearly ascribed to *Sodalis*, but others not (Rosenblueth et al. 2012). While the phylogeny of Flavobacteria reflects the phylogenetic relationships of the host (Rosenblueth et al. 2012), enterobacterial phylogenies follow different patterns, indicating they may shift hosts, without a committed interaction that leads to cospeciation.

The majority of gammaproteobacterial endosymbionts of insects are different from *Sodalis* because they exhibit considerable or extreme genome reduction, which prevents their cultivation in the lab (Dale et al. 2006, Burke et al. 2011, Clayton et al. 2012). With few exceptions, genome reduction is associated with a bias in DNA composition that is enriched in AT bases (Wernegreen 2002). One example of genome reduction with AT enrichment is provided by *Ca.* Blochmannia, an endosymbiont that has been described in all carpenter ants from the genus *Componotus*, which includes more than 1000 species (Feldhaar et al. 2007, Sauer et al. 2002, Williams and Wernegreen 2015). Interestingly, *Ca.* Blochmannia symbionts have been reported to improve colony growth and immune defense in the ant hosts (de Souza 2009).

The genus *Arsenophonus* provides a different example of gammaproteobacterial endosymbiosis with insects (Dale et al. 2006, Darby et al. 2010, Bressan et al. 2012, Michalik et al. 2018). *Arsenophonus* species have genomes around 3.5 Mb (Darby et al. 2010, Wilkes et al. 2010) and belong to the family of Moraxellaceae, which is characterized by the presence of multiple terminal oxidases that have been subsequently lost in other members of the Enterobacterales (Degli Esposti and Martinez-Romero 2017). *Arsenophonus* strains are widely distributed among insects (Novakova et al. 2009), including triatomine bugs that cause Chagas disease in the Americas (Hypsa and Dale 1997), as well as in ticks (Edouard et al. 2013, Bohacsova et al. 2016).

Experimental and natural replacements of obligate insect symbionts have been reported (Husnik and McCutcheon 2016, Chong and Moran 2018) indicating symbiont flexibility. These findings support the hypothesis of "It is the song, not the singer" (Doolittle et al. 2017) even for obligate symbioses. In view of the large unexplored insect diversity, the seemingly long list of insect symbionts will certainly become even longer in the future, confirming and expanding the insect affinities of gammaproteobacteria (Moran et al. 2003).

In the context of this book, genomic features well studied in gammaproteobacterial symbionts of insects constitute paradigms for the progressive reduction and functional streamlining of the genomes of intracellular parasites and endosymbionts, including the progenitors of mitochondria (Chapter seven). AT-rich genomes and rapid rates of evolution characterize gammaproteobacterial symbionts (Wernegreen 2002) and are equally present in obligate endocellular parasites of the alphaproteobacteria class, as well as in mitochondrial genomes (see Chapter seven). Most likely, these common features reflect general trends associated with the establishment of vertically transmitted endosymbiosis, as discussed earlier in this chapter with regard to sulfur-oxidizing symbionts. Intriguingly, symbionts of the *Sodalis* group have also been found inside betaproteobacterial endosymbionts of mealybug insects, thereby generating speculations about their possible organelle-like condition similar to mitochondrial organelles (Husnik and McCutcheon 2016). However, the functional similarities with mitochondria appear to be rather weak and are phylogenetically improbable, given that *Sodalis* belongs to a later branching group of gammaproteobacteria while the progenitors of mitochondria originated from an early branching lineage of alphaproteobacteria, as discussed in Chapter seven.

Xylella, the Killer of Olive Trees

Xylella fastidiosa, a broad spectrum plant pathogen that is difficult to cultivate (Wells et al. 1987), can also be considered among insect-associated gammaproteobacteria because it spends part of its lifecycle within insect vectors, even if its association with insects is semi-persistent (Perilla-Henao et al. 2016). *Xylella* belongs to the vast order of Xanthomonadales and is considered to be the only known vector-borne plant pathogen living as a xylem specialist (Chatterjee et al. 2008, Perilla-Henao et al. 2016, Saponari et al. 2017).

Xylella strains can infect and damage hundreds of diverse species of plants (Saponari et al. 2017), an ability that may be due to its semi-persistent relationship with insect vectors. The close relative *Xanthomonas* (Wells et al. 1987), which includes several plant pathogens, instead appears to establish persistent associations with its insect vectors (Perilla-Henao et al. 2016). Persistence may be due to the intracellular relationship that these pathogens share with plant and insect hosts. *X. fastidiosa* strains may have lost the traits promoting such relationships in evolving growth and survival in the plant host, also in dead cells of the xylem (Perilla-Henao et al. 2016). *X. fastidiosa* colonization of insect vectors requires adhesins and other traits that are not required for maximal infection of plants, thereby producing a conflict between adaptation to parasitic life in the insect vectors and plant host colonization (Chatterjee et al. 2008). Transmission of *Xylella* between plants of the same species, or among diverse plant species, is mediated by insects of the Hemiptera order which naturally feed on xylem sap (Chatterjee et al. 2008, Perilla-Henao et al. 2016). In the Americas, these insect vectors include leafhoppers and spittlebugs, while a single species of meadow spittlebugs is responsible for the spread of *Xylella* among olive trees and other cultivated plants in Europe (Saponari et al. 2017, and references therein). *Xylella* bacteria colonize only the foregut of the insect vectors and are not beneficial to their host, which often contain a mixture of commensals and endosymbionts in the midgut or other organs (see previous section on insect endosymbionts and also Chapter six). Notably, *Xylella* strains have been found in the gut microbiome of American ants and termites, as well as in the microbiome of red palm beetles that have recently invaded the Mediterranean basin (Degli Esposti and Martinez-Romero 2017, and references therein).

Since their original description under the species name *Xylella fastidiosa*, plant pathogens, now recognized as subspecies or local variants of *Xylella*, have been described as close relatives of species from the genus *Xanthomonas*, which are exclusively plant-associated and frequently pathogenic (Wells et al. 1987, Simpson et al. 2000). However, *Xylella* strains show several differences with *Xanthomonas* strains in their pathogenic traits, as well as in host and vector interactions; genomic analysis has provided various clues to explain such differences (Chatterjee et al. 2008). One difference vs. *Xanthomonas* is of particular interest to the bioenergetic leitmotif of this book: *Xylella* has the phenotypic character of 'cytochrome oxidase negative' (Wells et al. 1987) because its genome does not code for cytochrome *c* oxidases (Bhattacharyya et al. 2002). The genome of all strains of *X. fastidiosa* possess the operon of another terminal oxidase, that of bo3 ubiquinol oxidase described in Chapter two (Bhattacharyya et al. 2002), as shown in Fig. 1A (top) and Fig. 5. Because bo3 ubiquinol oxidase is frequently found in environmental bacteria which associate with insects (Chouaia et al. 2014, see previous section of this chapter and Fig. 5A), one could speculate that the presence of this enzyme in *X. fastidiosa* may also help the colonization and permanence inside the foregut of Hemipteran vectors. However, the role of bo3 ubiquinol oxidase in the colonization and infection of xylem tissue is unclear, unless it provides some functional advantage in dead xylem cells where the flux of oxygen may be limited.

Ubiquinol oxidases are often over-expressed under micro-oxic and partially anaerobic conditions but *Xylella* genomes do not have anaerobic traits which normally are found under the same conditions, for instance nitrate and nitrite reductases (Fig. 1A, cf. Bhattacharyya et al. 2002). However, the gene cluster of adenylyltransferase (APS reductase) is present in the genome of *X. fastidiosa* strains (Bhattacharyya et al. 2002), suggesting a possible anaerobic pathway of dissimilatory sulfate reduction. Such a pathway would be different from that previously described in Chapter four for sulfate-reducing deltaproteobacteria, since the donor for APS reductase, the MQ-oxidizing Qmo enzyme, is present only among deltaproteobacteria. Indeed, the sequence of the APS reductase subunits of *Xylella* is very distant from those of either sulfate-reducing organisms or those of sulfur-oxidizing gammaproteobacteria previously discussed in this chapter.

The unique presence of the bo3 ubiquinol oxidase in the respiratory chain of pathogenic *Xylella* has previously suggested that this enzyme could provide a suitable pharmacological target for specific anti-pathogen treatment (Bhattacharyya et al. 2002). In this regard, there are quinone-like inhibitors that specifically block bacterial bo3 oxidases and do not affect the two terminal oxidases present in plant cells, cytochrome *c* oxidase and alternative oxidase (AOX, see Chapter two for their description). This pharmacological lead should be probably revived for designing new intervention strategies against the plant disease recently transmitted by *Xylella* in Southern Europe, for which there is no cure at the moment (Abbott 2017).

Xylella has recently gained a surge of scientific interest, as well as media popularity—see for instance the article in Newsweek magazine by Wapner (2017)—because of the sudden and devastating appearance of a disease that this bacterium spreads among the olive groves of Southern Italy (Saponari et al. 2017, Abbott 2017). It all started in 2013, when ancient olive trees in the southernmost part of the Italian region of Apulia (Salento) manifested a progressive disease in their leaves and branches, leading to rapid decline of the plants (Saponari et al. 2017 and references therein). Until then, *X. fastidiosa* was endemic in the American continent and had a vast phytosanitary history there, where it infects several crops such as grapevine (in Northern America) and orange (predominantly in South America, Simpson et al. 2000). It appears that the spread of the pathogen in Southern Italy followed the import of ornamental plants containing *Xylella* from Central America by garden centers that are present in the Salento area. From one or probably more of these garden centers, *Xylella* begun to colonize the local species of meadow spittlebug that is not only common in the region, but had a population outburst from 2013 onwards due to milder winter conditions and hot summers following recent climate changes. These spittlebugs usually feed on the xylem sap of young branches of olive trees, thereby producing the rapid spread of a new disease now called Olive quick decline syndrome, OQDS (Saponari et al. 2017).

Nowadays, OQDS has spread to a large part of Apulia, as well as to Corsica, Baleari Islands and Southern France, posing a serious threat to all Southern Europe, since olive trees form the basis of traditional Mediterranean agriculture (Abbott 2017). Similar to the black death pandemic—incidentally produced by another gammaproteobacterium, *Yersinia pestis*, and also started in Italy—the fear of a devastating disease affecting the olive groves of the whole Mediterranean basin resonates with the media, and of course reverberates at the political level, local and Europe-wide (Abbott 2017). The major problem is that there is no cure for the devastating effects of *Xylella* on olive trees, other than eradication of entire, often secular groves, and spreading vast amounts of insecticides to control the insect vectors. Inevitably, these drastic measures produce problems in local communities and on the regional economy, with complicated and often contradictory political interventions that have hampered both basic and field research on this new disease (Abbott 2017). Fortunately, scientific knowledge on *Xylella* killers of olive trees is growing, not only at the genomic level, but also at the level of experimental agriculture, with the identification of the most susceptible varieties of the olive tree (Saponari et al. 2017).

Clearly, the story of transcontinental expansion of *Xylella* from Central America to Southern Italy will continue to develop for years to come, providing a unique instance in which basic knowledge of bacterial genomics and physiology may ultimately help finding a cure to an otherwise deadly disease, devastating Mediterranean economy and landscapes. Soon, *Xylella*-resistant variants of olive trees will be found, and a new twist of the story will start.

Acknowledgments

This work was sponsored by grants CONACyT Basic Science 253116 and PAPIIT (UNAM) IN207718 to E.M-R. We thank Dr. Dave W. Ussery (University of Arkansas, USA) for providing unpublished material and insightful comments on the manuscript of this chapter. We also thank Julio Martínez for his help in formatting parts of the manuscript.

References

Abbott, A. 2017. Italy rebuked for failure to prevent olive-tree tragedy. Nature 546: 193–194. doi:10.1038/546193a.

Alberts, B., A. Johnson, J. Lewis, M. Raff, K. Roberts and P. Walter. 2002. Molecular Biology of the Cell. 4th edn. New York: Garland Science.

Anantharaman, K., J.A. Breier, C.S. Sheik and G.J. Dick. 2013. Evidence for hydrogen oxidation and metabolic plasticity in widespread deep-sea sulfur-oxidizing bacteria. Proc. Natl. Acad. Sci. USA 110: 330–335. doi:10.1073/pnas.1215340110.

Bhattacharyya, A., S. Stilwagen, G. Reznik, H. Feil, W.S. Feil, I. Anderson et al. 2002. Draft sequencing and comparative genomics of Xylella fastidiosa strains reveal novel biological insights. Genome Res. 10: 1556–1563.

Baumann, P. 2005. Biology bacteriocyte-associated endosymbionts of plant sap-sucking insects. Annu. Rev. Microbiol. 59: 155–189.

Bazylinski, D.A., T.J. Williams, C.T. Lefèvre, R.J. Berg, C.L. Zhang, S.S. Bowser et al. 2013. *Magnetococcus marinus gen.* nov., sp. nov., a marine, magnetotactic bacterium that represents a novel lineage (*Magnetococcaceae* fam. nov., *Magnetococcales* ord. nov.) at the base of the Alphaproteobacteria. Int. J. Syst. Evol. Microbiol. 63: 801–808. doi:10.1099/ijs.0.038927-0.

Belda, E., A. Moya and F.J. Silva. 2005. Genome rearrangement distances and gene order phylogeny in gamma-Proteobacteria. Mol. Biol. Evol. 22: 1456–1467.

Bergin, C., C. Wentrup, N. Brewig, A. Blazejak, C. Erséus, O. Giere et al. 2018. Acquisition of a novel sulfur-oxidizing symbiont in the gutless marine worm *Inanidrilus exumae*. Appl. Environ. Microbiol. pii: AEM.02267-17. doi:10.1128/AEM.02267-17.

Bohacsova, M., O. Mediannikov, M. Kazimirova, D. Raoult and Z. Sekeyova. 2016. Arsenophonus nasoniae and rickettsiae infection of ixodes ricinus due to parasitic wasp ixodiphagus hookeri. PLoS One 11: e0149950. doi:10.1371/journal.pone.0149950.

Boden, R., L.P. Hutt, M. Huntemann, A. Clum, M. Pillay, K. Palaniappan et al. 2016. Permanent draft genome of *Thermithiobacillus tepidarius* DSM 3134(T), a moderately thermophilic, obligately chemolithoautotrophic member of the Acidithiobacillia. Stand. Genomic Sci. 11: 74.

Boden, R., K.M. Scott, J. Williams, S. Russel, K. Antonen, A.W. Rae and L.P. Hutt. 2017. An evaluation of *Thiomicrospira, Hydrogenovibrio* and *Thioalkalimicrobium*: reclassification of four species of Thiomicrospira to each *Thiomicrorhabdus gen.* nov. and *Hydrogenovibrio*, and reclassification of all four species of *Thioalkalimicrobium* to *Thiomicrospira*. Int. J. Syst. Evol. Microbiol. 67: 1140–1151. doi:10.1099/ijsem.0.001855.

Bodilis, J., S. Nsigue Meilo, P. Cornelis, P. De Vos and S. Barray. 2011. A long-branch attraction artifact reveals an adaptive radiation in pseudomonas. Mol. Biol. Evol. 28: 2723–2726. doi:10.1093/molbev/msr099.

Borisov, V.B. and M.I. Verkhovsky. 2015. Oxygen as acceptor. Eco. Sal. Plus. 6(2). doi:10.1128/ecosalplus.ESP-0012-2015.

Bressan, A., F. Terlizzi and R. Credi. 2012. Independent origins of vectored plant pathogenic bacteria from arthropod-associated *Arsenophonus* endosymbionts. Microb. Ecol. 63: 628–638. doi:10.1007/s00248-011-9933-5.

Buchner, P. 1965. Endosymbiosis of animals with plant microorganisms. New York: John Wiley & Sons, Inc.: Interscience Publ.

Burke, G.R. and N.A. Moran. 2011. Massive genomic decay in *Serratia symbiotica*, a recently evolved symbiont of aphids. Genome Biol. Evol. 3: 195–208. doi:10.1093/gbe/evr002.

Cavanaugh, C.M., S.L. Gardiner, M.L. Jones, H.W. Jannasch and J.B. Waterbury. 1981. Prokaryotic cells in the hydrothermal vent tube worm *Riftia pachyptila* Jones: Possible chemoautotrophic symbionts. Science 213: 340–342.

Chatterjee, S., R.P. Almeida and S. Lindow. 2008. Living in two worlds: The plant and insect lifestyles of *Xylella fastidiosa*. Annu. Rev. Phytopathol. 46: 243–71.

Chong, R.A. and N.A. Moran. 2018. Evolutionary loss and replacement of *Buchnera*, the obligate endosymbiont of aphids. ISME J. doi:10.1038/s41396-017-0024-6.

Chouaia, B., S. Gaiarsa, E. Crotti, F. Comandatore, M. Degli Esposti, I. Ricci et al. 2014. Acetic acid bacteria genomes reveal functional traits for adaptation to life in insect guts. Genome Biol. Evol. 6: 912–920. doi:10.1093/gbe/evu062.

Clark, M.A., N.A. Moran, P. Bauman and J.J. Wernegreen. 2000. Cospeciation between bacterial endosymbionts (*Buchnera*) and a recent radiation of aphids (*Uroleucon*) and pitfalls of testing for phylogenetic congruence. Evolution. 54: 517–525.

Crespo-Medina, M., A. Chatziefthimiou, R. Cruz-Matos, I. Pérez-Rodríguez, T. Barkay, R.A. Lutz et al. 2009. *Salinisphaera hydrothermalis* sp. nov., a mesophilic, halotolerant, facultatively autotrophic, thiosulfate-oxidizing gammaproteobacterium from deep-sea hydrothermal vents, and emended description of the genus *Salinisphaera*. Int. J. Syst. Evol. Microbiol. 59: 1497–1503.

Dahl, C. 2015. Cytoplasmic sulfur trafficking in sulfur-oxidizing prokaryotes. IUBMB Life. 67: 268–274. doi:10.1002/iub.1371.

Dale, C., M. Beeton, C. Harbison, T. Jones and M. Pontes. 2006. Isolation, pure culture, and characterization of "*Candidatus* Arsenophonus arthropodicus," an intracellular secondary endosymbiont from the hippoboscid louse fly *Pseudolynchia canariensis*. Appl. Environ. Microbiol. 72: 2997–3004.

Darby, A.C., J.H. Choi, T. Wilkes, M.A. Hughes, J.H. Werren, G.D. Hurst et al. 2010. Characteristics of the genome of *Arsenophonus nasoniae*, son-killer bacterium of the wasp *Nasonia*. Insect Mol. Biol. 19: 75–89. doi:10.1111/j.1365-2583.2009.00950.x.

Degli Esposti, M. 2014. Bioenergetic evolution in proteobacteria and mitochondria. Genome Biol. Evol. 6: 3238–3251. doi:10.1093/gbe/evu257.

Degli Esposti, M. 2015. Genome analysis of structure-function relationships in Respiratory Complex I, an ancient bioenergetic enzyme. Genome Biol. Evol. 8: 126–147. doi:10.1093/gbe/evv239.

Degli Esposti, M., T. Rosas-Pérez, L.E. Servín-Garcidueñas, L.M. Bolaños, M. Rosenblueth and E. Martínez-Romero. 2015. Molecular evolution of cytochrome bd oxidases across proteobacterial genomes. Genome Biol. Evol. 7: 801–820. doi:10.1093/gbe/evv032.

Degli Esposti, M. and E. Martinez Romero. 2016. A survey of the energy metabolism of nodulating symbionts reveals a new form of respiratory complex I. FEMS Microbiol. Ecol. 92: fiw084. doi:10.1093/femsec/fiw084.

Degli Esposti, M., D. Cortez, L. Lozano, S. Rasmussen, H.B. Nielsen and E. Martinez Romero. 2016. Alpha proteobacterial ancestry of the [Fe-Fe]-hydrogenases in anaerobic eukaryotes. Biol. Direct. 11: 34. doi:10.1186/s13062-016-0136-3.

Degli Esposti, M. 2017. A journey across genomes uncovers the origin of ubiquinone in cyanobacteria. Genome Biol. Evol. 9: 3039–3053. doi:10.1093/gbe/evx225.

Degli Esposti, M. and E. Martinez Romero. 2017. The functional microbiome of arthropods. PLoS One 12(5): e0176573. doi:10.1371/journal.pone.0176573.

de Souza, D.J., A. Bézier, D. Depoix, J.M. Drezen and A. Lenoir. 2009. *Blochmannia* endosymbionts improve colony growth and immune defence in the ant *Camponotus fellah*. BMC Microbiol. 9: 29. doi:10.1186/1471-2180-9-29.

Doolittle, W.F. and A. Booth. 2017. It's the song, not the singer: an exploration of holobiosis and evolutionary theory. Biol. Philos. 32: 5–24.

Dmytrenko, O., S.L. Russell, W.T. Loo, K.M. Fontanez, L. Liao, G. Roeselers et al. 2014. The genome of the intracellular bacterium of the coastal bivalve, *Solemya velum*: a blueprint for thriving in and out of symbiosis. BMC Genomics 15: 924. doi:10.1186/1471-2164-15-924.

Dubilier, N., C. Bergin and C. Lott. 2008. Symbiotic diversity in marine animals: The art of harnessing chemosynthesis. Nat. Rev. Microbiol. 6: 725–740. doi:10.1038/nrmicro1992.

Dufour, S.C., J.R. Laurich, R.T. Batstone, B. McCuaig, A. Elliott and K.M. Poduska. 2014. Magnetosome-containing bacteria living as symbionts of bivalves. ISME J. 8: 2453–2462. doi:10.1038/ismej.2014.93.

Edouard, S., G. Subramanian, B. Lefevre, A. Dos Santos, P. Pouedras, Y. Poinsignon et al. 2013. Co-infection with *Arsenophonus nasoniae* and *Orientia tsutsugamushi* in a traveler. Vector Borne Zoonotic Dis. 13: 565–571. doi:10.1089/vbz.2012.1083.

Feldhaar, H., J. Straka, M. Krischke, K. Berthold, S. Stoll, M.J. Mueller et al. 2007. Nutritional upgrading for omnivorous carpenter ants by the endosymbiont *Blochmannia*. BMC Biol. 5: 48.

Flood, B.E., D.S. Jones and J.V. Bailey. 2005. *Sedimenticola thiotaurini* sp. nov., a sulfur-oxidizing bacterium isolated from salt marsh sediments, and emended descriptions of the genus *Sedimenticola* and *Sedimenticola selenatireducens*. Int. J. Syst. Evol. Microbiol. 65: 2522–2530. doi:10.1099/ijs.0.000295.

Flood, B.E., D.S. Jones and J.V. Bailey. 2015. Complete genome sequence of *Sedimenticola thiotaurini* strain SIP-G1, a polyphosphate- and polyhydroxyalkanoate-accumulating sulfur-oxidizing Gammaproteobacterium isolated from salt marsh sediments. Genome Announc. 3(3). pii: e00671-15. doi:10.1128/genomeA.00671-15.

Friedrich, C.G., F. Bardischewsky, D. Rother, A. Quentmeier and J. Fischer. 2005. Prokaryotic sulfur oxidation. Curr. Opin. Microbiol. 8: 253–259.

Garrity, G.M., J.A. Bell and T.G. Lilburn. 2005. Class III. *Gammaproteobacteria* class. nov., p. 1. *In*: Brenner, D.J., N.R. Krieg, J.T. Staley and G.M. Garrity (eds.). Bergey's Manual of Systematic Bacteriology, 2nd ed., vol. 2. Springer, New York, NY.

Gennis, R.B. 1991. Some recent advances relating to prokaryotic cytochrome c reductases and cytochrome c oxidases. Biochim. Biophys. Acta. 1058(1): 21–24.

Gil, R., E. Belda, M.J. Gosalbes, L. Delaye, A. Vallier, C. Vincent-Monégat et al. 2008. Massive presence of insertion sequences in the genome of SOPE, the primary endosymbiont of the rice weevil *Sitophilus oryzae*. Int. Microbiol. 11: 41–48.

González, C., M. Lazcano, J. Valdés and D.S. Holmes. 2016. Bioinformatic analyses of unique (orphan) core genes of the genus *Acidithiobacillus*: Functional inferences and use as molecular probes for genomic and metagenomic/transcriptomic interrogation. Front. Microbiol. 7: 2035. doi:10.3389/fmicb.2016.02035.

Gosalbes, M.J., A. Latorre, A. Lamelas and A. Moya. 2010. Genomics of intracellular symbionts in insects. Int. J. Med. Microbiol. 300: 271–278. doi:10.1016/j.ijmm.2009.12.001.

Greening, C., A. Biswas, C.R. Carere, C.J. Jackson, M.C. Taylor, M.B. Stott et al. 2016. Genomic and metagenomic surveys of hydrogenase distribution indicate H_2 is a widely utilised energy source for microbial growth and survival. ISME J. 10: 761–77. doi:10.1038/ismej.2015.153.

Hansen, A.K. and N.A. Moran. 2011. Aphid genome expression reveals host-symbiont cooperation in the production of amino acids. Proc. Natl. Acad. Sci. USA 108: 2849–2854. doi:10.1073/pnas.1013465108.

Hansen, A.K. and N.A. Moran. 2014. The impact of microbial symbionts on host plant utilization by herbivorous insects. Mol. Ecol. 23: 1473–96. doi:10.1111/mec.12421.

Heddi, A., A.M. Grenier, C. Khatchadourian, H. Charles and P. Nardon. 1999. Four intracellular genomes direct weevil biology: nuclear, mitochondrial, principal endosymbiont, and *Wolbachia*. Proc. Natl. Acad. Sci. USA 96: 6814–6819.

Husník, F., T. Chrudimský and V. Hypša. 2011. Multiple origins of endosymbiosis within the Enterobacteriaceae (γ-Proteobacteria): convergence of complex phylogenetic approaches. BMC Biol. 9: 87. doi:10.1186/1741-7007-9-87.

Husník, F. and J.P. McCutcheon. 2016. Repeated replacement of an intrabacterial symbiont in the tripartite nested mealybug symbiosis. Proc. Natl. Acad. Sci. USA 113: E5416–E5424. doi:10.1073/pnas.1603910113.

Hypsa, V. and C. Dale. 1997. In vitro culture and phylogenetic analysis of "*Candidatus* Arsenophonus triatominarum," an intracellular bacterium from the triatomine bug, *Triatoma infestans*. Int. J. Syst. Bacteriol. 47: 1140–1144.

Ikuta, T., Y. Takaki, Y. Nagai, S. Shimamura, M. Tsuda, S. Kawagucci et al. 2016. Heterogeneous composition of key metabolic gene clusters in a vent mussel symbiont population. ISME J. 10: 990–1001. doi:10.1038/ismej.2015.176.

Jiang, Z.F., F. Xia, K.W. Johnson, C.D. Brown, E. Bartom, J.H. Tuteja et al. 2013. Comparison of the genome sequences of "*Candidatus* Portiera aleyrodidarum" primary endosymbionts of the whitefly *Bemisia tabaci* B and Q biotypes. Appl. Environ-Microbiol. 79: 1757–1759. doi:10.1128/AEM.02976-12.

Johnson, D.B. 2014. Biomining-biotechnologies for extracting and recovering metals from ores and waste materials. Curr. Opin. Biotechnol. 30: 24–31. doi:10.1016/j.copbio.2014.04.008.

Kleiner, M., J.M. Petersen and N. Dubilier. 2012. Convergent and divergent evolution of metabolism in sulfur-oxidizing symbionts and the role of horizontal gene transfer. Curr. Opin. Microbiol. 15: 621–631. doi:10.1016/j.mib.2012.09.003.

König, S., O. Gros, S.E. Heiden, T. Hinzke, A. Thürmer, A. Poehlein et al. 2016. Nitrogen fixation in a chemoautotrophic lucinid symbiosis. Nat. Microbiol. 2: 16193. doi:10.1038/nmicrobiol.2016.193.

Ku, C., S. Nelson-Sathi, M. Roettger, S. Garg, E. Hazkani-Covo and W.F. Martin. 2015. Endosymbiotic gene transfer from prokaryotic pangenomes: Inherited chimerism in eukaryotes. Proc. Natl. Acad. Sci. USA 112: 10139–10146. doi:10.1073/pnas.1421385112.

Kuwahara, H., Y. Takaki, S. Shimamura, T. Yoshida, T. Maeda, T. Kunieda et al. 2011. Loss of genes for DNA recombination and repair in the reductive genome evolution of thioautotrophic symbionts of *Calyptogena* clams. BMC Evol. Biol. 11: 285. doi:10.1186/1471-2148-11-285.

Land, M., L. Hauser, S.R. Jun, I. Nookaew, M.R. Leuze, T.H. Ahn et al. 2015. Insights from 20 years of bacterial genome sequencing. Funct. Integr. Genomics 15: 141–61. doi:10.1007/s10142-015-0433-4.

Lo, N., C. Bandi, H. Watanabe, C. Nalepa and T. Beninati. 2003. Evidence for cocladogenesis between diverse dictyopteran lineages and their intracellular endosymbionts. Mol. Biol. Evol. 20: 907–913.

Marshall, K.T. and R.M. Morris. 2013. Isolation of an aerobic sulfur oxidizer from the SUP05/Arctic96BD-19 clade. ISME J. 7: 452–455. doi:10.1038/ismej.2012.78.

McCutcheon, J.P., B.R. McDonald and N.A. Moran. 2009. Convergent evolution of metabolic roles in bacterial co-symbionts of insects. Proc. Natl. Acad. Sci. USA 106: 15394–15399. doi:10.1073/pnas.0906424106.

McGill, S.E. and D. Barker. 2017. Comparison of the protein-coding genomes of three deep-sea, sulfur-oxidising bacteria: "*Candidatus* Ruthia magnifica", "*Candidatus* Vesicomyosocius okutanii" and *Thiomicrospira crunogena*. BMC Res. Notes 10: 296. doi:10.1186/s13104-017-2598-5.

Michalik, A., F. Schulz, K. Michalik, F. Wascher, M. Horn and T. Szklarzewicz. 2018. Coexistence of novel gammaproteobacterial and *Arsenophonus* symbionts in the scale insect *Greenisca brachypodii* (Hemiptera, Coccomorpha: Eriococcidae). Environ. Microbiol. doi:10.1111/1462-2920.14057.

Moran, N.A., M.A. Munson, P. Baumann and H. Ishikawa. 1993. A molecular clock in endosymbiotic bacteria is calibrated using the insect hosts. Proc. R. Soc. Lond. B. 253: 167–171. doi:10.1098/rspb.1993.0098.

Moran, N.A. and P. Baumann. 2000. Bacterial endosymbionts in animals. Curr. Opin. Microbiol. 3: 270–275.

Moran, N.A., C. Dale, H. Dunbar, W.A. Smith and H. Ochman. 2003. Intracellular symbionts of sharpshooters (Insecta: Hemiptera: Cicadellinae) form a distinct clade with a small genome. Environ. Microbiol. 5: 116–126.

Moran, N.A., J.P. McCutcheon and A. Nakabachi. 2008. Genomics and evolution of heritable bacterial symbionts. Annu. Rev. Genet. 42: 165–190. doi:10.1146/annurev.genet.41.110306.130119.

Nakabachi, A., A. Yamashita, H. Toh, H. Ishikawa, H.E. Dunbar, N.A. Moran et al. 2006. The 160-kilobase genome of the bacterial endosymbiont *Carsonella*. Science 314: 267.

Nakagawa, S., S. Shimamura, Y. Takaki, Y. Suzuki, S. Murakami, T. Watanabe et al. 2014. Allying with armored snails: the complete genome of gammaproteobacterial endosymbiont. ISME J. 8: 40–51. doi:10.1038/ismej.2013.131.

Neave, M.J, A. Apprill, C. Ferrier-Pagès and C.R. Voolstra. 2016. Diversity and function of prevalent symbiotic marine bacteria in the genus Endozoicomonas. Appl. Microbiol.Biotechnol. 100: 8315–8324. doi: 10.1007/s00253-016-7777-0.

Newton, I.L., T. Woyke, T.A. Auchtung, G.F. Dilly, R.J. Dutton, M.C. Fisher et al. 2007. The *Calyptogena magnifica* chemoautotrophic symbiont genome. Science 315: 998–1000.

Nováková, E., V. Hypsa and N.A. Moran. 2009. *Arsenophonus*, an emerging clade of intracellular symbionts with a broad host distribution. BMC Microbiol. 9: 143. doi:10.1186/1471-2180-9-143.

Nunoura, T., Y. Takaki, H. Kazama, J. Kakuta, S. Shimamura, H. Makita et al. 2014. Physiological and genomic features of a novel sulfur-oxidizing gammaproteobacterium belonging to a previously uncultivated symbiotic lineage isolated from a hydrothermal vent. PLoS One. 9(8): e104959. doi:10.1371/journal.pone.0104959.

Oakeson, K.F., R. Gil, A.L. Clayton, D.M.Dunn, A.C. von Niederhausern, C. Hamil et al. 2014. Genome degeneration and adaptation in a nascent stage of symbiosis. Genome Biol. Evol. 6: 76–93. doi:10.1093/gbe/evt210.

Oliver, K.M., J.A. Russell, N.A. Moran and M.S. Hunter. 2003. Facultative bacterial symbionts in aphids confer resistance to parasitic wasps. Proc. Natl. Acad. Sci. USA 100: 1803–1807.

Oshiki, M., T. Fukushima, S. Kawano and J. Nakagawa. 2017. Draft genome sequence of *Thiohalobacter thiocyanaticus* strain FOKN1, a neutrophilic halophile capable of thiocyanate degradation. Genome Announc. 5(32): e00799–17. doi:10.1128/genomeA.00799-17.

Pandelia, M.E., W. Lubitz and W. Nitschke. 2012. Evolution and diversification of Group 1 [NiFe] hydrogenases. Is there a phylogenetic marker for O_2-tolerance? Biochim. Biophys. Acta, 1817: 1565–1575.

Peix, A., M.H. Ramírez-Bahena and E. Velázquez. 2009. Historical evolution and current status of the taxonomy of genus *Pseudomonas*. Infect. Genet. Evol. 9: 1132–1147. doi:10.1016/j. meegid.2009.08.001.

Peix, A., M.H. Ramírez-Bahena and E. Velázquez. 2018. The current status on the taxonomy of *Pseudomonas* revisited: An update. Infect. Genet. Evol. 57: 106–116. doi:10.1016/j. meegid.2017.10.026.

Perilla-Henao, L.M. and C.L. Casteel. 2016. Vector-Borne Bacterial Plant Pathogens: Interactions with Hemipteran Insects and Plants. Front. Plant. Sci. 7: 1163. doi:10.3389/fpls.2016.01163.

Petersen, J.M., F.U. Zielinski, T. Pape, R. Seifert, C. Moraru, R. Amann et al. 2011. Hydrogen is an energy source for hydrothermal vent symbioses. Nature 476: 176–180. doi:10.1038/nature10325.

Petersen, J.M., C. Wentrup, C. Verna, K. Knittel and N. Dubilier. 2012. Origins and evolutionary flexibility of chemosynthetic symbionts from deep-sea animals. Biol. Bull. 223: 123–137.

Philippe, H. and P. Forterre. 1999. The rooting of the universal tree of life is not reliable. J. Mol. Evol. 49: 509–523.

Ponnudurai, R., L. Sayavedra, M. Kleiner, S.E. Heiden, A. Thürmer, H. Felbeck et al. 2017. Genome sequence of the sulfur-oxidizing *Bathymodiolus thermophilus* gill endosymbiont. Stand. Genomic Sci. 12: 50. doi:10.1186/s40793-017-0266-y.

Rao, Q., P.A. Rollat-Farnier, D.T. Zhu, D. Santos-Garcia, F.J. Silva, A. Moya et al. 2015. Genome reduction and potential metabolic complementation of the dual endosymbionts in the whitefly *Bemisia tabaci*. BMC Genomics 16: 226. doi:10.1186/s12864-015-1379-6.

Reveillaud, J., R. Anderson, S. Reves-Sohn, C. Cavanaugh and J.A. Huber. Metagenomic investigation of *Vestimentiferan* tubeworm endosymbionts from Mid-Cayman Rise reveals new insights into metabolism and diversity. Microbiome 6(1): 19. doi:10.1186/s40168-018-0411-x.

Rosas-Pérez, T., A. Vera-Ponce de León, M. Rosenblueth, S.T. Ramírez-Puebla, R. Rincón-Rosales, J. Martínez-Romero et al. 2017. The symbiome of Llaveia Cochineals (Hemiptera: Coccoidea: Monophlebidae) includes a Gammaproteobacterial cosymbiont Sodalis TME1 and the known Candidatus Walczuchella monophlebidarum. *In*: Shields, V.D.C. (ed.). Agricultural and Biological Sciences "Insect Physiology and Ecology". Elsevier. doi:10.5772/66442.

Rosenblueth, M., J. Martínez-Romero, S.T. Ramírez-Puebla, A. Vera-Ponce de León, T. Rosas-Pérez, R. Bustamante-Brito et al. 2018. Endosymbiotic microorganisms of scale insects. TIP Revista Especializada en Ciencias Químico-Biológicas 21: 53–69. doi:10.1016/j.recqb.2017.08.006.

Rosenblueth, M., L. Sayavedra, H. Sámano-Sánchez, A. Roth and E. Martínez-Romero. 2012. Evolutionary relationships of flavobacterial and enterobacterial endosymbionts with their scale insect hosts (Hemiptera: Coccoidea). J. Evol. Biol. 25: 2357–2368. doi:10.1111/j.1420-9101.2012.02611.x.

Russell, S.L., R.B. Corbett-Detig and C.M. Cavanaugh. 2017. Mixed transmission modes and dynamic genome evolution in an obligate animal-bacterial symbiosis. ISME J. 11: 1359–1371. doi:10.1038/ismej.2017.10.

Santos-Garcia, D., P.A. Farnier, F. Beitia, E. Zchori-Fein, F. Vavre, L. Mouton et al. 2012. Complete genome sequence of "*Candidatus* Portiera aleyrodidarum" BT-QVLC, an obligate symbiont that supplies amino acids and carotenoids to *Bemisia tabaci*. J. Bacteriol. 194: 6654–6655. doi:10.1128/JB.01793-12.

Saponari, M., D. Boscia, G. Altamura, G. Loconsole, S. Zicca, G. D'Attoma et al. 2017. Isolation and pathogenicity of *Xylella fastidiosa* associated to the olive quick decline syndrome in southern Italy. Sci. Rep. 7: 17723. doi:10.1038/s41598-017-17957-z.

Sargent, F. 2016. The Model [NiFe]-Hydrogenases of *Escherichia coli*. Adv. Microb-Physiol. 68: 433–507. doi:0.1016/bs.ampbs.2016.02.008.

Sauer, C., D. Dudaczek, B. Hölldobler and R. Gross. 2002. Tissue localization of the endosymbiotic bacterium "*Candidatus* Blochmannia floridanus" in adults and larvae of the carpenter ant *Camponotus floridanus*. Appl. Environ. Microbiol. 68: 4187–4193.

Schulz, F., E.A. Eloe-Fadrosh, R.M. Bowers, J. Jarett, T. Nielsen, N.N. Ivanova et al. 2017. Towards a balanced view of the bacterial tree of life. Microbiome 5(1): 140. doi:10.1186/s40168-017-0360-9.

Scott, K.M., S.M. Sievert, F.N. Abril, L.A. Ball, C.J. Barrett, R.A. Blake et al. 2006. The genome of deep-sea vent chemolithoautotroph *Thiomicrospira crunogena* XCL-2. PLoS Biol. 4(12): e383.

Segata, N., D. Börnigen, X.C. Morgan and C. Huttenhower. 2013. PhyloPhlAn is a new method for improved phylogenetic and taxonomic placement of microbes. Nat. Commun. 4: 2304.

Shah, V. and R.M. Morris. 2015. Genome sequence of "*Candidatus* Thioglobus autotrophica" strain EF1, a chemoautotroph from the SUP05 clade of marine Gammaproteobacteria. Genome Announc. 3(5). pii: e01156-15. doi:10.1128/genomeA.01156-15.

Shigenobu, S., H. Watanabe, M. Hattori, Y. Sakaki and H. Ishikawa. 2000. Genome sequence of the endocellular bacterial symbiont of aphids *Buchnera* sp. APS. Nature 407: 81–86.

Sloan, D.B. and N.A. Moran. 2012. Endosymbiotic bacteria as a source of carotenoids in whiteflies. Biol. Lett. 8: 986–989. doi:10.1098/rsbl.2012.0664.

Simon, J. and M.G. Klotz. 2013. Diversity and evolution of bioenergetic systems involved in microbial nitrogen compound transformations. Biochim. Biophys. Acta. 1827(2): 114–135. doi:10.1016/j.bbabio.2012.07.005.

Simpson, A.J., F.C. Reinach, P. Arruda, F.A. Abreu, M. Acencio, R. Alvarenga et al. 2000. The genome sequence of the plant pathogen *Xylella fastidiosa*. Nature 406: 151–159.

Sorokin, D.Y., T.P. Tourova, E.A. Galinski, G. Muyzer and J.G. Kuenen. 2008. *Thiohalorhabdus denitrificans gen. nov., sp. nov.*, an extremely halophilic, sulfur-oxidizing, deep-lineage gammaproteobacterium from hypersaline habitats. Int. J. Syst. Evol. Microbiol. 58: 2890–2897. doi:10.1099/ijs.0.2008/000166-0.

Stackebrandt, E. 1985. The phylogeny of purple bacteria: The gamma subdivision. Syst. Appl. Microbiol. 6: 25–33.

Tran, T.T.T., S. Mangenot, G. Magdelenat, E. Payen, Z. Rouy, H. Belahbib et al. 2017. Comparative genome analysis provides insights into both the lifestyle of *Acidithiobacillus ferrivorans* strain CF27 and the chimeric nature of the iron-oxidizing acidithiobacilli genomes. Front. Microbiol. 8: 1009. doi:10.3389/fmicb.2017.01009.

Walsh, D.A., E. Zaikova, C.G. Howes, Y.C. Song, J.J. Wright, S.G. Tringe et al. 2009. Metagenome of a versatile chemolithoautotroph from expanding oceanic dead zones. Science 326: 578–582. doi:10.1126/science.1175309.

Wapner, J. 2017. Half of world's olive oil threatened by deadly bacteria in Spain. Newsweek 6 June 2017.

Wells, J.M., B.C. Raju, H.Y. Hung, W.G. Weisburg, L.M. Parl and D. Beemer. 1987. *Xylella fastidiosa* gen. nov., sp. nov.: Gram-negative, xylem-limited, fastidious plant bacteria related to *Xanthomonas* spp. Int. J. Syst. Bacteriol. 37: 136–143.

Wernegreen, J.J. 2002. Genome evolution in bacterial endosymbionts of insects. Nat. Rev. Genet. 3: 850–861.

Wilkes, T.E., A.C. Darby, J.H. Choi, J.K. Colbourne, J.H. Werren and G.D. Hurst. 2010. The draft genome sequence of Arsenophonus nasoniae, son-killer bacterium of *Nasonia vitripennis*, reveals genes associated with virulence and symbiosis. Insect Mol. Biol. 19 Suppl. 1: 59–73. doi:10.1111/j.1365-2583.2009.00963.x.

Williams, K.P., J.J. Gillespie, B.W. Sobral, E.K. Nordberg, E.E. Snyder, J.M. Shallom et al. 2010. Phylogeny of gammaproteobacteria. J. Bacteriol. 192: 2305–2314.

Williams, K.P. and D.P. Kelly. 2013. Proposal for a new class within the phylum Proteobacteria, Acidithiobacillia *classis nov.*, with the type order Acidithiobacillales, and emended description of the class Gammaproteobacteria. Int. J. Syst. Evol. Microbiol. 63(Pt 8): 2901–2906. doi:10.1099/ijs.0.049270-0.

Williams, L.E. and J.J. Wernegreen. 2015. Genome evolution in an ancient bacteria-ant symbiosis: parallel gene loss among Blochmannia spanning the origin of the ant tribe Camponotini. Peer J. 3: e881.

Woese, C.R. 1987. Bacterial evolution. Microbiol. Rev. 51: 221–271.

6

Betaproteobacteria

Mauro Degli Esposti, Paola Bonfante, Mónica Rosenblueth and *Esperanza Martínez-Romero*

This chapter is dedicated to the class of betaproteobacteria and is structured in three parts. **Part 1**, written by Mauro Degli Esposti, describes general features of betaproteobacteria and their phylogeny. **Part 2**, written by Paola Bonfante, presents the interactions between betaproteobacteria and fungi. **Part 3** is dedicated to betaproteobacteria that are endosymbionts of plants and animals, and is written by Esperanza Martínez-Romero together with Mónica Rosenblueth.

Center for Genomic Sciences, UNAM Campus de Cuernavaca, Cuernavaca, 62130 Morelos, Mexico.
Email: mauro1italia@gmail.com

PART 1

General Features of Betaproteobacteria and Their Phylogeny

Mauro Degli Esposti

General Features of Betaproteobacteria

Betaproteobacteria, the third largest class of proteobacteria, were originally named by Woese et al. (1984) to distinguish them from alphaproteobacteria, which previously were classified together under the broad name of purple non sulfur bacteria. Despite their widespread distribution in the environment, betaproteobacteria are often associated with metazoan organisms and can also be pathogenic. Strains of the genera *Neisseria* and *Bordetella*, which belong to the order of Neisseriales (Li et al. 2017a), are responsible for common human diseases such as meningitis and whooping cough. There are betaproteobacteria from the ocean as well, for example clade OM43 that forms a group of methylotrophic marine bacteria with small genomes (Huggett et al. 2012). However, the focus of this chapter is on betaproteobacteria that form mutualistic associations with fungi, plants and animals, which are much more common than pathogens and belong to two of the four major orders in which the class is subdivided: the Burkholderiales and the Rhodocyclales. According to a recent re-classification (Boden et al. 2017), the other major orders of betaproteobacteria are the above mentioned Neisseriales and the Nitrosomonadales. The latter constitutes a large group that includes many chemolithotrophic organisms that were previously classified into separate groups, such as Fe^{II}-oxidizing Gallionellaceae and methylotrophic Methylophilales (Chistoserdova et al. 2009). The previous order of Hydrogenophilales has now been removed from the class and proposed as a novel class of the Proteobacteria, the Hydrogenophilalia (Boden et al. 2017), which includes thermophilic organisms such as *Tepidiphilus* (Poddar et al. 2014). The depth and consistency of this newly proposed class of proteobacteria awaits further verification.

Betaproteobacteria include species that share key metabolic traits with alphaproteobacteria such as methylotrophy, nitrogen fixation in association with plants and the presence of rhodoquinone in the membrane (see Chapters four and seven). However, betaproteobacteria normally form a sister clade of gammaproteobacteria, especially in phylogenetic trees of bioenergetic proteins (Degli Esposti 2014; see also Fig. 6 in Chapter four). The phylogeny of betaproteobacteria is relatively simple in comparison to that of gammaproteobacteria discussed in Chapter five; it is dominated by the order of Burkholderiales, which effectively is a super-order for it includes more than half of the known species of the class and the majority of their phenotypic traits, including endosymbiosis and N-fixing nodulation. Several members of this order will be described in the following sections of the chapter. In phylogenetic trees, proteins from Rhodocyclales often occupy the deepest branches while their homologs from *Burkholderia* organisms occupy the latest divergent branches (Degli Esposti and Martinez-Romero 2017).

Of note is that more than 20% of currently available genomes of betaproteobacteria belong to taxa that remain unclassified, thereby indicating that the phylogenetic depth of the class may be larger than the current classification embraces. On the other hand, there are phenotypically diverse organisms such as the Sutterellaceae which remain classified under the order of Burkholderiales, even if their strictly anaerobic lifestyle and possession of menaquinone (Nagai et al. 2009) clearly suggest an ancestral origin within the class (Degli Esposti and Martinez-Romero 2017). Indeed, Sutterellaceae is the only group of betaproteobacteria lacking terminal oxidases of the superfamily of heme copper oxygen reductases (HCO), but containing anaerobic bioenergetic systems such as [FeFe]-hydrogenases (Degli Esposti and Martinez-Romero 2017–see also Chapter two and seven).

Conversely, *Candidatus* Accumulibacter strains might be among the most ancestral betaproteobacteria. While *Sutterella* and its relatives have been isolated from human microbiomes (Wexler et al. 1996, Nagai et al. 2009, Dione et al. 2017), *Ca.* Accumulibacter has been isolated from the sludge of water processing plants and has an almost unique metabolism of accumulating polyphosphates and other storage material under anaerobic conditions (Hesselmann et al. 1999, Oyserman et al. 2016). *Ca.* Accumulibacter is often associated with the family Rhodocyclaceae (Oyserman et al. 2016), but in NCBI resources is listed among unclassified betaproteobacteria. The respiratory chain of *Ca.* Accumulibacter contains several terminal oxidases of the HCO super-family that render this organism a facultatively anaerobe, well adapted to the fluctuations of oxygen levels existing in water treatment systems. These oxidases, together with other respiratory systems such as complex I, are considered ancestral components of the genome of *Ca.* Accumulibacter, with the exception of the genes of a ba3 oxidase that appears to have been acquired via LGT (Oyserman et al. 2016). So, the respiratory chain of *Ca.* Accumulibacter can be considered a general example for the aerobic metabolism of betaproteobacteria.

Before introducing the various betaproteobacteria associated with metazoans in the following sections, it is worth mentioning that the class also includes symbionts of ancestral protists such as Kinetoplastidia (Alves et al. 2013). Of note is that all these endosymbionts are heterotrophs like the gammaproteobacterial symbionts of insects presented in previous Chapter five of the book.

PART 2

Betaproteobacteria and Fungi: a Look at their Interactions

Paola Bonfante

Introduction

As relevant components of animal and plant microbiota, bacteria and fungi often share the same ecological niche. For a long time, the capacity of fungi and bacteria to produce antibacterial and antifungal molecules, respectively, has supported the concept that the interactions between the two microbial groups are mostly of antagonistic nature. Strategies for producing antifungal metabolites are very diverse, also including mycophagy: thanks to the production of chitinolytic enzymes, some bacteria grow by using fungi as a source of nutrients and energy (Leveau and Preston 2008). By contrast, the most recent data report on cooperative activities of fungi which positively interact with bacteria (Deveau et al. 2018), indicating that such inter-domain associations are more common than previously expected (Olsson et al. 2017).

   Among the major classes of proteobacteria, betaproteobacteria play a relevant role in bacterial-fungal interactions. They predominantly belong to the dominant order of Burkholderiales, which have large genomes and are characterized by a relevant phenotypic plasticity, i.e., the capacity to interact with a wide range of eukaryotes, from fungi, as discussed in this section, to plants and animals, as presented in next section. With fungi, in particular, betaproteobacteria can cause hyphal disintegration thanks to the expression of a prophage tail-like protein (Swain et al. 2017), but can also lead to the establishment of endosymbiotic events (Bonfante and Desirò 2017). The aim of this part of the chapter is to illustrate how betaproteobacteria interact with fungi, identifying two main issues:

Department of Life Sciences and Systems Biology, Viale Mattioli 25 - 10125 Torino, Italy.
Email: paola.bonfante@unito.it

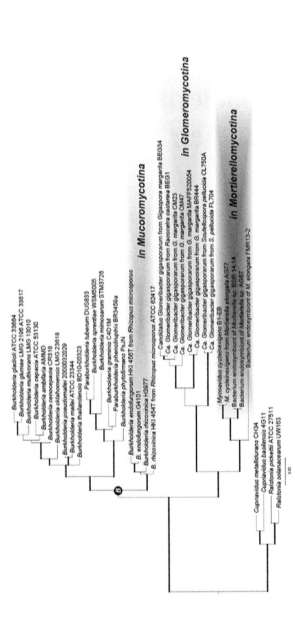

Fig. 1. Phylogenetic placement of Betaproteobacteria endosymbionts identified within Mucoromycota fungi based on 16S rRNA gene sequences. Three lineages of Burkholderia-related symbionts can live as endobacteria inside several taxa of the early diverging *phylum* Mucoromycota. *Burkholderia endofungorum* and *Burkholderia rhizoxinica* live inside the fungus *Rhizopus microsporus* (Mucoromycotina). These two endobacterial lineages cluster within the Burkholderia clade (B) and represent the sister group to a clade that encompasses several free-living *Burkholderia* spp. The *Burkholderia* clade (B), in turn, is sister to a clade that includes the other two lineages of betaproteobacterial endosymbionts. *Candidatus* Glomeribacter gigasporarum (CaGg), which resides in spores and mycelia of numerous species of the family Gigasporaceae (Glomeromycotina), clusters within a clade that is sister to the one that includes *Mycoavidus cysteinexigens* and other *Mycoavidus*-related endobacteria that dwell inside the fungus *Mortierella* (Mortierellomycotina). *Cupriavidus* and *Ralsonia* form an outgroup rooting the tree. The phylogenetic reconstruction was carried out with RAxML v.8.2.4. Maximum likelihood (ML) analysis was conducted under the GTRCAT nucleotide substitution model. Branches with bootstrap support values ≥ 75 are thickened. Figure by courtesy of Alessandro Desirò. See Fig. 5 for more phylogenetic relationships between these and other endosymbionts of betaproteobacteria.

betaproteobacteria as components of the fungal microbiota, including lichens, and as endosymbionts of basal fungi.

Betaproteobacteria as Components of Fungal Microbiota

Before discussing the interaction of betaproteobacteria with fungi, some basic information needs to be introduced about this highly diverse group of eukaryotes. Fungi include unicellular yeasts as well as the giant *Armillariella* (Spatafora et al. 2016), from beneficial endophytes to dangerous pests. Fungi occur in all ecosystems where they colonize the most diverse substrates thanks to their hyphae, a network of interconnected filaments. On one hand, fungi represent a large proportion of the genetic diversity on Earth and a relevant component of animal and plant microbiota (Peay et al. 2016), while on the other hand they possess their own microbiota. The wide surface offered by hyphae represents a good niche for many microbial colonies. Interestingly, attention has been mostly given to bacteria which thrive associated to mycorrhizal fungi, i.e., the fungi which live in symbiosis with most of land plants (Tedersoo 2017), giving rise to three-partite interactions (Bonfante and Anca 2009). Among mycorrhizal fungi, Basidio- and Ascomycota, which produce conspicuous fruit bodies, are of interest to humans as food source or active molecules. For many years, data on the microbiota of such fungi have been limited to bacterial taxa which grow in pure culture (among many references, see Rangel-Castro et al. 2002). Only the advent of next generation sequencing (NGS) has deeply changed our views. In one of the first comprehensive analysis of the

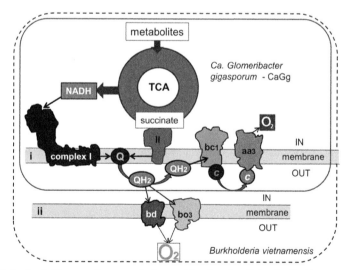

Fig. 2. Reduction in bioenergetic systems in the transition from free-living *Burkholderia vietnamensis* to the endosymbiont *Ca.* Glomeribacter. Analysis of bioenergetic systems has been conducted as described before (Degli Esposti 2014) and graphically rendered as in previous chapters of the book. The dashed box represents the respiratory chain of free-living *Burkholderia* with additional ubiquinol oxidases working with lower levels of O_2.

structure of bacterial communities in forest mushrooms, Pent and colleagues (Pent et al. 2017) recently found that microbial communities across fruit bodies from eight ectomycorrhizal genera were primarily affected by soil pH and fungal identity. Interestingly, betaproteobacteria, together with gammaproteobacteria, were the dominant taxa among the samples. At lower taxonomic levels, members of the family Burkholderiaceae were found in 42% of fruit bodies, including: *Burkholderia phytofirmans* (in 10%), *B. phenazinium* (8%), *B. xenovorans* (6%), *B. bryophila* (6%) and *B. graminis* (5%). These data confirm the general concept that soil is one of the major drivers of bacterial communities (Jeanbille et al. 2016). On the other hand, the same data suggest that the effect of host identity is significant. Thus, bacteria inhabiting fungal fruit bodies may be non-randomly selected from the environment, probably thanks to their specific capacities to interact with fungi (Pent et al. 2017), or from their metabolic capacities to take advantage of nutrient-rich niches.

An interesting specific function that differs from mycorrhiza-helper bacteria (Garbaye 1994) has been assigned to some betaproteobacteria living inside the fruit bodies of truffles. White and black truffles are symbiotic Ascomycetes, well known for the aroma of their fruiting bodies (Martin et al. 2010), which host bacterial communities among their hyphae and asci. Splivallo et al. (2015) found that thiophene derivatives characteristic of *T. borchii* fruiting bodies resulted from the biotransformation of non-volatile precursor(s) into volatile compounds by bacteria, among which betaproteobacteria such as *Comamonas* were dominant. Treatment of truffles with antibacterial agents fully suppressed the production of thiophene volatiles, while fungicides had no inhibitory effect (Splivallo et al. 2015). Since truffle dispersion is strongly dependent on their aroma and its capacity to attract insects and small animals with pheromone-like activity, we can suggest that this specific bacterial-fungal interaction may have a crucial ecological relevance for the success of truffles.

Among mycorrhizal fungi, the arbuscular mycorrhizal ones (AM) are the most widespread, being present both in soils as mycelia and spores, and inside plant tissues as endophytes (Bonfante and Genre 2010). Bacterial communities associated to AM fungi during their extra radical phase have been widely investigated to detect both cultivable and un-cultivable microbes (Bonfante and Anca 2009). Burkholderiales resulted to be among the most common bacterial taxa living on the surface of AM hyphae (Bonfante and Anca 2009) or of AM spores (Agnolucci et al. 2015, Long et al. 2017). The most abundant of these betaproteobacteria were: *Methylibium* and *Ideonella* (Burkholderiales genera *incertae sedis*), and *Duganella, Massilia* and *Janthinobacterium* of the Oxalobacteriaceae family (see later Fig. 5). Members of this family have been reported to be particularly abundant on AMF hyphae (Scheublin et al. 2010), as well as in the truffle grounds (Zampieri et al. 2016). AM fungi living in petroleum hydrocarbon-polluted environment have been recently reported to harbor bacterial Operational Taxonomic Units (OTUs) belonging to many other groups, besides betaproteobacterial *Massilia* (Iffis et al. 2014). These data open the question whether AM-associated betaproteobacteria

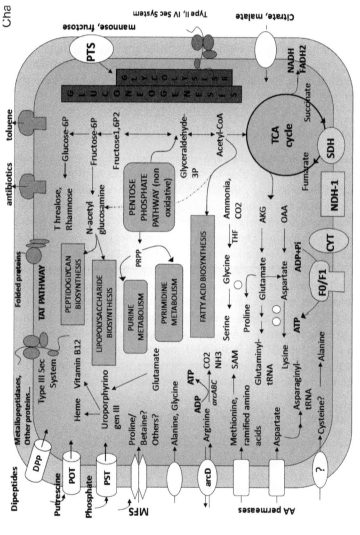

Fig. 3. Reconstruction of the main metabolic pathways in CaGg. White circles represent enzymes whose corresponding genes have not been found in the genome for a given pathway. Partly represented pathways are not illustrated. arcD, arginine/ornithine antiporter; CYT, cytochrome bc1 complex (see Fig. 2); DPP, dipeptide transporter; F0/F1, ATP synthase; MFS, major facilitator superfamily transporters; NDH-1, NAD dehydrogenase or complex I (see Fig. 2); POT, putrescine transporter; PST, phosphate transporter; PTS, sugar transporter; SDH, succinate dehydrogenase (complex II in Fig. 2); TAT, twin-Arginine translocation pathway; TCA, tricarboxylic acid cycle as in Fig. 2. Modified from Ghignone et al. (2012).

are negatively or positively selected in polluted soils. Interestingly, some NGS investigations have reported slight changes in the bacterial biodiversity profile in the presence or absence of *Funnelliformis mosseae*, a widespread AM fungus, while detecting a decrease in the abundance of betaproteobacteria, as well as gammaproteobacteria, indicating that there was a rhizosphere competition from other fast-growing bacteria such as Actinobacteria (Gui et al. 2017).

Taken together, these sometimes contrasting results suggest that the profile of betaproteobacteria associated to mycorrhizal fungi is not well defined yet; consequently, their role remains obscure. It will be crucial to associate NGS analyses with culture-dependent approaches, to sequence the more consistent and frequent bacterial species, in order to better understand their potential activities when interacting with symbiotic fungi.

Betaproteobacteria as Components of Lichen Microbiota

As a stable symbiosis between fungi and algae, lichens are often a dominating feature of many environments, also those hostile such as high mountains or subpolar regions (Cardinale et al. 2012). NGS studies have revealed that bacterial communities are a constant feature of lichenic thalli, and the combined use of molecular and *in situ* fluorescent techniques have allowed to identify and to locate a huge number of prokaryotes. This emerging evidence suggests that lichens can be described as ecosystems in miniature, where algae are the primary producers, fungi the consumers, while bacteria contribute to nutrient acquisition, recycling and antagonism (Cardinale et al. 2012). Alphaproteobacteria dominate the lichen-associated communities, while betaproteobacteria seem to be limited to some lichen taxon, such as *Cladonia*. In addition, their presence changes depending on the thallus age. Interestingly, similar results were obtained when lichens from Artic or Antarctic sites were sampled (Lee et al. 2014), suggesting some specific selection mechanism for bacterial association. It seems, therefore, that the presence and role of betaproteobacteria in lichen have still to be clarified.

Betaproteobacteria as Endosymbionts of Basal Fungi

While many of the fungal-bacterial associations previously discussed are non-heritable (i.e., bacterial communities are *de-novo* assembled by each generation of the host from the environment), heritable bacterial symbionts are transmitted vertically from the host fungal parent to the next generation of the fungal-bacterial association. According to Olsson et al. (2017), heritable bacterial symbionts can be confined to their eukaryotic host's intracellular environment and have no extracellular state (obligate endobacteria), or are able to live both in fungal cells and in extracellular environments (facultative endobacteria). Finally, depending on the impact on their fungal host, they can be divided into essential and nonessential. The mechanisms that allow a free independent bacterium to be hosted inside a fungus as a heritable endosymbiont are largely unknown: however, and interestingly, the most investigated examples of fungal endosymbionts belong to betaproteobacteria, while their fungal host belongs to a basal group of fungi, the Mucoromycota. In this context, three fungi which belong to the three subphyla of Mucoromycota (Mucoromycotina, Morterellomycotina and Glomeromycotina) have very different lifestyles. Yet, all of their symbionts are phylogenetically strictly related (see later Fig. 5), being members of the *Burkholderia* group (Bonfante and Desirò 2017).

The study of the rice seedling blight fungus *Rhizopus microsporus* (Mucoromycotina) and its betaproteobacterial endosymbiont *Burkholderia rhizoxinica* gives many hints to understand how a microbe may evolve antagonism into mutualism (Partida-Martinez and Hertweck 2005, Lackner et al. 2011). The endobacteria have been found in 8 out of 22 (36%) strains of *R. microsporus* (Lackner and Nunnari 2009, Partida-Martinez 2013), while a wider screen of 64 strains from the Centraalbureau voor Schimmelcultures (CBS-KNAW Fungal Biodiversity Centre) confirmed their presence in seven (Dolatabadi et al. 2016). Such findings suggest that the spread of endobacteria is relatively relevant, also in cultured collections of *Rhizopus*. *B. rhizoxinica* directly resides within the fungal cytoplasm (Partida-Martinez et al. 2007a), even if an exopolysaccharide matrix can be detected at the bacterial surface (Uzum et al. 2015). The bacteria are transmitted vertically along the spore generations, confirming their heritable status (Partida Martinez 2017).

Since the fungi could be cured of their endobacteria with antibiotic treatment (Partida-Martinez et al. 2007b), important functional features emerged. First, *B. rhizoxinica* and not the host *R. microsporus* is the true producer of the toxins 'rhizoxin' and 'rhizonin' (Partida-Martinez and Hertweck 2005, Partida-Martinez et al. 2007c), which causes disease in rice plants. As a second consequence, bacteria are required for the asexual development of the fungal host (Partida-Martinez 2017). Thanks to genome sequencing, some molecular mechanisms of bacterial-fungal interactions have been clarified: type III secretion system is required for eliciting asexual development (Lackner et al. 2011); type II secretion system and the production of chitinase are needed for the active invasion of fungal cells (Moebius et al. 2014); exopolysaccharides may mediate interactions with the fungal environment (Uzum et al. 2015), while diacylglycerol kinase enzymes regulate the fungal lipid profile, being the most differentially expressed gene during the interaction with endobacteria (Lastovetsky et al. 2016). On the basis of the last observation, in addition to the mutualistic interaction, an antagonistic one was detected in which bacteria do not populate fungal hyphae of a different *R. microsporus* isolate and cause phenotypic alterations in fungal growth.

Heritable endosymbiotic betaproteobacteria have also been found in *Mortierella elongata* (Morteriellomycotina), a common soil saprotrophic fungus (Sato et al. 2010). This has been confirmed only recently thanks to the use of NGS approaches (Uehling et al. 2017). Based on a draft genome and the property of growing in pure culture after the addition of some amino acids for which it is auxotroph, this endosymbiont has been called *Mycoavidus cysteinexigens* (Fujimura et al. 2014). The analysis of its full genome (2.6 Mb), together with that of its fungal host, allowed

Uehling and collaborators (2017) to trace back the history of their association, which was likely established 350 Mya. The fungal and bacterial couple has an excellent capacity to produce fatty acids, thanks to the presence of both fungal and bacterial fatty acid synthases (FAS). Accordingly to its saprotrophic lifestyle, the host *M. elongata* does not degrade complex polymers like lignins, while the bacterium fully depends on the fungus for many amino acids. A proteomic analysis (Li et al. 2017b) revealed interconnections between the carbon and nitrogen metabolism of this fungal-bacterial couple. Under low nitrogen availability, the bacterial population flourishes, in contrast with the fungus which has a more limited growth under the same conditions. Uehling et al. (2017) suggested that environmental conditions probably mirror nitrogen deprivation, leading to situations which favor bacterial survival. Indeed, under laboratory conditions, with no nutritional stress, *M. elongata* grows rapidly without the bacterium (Uehling et al. 2017)).

The third example of betaproteobacteria living inside fungi has been investigated in *Gigaspora margarita* (Glomeromycotina), a member of arbuscular mycorrhizal fungi (Bianciotto et al. 1996). This is part of a group of fungal symbionts which are defined as obligate biotrophs, since they depend on the plant host to accomplish their life cycle (Bonfante and Genre 2010). Their endosymbionts, which are vertically transmitted and thus heritable, have been named *Candidatus* Glomeribacter gigasporarum (hereafter referred to as CaGg) and are fully dependent on their fungal host for thriving, even if they are not required for the fungal vitality and its capacity to colonize plants (Lumini et al. 2007). *Ca.* Glomeribacter strains have been found in many diverse fungal isolates belonging to the Gigasporales (11 out of 13 isolates of *Gigaspora margarita*), with 16S rRNA gene sequences resulting relatively conserved, irrespective of the geographic origin of the fungal host (Desirò et al. 2014). The genome draft of CaGg (1.8 Mb) has revealed how this betaproteobacterium fully depends on its fungal host for many amino acids, as well as for many mineral nutrients (Ghignone et al. 2012). There is, therefore, a complex energetic flux within the tripartite association, plant-fungus-endosymbiont, since the photosynthetic plant provides sugars which fuel both the fungus and bacterium, while the fungus releases both phosphate and nitrogen compounds to its plant and bacterial partners. The endosymbionts of Mucoromycotina cluster within the *Burkholderia* clade, but those of Mortierellomycotina and Glomeromycotina belong to a separate clade (see later Fig. 5). Indeed, many of their primary metabolic pathways are very similar among these betaproteobacteria. However, the respiratory chain of CaGg is further streamlined from the free-living relatives of the Burkholderiales such as *B. thailandensis* and *B. vietnamensis*, which possess alternative ubiquinol oxidases of bd and bo3 type (Fig. 2, cf. Degli Esposti 2014). Vestiges of the bd oxidases are still present in the genome of CaGg but are non-functional, thereby rendering the central metabolism of this endosymbiont as aerobic as that of mitochondrial organelles (Fig. 2, cf. Degli Esposti 2014). In contrast, the genome of *B. rhizoxinica* still retains both types of bd oxidases that are present in free-living relatives (Degli Esposti 2014), thereby indicating a latent facultatively anaerobic metabolism that may sustain the endosymbiont under conditions of low ambient oxygen, as in rice paddy soils.

Other common features have been detected in the primary metabolism of CaGg, *Mycoavidus* and *B. rhizoxinica*: all these endobacteria cannot degrade complex sugars since their glycolytic pathway is not fully efficient. Their genome lacks genes for key glycolytic enzymes such as phosphofructokinase, while malate and citrate are imported via specific transporters to fuel the TCA cycle. On the other hand, sugar synthesis capacities are maintained, and indeed a single copy of a sugar phosphate transporter gene (pts) was identified in CaGg. All the above endobacteria have a restricted biosynthetic capacity for various amino acids, while possessing a large number of amino acid transporters. All three endobacteria encode type II, III, and IV secretions systems (SS), which probably allow them to interact with their fungal partner and help to establish endosymbiosis. However, type III SS of CaGg and *Mycoavidus* shows homology with the type III SS of gammaproteobacteria rather than with that of other betaproteobacteria, suggesting a deep divergence between *B. rhizoxinica* and the clade Mycoavidus-CaGg. Interestingly, both *B. rhizoxinica* and CaGg possess toxin-antitoxin systems; in CaGg, this system is expressed both during the pre- and symbiotic phase (Lackner et al. 2011, Salvioli et al. 2017). Other functions seem to be differently expressed in the three endosymbionts above: while *B. rhizoxinica* and *M. cysteinexigens* fully express the metabolic pathway leading to fatty acids, and some enzymes, as diacylglycerol kinase, modify the lipid profile of the *Rhizopus* fungal host (Lastovetsky et al. 2016), the situation is less clear-cut for CaGg. AM fungi have been recently demonstrated to miss many genes and transcripts for the biosynthesis of fatty acids (Wewer et al. 2014, Vannini et al. 2016), while plants have been shown to release lipids to the AM fungi (Keymer et al. 2017, Jiang et al. 2017, Luginbuehl et al. 2017). CaGg presents only some genes for fatty acid biosynthesis, suggesting that it cannot complement the missing genes in the fungal genome. Therefore, it is likely that there is also a flow of lipids from the plant to the fungus and ultimately to the endobacterium.

Genomic and functional data are summarized in Fig. 3 and clearly illustrate how the above examples of betaendobacteria depend on their fungal hosts, also on the basis of a long common evolutionary history (Mondo et al. 2012). However, the opposite question—what is the impact of the bacterium on the fungal physiology—only has a clear response for *B. rhizoxinica*. Since the bacterium produces toxins, the fungus, as a pathogen, has a clear evolutionary advantage. The adaptive advantages that *M. cysteinexigens* may provide to its fungal host have not yet been identified, also due to the contrasting physiological data (the fungi sampled in nature have been found to contain the bacterium, but under laboratory conditions, the cured line performs better). On the other hand, extensive transcriptomic and proteomics investigations have provided valuable information on CaGg interaction with its fungal host. The presence of this bacterium raises the bioenergetic capacity of the fungus, which reveals more sustained respiratory activities, higher ATP production, and more intense ROS-scavenging activities (Salvioli et al. 2016, Vannini et al. 2016). In the absence of the bacterium, the fungus shifts to alternative metabolic pathways (for example, towards the pentose phosphate pathway in order to get reducing power). The higher antioxidant capacity of the fungus with its endobacterium does not seem to provide further fitness in the presence of environmental oxidative stress, since also the cured line efficiently counteracts treatment with radical-producing H_2O_2, thereby suggesting a strong adaptive metabolism for the fungus in the absence of its endobacterium (Venice et al. 2017). Surprisingly, however, host plants colonized by the fungus with and without the bacterium revealed a higher level of carbonylated proteins, a biological marker of oxidative stress, in the plants colonized by the cured line (Vannini et al. 2016). These data could suggest that there is a long-term benefit exerted by the betaproteobacterium on the mycorrhizal plant also, which is the third partner of the symbiosis.

In conclusion, betaproteobacteria related to *Burkholderia* seem to be particularly prone to interact with basal fungi as specific endobacteria. Their interactions, which date back to hundreds of million years ago, lead to interesting evolutionary hypotheses (Bonfante and Desirò 2017), in which these bacteria are viewed as evolutionary markers of speciation in the basal fungal groups belonging to Mucoromycota.

PART 3

Betaproteobacteria Endosymbionts in Plants and Animals

Mónica Rosenblueth and *Esperanza Martínez-Romero**

Betaproteobacteria Endosymbionts of Plants

There is a natural symbiotic association of nitrogen-fixing bacteria and plants, in which partners exchange C for N, the latter element being normally limiting for plant growth in agricultural fields and diverse natural environments. Thus, N fixation is considered as an ecological service that benefits the community (Morris et al. 2012). However, Martínez-Romero (2012) suggested a novel perspective for nitrogen fixation considering that this process may be a bacterial strategy to enhance the carrying capacity of their habitats to the benefit of bacteria.

Nitrogen-fixing capabilities are unequally distributed among bacterial phyla, genera, species or strains (Ormeño-Orrillo et al. 2013), perhaps in relation to the Black Queen hypothesis. Such a hypothesis considers that the costly process of nitrogen fixation may be a disadvantage for a species or individual bacterium, but necessary for a community (Morris et al. 2012). Consequently, nitrogen-fixing bacteria are normally minor components in bacterial communities. Within betaproteobacteria, there are many diazotrophs that have been isolated from soil, water or associated with plants (Balandreau et al. 2001, Estrada de los Santos et al. 2001, Reis et al. 2004, Caballero-Mellado et al. 2007). Betaproteobacteria may have independently acquired *nif* genes for the nitrogenase enzyme on several occasions. For example, in Burkholderiales *nif* genes seem to have a single origin, except for *Herbaspirillum* and *Rubrivivax*. Rhodocyclales also seem to have a single origin, except for *Azoarcus* (Fig. 4).

Center for Genomic Sciences, UNAM Campus de Cuernavaca, Cuernavaca, 62130 Morelos, Mexico.
*Corresponding author: esperanza.academica@yahoo.com

Many nitrogen-fixing betaproteobacteria may colonize the rhizosphere (the soil tightly adjacent to roots) or the inside of plant tissues (thus designated endophytes–Rosenblueth and Martinez-Romero 2006). For instance, there was a prevalence of betaproteobacterial sequences of *nifH* genes recovered from modern rice roots (Wu et al. 2009). *Burkholderia* is a genus rich in nitrogen-fixing and plant-associated bacteria (Estrada de los Santos et al. 2001). *Burkholderia* strains are versatile and are also found in humans as pathogens or opportunistic pathogens, as symbionts of insects, plants and fungi, as well as soil residents; they also produce a diversity of products (Vial et al. 2007, 2011). Genetically distinct *Burkholderia* organisms may be found differentially distributed inside seeds of maize plants and in variable numbers (Rosenblueth et al. 2012). Notably, some betaproteobacteria are capable of

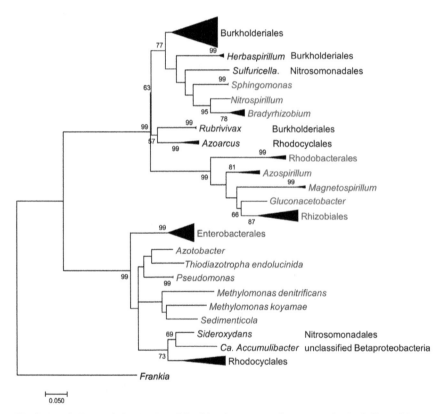

Fig. 4. ML phylogeny (LG + G + I model) of the nitrogenase reductase proteins (coded by *nifH* gene) was obtained with 500 bootstraps with the program MEGA7. Nodes with bootstrap values above 50 are shown. Betaproteobacteria are shown in blue, alphaproteobacteria in green and gammaproteobacteria in brown. The outgroup *Frankia* is a member of the Firmicutes class.

forming root nodules in legumes (see below), where nitrogen fixation takes place (Moulin et al. 2001, Chen et al. 2003).

In free-living conditions, several nitrogen-fixing *Burkholderia* strains are capable of reducing acetylene—the biochemical hallmark for nitrogenase activity —in culture media under microaerobic conditions (Reis et al. 2004). Most of these bacteria were originally isolated from plants, but some of them turned out to belong to the species *Burkholderia cepacia*, which is an aggressive human opportunistic pathogen infecting patients with cystic fibrosis. Even if nitrogen-fixing *B. cepacia* and the related *B. vietnamiensis* strains are plant-growth promotors (Tran Van et al. 2000, Peix et al. 2001), their use as inoculants in agriculture is obviously not recommended. Genes involved in pathogenicity may allow the identification of pathogenic *Burkholderia* (Eberl and Vandamme 2016).

Most plant isolates have been recently reclassified as *Paraburkholderia* and *Caballeronia* (Sawana et al. 2014, Dobritsa and Samadpour 2016), genera that include only a few strains from hospital samples (Figs. 4 and 5). Plant-associated *Paraburkholderia* organisms have different traits that stimulate plant growth. They may solubilize phosphate (Estrada et al. 2013), produce phytohormones (Joo et al. 2009) and ACC (1-aminocyclopropane-1-carboxylic acid, the immediate precursor of ethylene in plants) deaminase which reduces ethylene levels in plants (Onofre-Lemus et al. 2009). These traits are found in many root-associated bacteria including alpha- and gammaproteobacteria.

Besides Burkholderiaceae, nitrogen-fixing betaproteobacteria are present in other families of the class such as Oxalobacteraceae, which includes *Herbaspirillum,* and Rhodocyclaceae. *Herbaspirillum nifH* genes (that code for nitrogenase reductase participating in the process of nitrogen fixation) were found to be expressed in plants (Roncato-Maccari et al. 2003), suggesting that these bacteria contribute fixed nitrogen to plants. *Herbaspirillum* is capable of solubilizing phosphate as well (Stephen and Jisha 2011, Estrada et al. 2013). Rhodocyclaceae includes other nitrogen-fixing species besides *Azoarcus,* in particular *Azovibrio* and *Azonexus. Azoarcus* was proposed as a model to study nitrogen-fixation in non-legumes (Hurek and Reinhold-Hurek 2003), since it fixes nitrogen in the important crop rice.

In contrast to what occurs in nodules (see Chapter seven for details), it is common that free-living nitrogen-fixing bacteria (diazotrophs) do not excrete nitrogen compounds to their host, or to their culture medium. Ammonium, the product of nitrogen-fixation, is assimilated and used for own bacterial growth. Ammonium excreting mutants of alpha- and gammaproteobacteria such as *Azospirillum, Kosakonia, Pseudomonas* and *Azotobacter* (Zhang et al. 2012, Setten et al. 2013, Geddes et al. 2015, Ambrosio et al. 2017, Bageshwar et al. 2017) are capable of stimulating plant growth. Nodule-forming *Burkholderia* strains from Mexico did not show the capacity to fix N_2 under free-living conditions (E.M.R. unpublished results) as those nitrogen-fixing but non-nodulating Burkholderias. In contrast, *B. phymatum* was found to be the first betaproteobacterium capable of fixing nitrogen under free-living conditions.

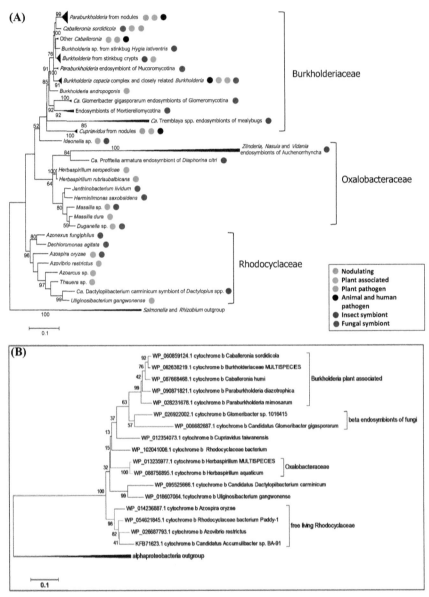

Fig. 5. Phylogenetic trees of betaproteobacteria associated with plants, fungi and insects. (A) ML tree of 16S rRNA sequences from a comprehensive set of betaproteobacterial symbionts of plants, fungi and insects. The tree was constructed with 500 bootstraps and only nodes with bootstrap values above 50 are shown. (B) Neighbor Joining (NJ) tree of the cytochrome *b* protein from a reduced set of the taxa in A. The tree was obtained with 500 bootstraps, the percentage value of which is reported by the nodes. Note that insect endosymbiont such as *Ca.* Tremblaya (shown in A) do not have cytochrome *b* because of their extremely reduced genomes. *Ca.* Dactylopiibacterium and *Uliginosibacterium* are closer to Rhodocyclaceae taxa than to plant associated *Burkholderia* taxa, as in A. Of note, a ML tree produced similar results.

As denoted before for mycorrhiza and other fungal associations with betaproteobacteria (Part 2 of this chapter), *Azoarcus* shows a remarkable affinity of association with fungi; in the fungal association, this betaproteobacterium shows stronger nitrogen fixation capability than in association with plants, forming peculiar structures called diazosomes (Dörr et al. 1998).

Betaproteobacteria in Legume Nodules

Plant specialized structures called nodules contain large numbers of symbiotic bacteria. Nitrogen-fixing nodules may contain 10^{12} nitrogen-fixing bacterial symbionts, whereas the concentrations of the same bacteria are around 10^9 per g in roots and 10^3–10^6 inside other plant tissues. For many years, only alphaproteobacteria were known to form nodules in legumes (see part on Rhizobia in next chapter of the book). After the discovery of betaproteobacteria within plant nodules (Moulin et al. 2001), a large number of studies have analysed their prevalence and nodulation process. These studies indicate that there are only some species within three genera of betaproteobacteria (*Burkholderia, Paraburkholderia* and *Cupriavidus*) that form nodules in legumes (Fig. 5), while there are at least fifteen distinct genera of alphaproteobacteria with hundreds of species that are capable of forming the same nodules (see Chapter seven). Furthermore, the number of legume species nodulated by betaproteobacteria is limited.

Seemingly, nodulation in betaproteobacteria arose later than nodulation in alphaproteobacteria (Bontemps et al. 2010, Martinez-Romero et al. 2010), even if some authors have claimed that the nodulation process evolved first in betaproteobacteria (Aoki et al. 2013). Zheng et al. (2017) have supported instead the origin of symbiotic genes from a common ancestor of beta- and alphaproteobacteria. In betaproteobacteria, symbiosis genes are located in plasmids, while in alphaproteobacteria they are distributed both in plasmids and in chromosomes or chromids (Harrison et al. 2010). Furthermore, there are more *nod* genes in the symbiosis regions of alphaproteobacteria than in those of betaproteobacteria. Thus, widespread distribution of nodulation capabilities and the larger diversity and number of *nod* genes would indicate that nodulation evolved first in alphaproteobacteria. Further, *Cupriavidus* strains (Platero et al. 2016) share a conserved symbiotic plasmid (Parker 2015); these strains may be adapted to soils containing metals (Chen et al. 2008).

Some nodulating betaproteobacteria easily lose nodulation capabilities after the loss of symbiotic plasmids (Ormeño-Orrillo et al. 2012, López-Guerrero et al. 2012), as also observed in some alphaproteobacterial rhizobia (see the section of Chapter seven that is dedicated to Rhizobia). Intriguingly, betaproteobacteria are the main nodule-forming symbionts of *Mimosa* plants in South America (De León Martínez et al. 2017), but not in North America (Bontemps et al. 2016). Betaproteobacteria may be competitive symbionts in soil with low pH or low nitrogen. Differences in *Mimosa* plants may dictate distinct affinities for nodule bacteria. In Africa, other legumes such as fymbos may be nodulated by *Burkholderia* strains (Beukes et al. 2013). Nevertheless, we are just beginning to appreciate the distribution and physiological role of betaproteobacterial strains associated with plant roots in tropical countries.

Gut and Crypt Symbionts in Arthropods

Betaproteobacteria have been detected in the gut of several arthropods and their contribution to insect microbiota has recently been analyzed using functional traits (Degli Esposti and Martinez-Romero 2017). Some gut symbionts belonging to betaproteobacteria may have contributed to the evolution of herbivory in insects (Russell et al. 2009) and, as it occurs in mammals, insect gut bacteria help in degrading plant polymers from the diet. Some phytophagous stinkbugs possess particular structures called crypts in the posterior region M4 of their midgut. These crypts contain extracellular symbiotic bacteria belonging to the genus *Burkholderia* (Kikuchi et al. 2005, 2007, 2008, 2011). *In situ* hybridization confirmed the presence of *Burkholderia* organisms in the lumen of the crypts (Boucias et al. 2012, Kikuchi et al. 2011, Xu et al. 2016a, Kuechler et al. 2016). These bacteria seem to exert a beneficial role in the insects because aposymbiotic stinkbugs, namely insects without gut crypts symbionts, display retarded development (Kikuchi et al. 2007, Boucias et al. 2012). However, the endosymbionts seem to be facultative, since insects are able to survive and reproduce without them (Kikuchi et al. 2007), while the bacteria can be cultured in standard microbiological media (Kikuchi et al. 2007, 2011, Xu et al. 2016b). Intriguingly, Kikuchi et al. (2012) have showed that insecticide-degrading *Burkholderia* strains were capable of colonizing the gut of the broad-headed bug *Riptortus pedestris* and conferred resistance to organophosphorus insecticides such as fenitrothion.

Crypt-isolated *Burkholderia* strains are closely related to soil and plant gall-forming bacteria of the same genus (Kikuchi et al. 2011, Kuechler et al. 2016). These free-living organisms have been found in the rhizosphere of soybean (Kikuchi et al. 2007) or as grass endophytes (Xu et al. 2016a). Notably, stinkbugs feed on these plants. Therefore, some *Burkholderia* organisms seem to have a dual lifestyle, for they may be found associated with plants and insects. Stinkbug *Burkholderia* phylogenies are not concordant with the host phylogeny and do not form separate clades, but are scattered along the phylogenetic tree of *Burkholderia* (Kikuchi et al. 2011, Kuechler et al. 2016). Such results indicate that even if this association seems to be ancient in evolutionary terms, bacterial symbionts are generally acquired from the environment (plant or soil), every generation (Kikuchi et al. 2011, Kuechler et al. 2016) following filtering mechanisms (Degli Esposti and Martinez-Romero 2017). However, in the stinkbug *Caveleius saccharivorus*, there is a mixed acquisition (from the environmental and maternally inherited) of the symbiotic *Burkholderia*, since 30% of the insects receive these symbionts vertically via egg smearing (Itoh et al. 2014).

Crypt *Burkholderia* strains are selected from a large number of bacteria populating the environment. Mechanisms involved in symbiont selection may be related to insect defence (Garcia et al. 2014); a constricted region found before M4 blocks the entry of other bacteria (Ohbayashi et al. 2015), the capacity of *Burkholderia* to move with flagella (Ohbayashi et al. 2015) and to resist osmotic pressure by forming biofilms (Kim et al. 2013, Kim et al. 2014, Kim and Lee 2015). These data show that both bacterial symbionts and host are involved in crypt symbiont colonization (Kim and Lee 2015). Aposymbiotic stinkbugs that do not have *Burkholderia* present a distinct morphology in the gut with atrophied regions (Futahashi et al. 2013). Kikuchi et al. (2011) also observed that the stinkbug species that were not infected with *Burkholderia* did not exhibit crypts in M4 region. This might indicate that *Burkholderia* infection is involved in the formation of the symbiotic structure in the insect, thereby assigning a role in gut development and plasticity to the bacterial symbionts. Interestingly, different *Burkholderia* strains occasionally coexist in the same insect (Kikuchi et al. 2011, Xu et al. 2016b), while species not belonging to the Burkholderiaceae family may also colonize the crypts (Boucias et al. 2012, Garcia et al. 2014).

Kikuchi et al. (2011) suggested that stinkbug-*Burkholderia* symbioses are analogous to legume-rhizobial symbiosis in the following aspects: symbionts are acquired from the environment in every generation, some symbiotic structures (nodules or crypts) can have mixed population of different bacterial species (Moawad and Schmidt 1987) and bacteria are needed to induce the formation of the symbiotic structures (Demont et al. 1993). Nevertheless, it is important to mention that crypt *Burkholderia* are not phylogenetically related to legume-nodule *Burkholderia*, as shown in Fig. 5. Hence, the above similarities appear to derive from instances of convergent evolution.

Betaproteobacterial Endosymbionts in Insects

Some insects possess bacterial endosymbionts that provide nutrients that are missing from their diets, such as essential amino acids and vitamins. They reside inside specialized cells called bacteriocytes (see previous Chapter five), or inside tissues such as the fat body, haemolymph or salivary glands. As already presented in Chapter five, endosymbionts are vertically transferred and have reduced genomes which are usually AT-rich (Baumann 2005, Moran et al. 2005). These bacteria have enabled insects to live on plants, which normally constitute poor diets in terms of nitrogen-containing nutrients.

A phylogeny of betaproteobacteria symbionts is presented in Fig. 5. Betaproteobacterial endosymbionts are found in sap-feeding hemipterans, which harbor *Ca.* Vidania fulgoroideae (Gonella et al. 2011), spittlebugs, which harbor *Ca.* Zinderia insecticola (McCutcheon and Moran 2010), and leafhoppers, which harbor *Ca.* Nasuia deltocephalinicola (Noda et al. 2012). All these endosymbionts are closely related to each other within the family Oxalobacteraceae. They are co-symbionts of the Flavobacteria organism, *Ca.* Sulcia muelleri (CFB phylum). Bennett and Moran (2013) proposed that these betaproteobacteria have the same evolutionary origin, and have been endosymbionts of spittlebugs since *Sulcia* associated with the same insects (> 260 million years, cf. Moran et al. 2005). It seems that some spittlebugs have lost both endosymbionts (the Flavobacteria and the betaproteobacteria), but others have specifically lost the betaproteobacteria, which seemingly were replaced by gammaproteobacteria (*Sodalis*-like strains, *Ca.* Purcelliella or *Ca.* Baumannia cicadellinicola) or by alphaproteobacteria (*Ca.* Hodgkinia cicadicola) (Bennett and Moran 2013, Koga et al. 2013).

The genomes of the secondary betaproteobacterial symbionts *Nasuia* and *Zinderia* are extremely reduced in size, being a little over 100 kb (McCutcheon and Moran 2010, Bennett and Moran 2013). These reduced genomes encode enzymes for the biosynthesis of essential amino acids that are not produced by the primary insect endosymbionts (McCutcheon and Moran 2010, Bennett and Moran 2013). Their small size makes these endosymbionts look like new organelles, perhaps reproducing organelle evolution (McCutcheon and Keeling 2014). However, the evolution of mitochondrial organelles occurred much earlier than the symbiosis of beta- or gammaproteobacteria with insects, and followed a unique process, as discussed in Chapters seven and eight of the book.

The Asian citrus psyllid, *Diaphorina citri*, causes the greening disease of Huanglongbing in citrus and harbors three endosymbionts: *Wolbachia* (alphaproteobacteria), *Ca.* Carsonella rudii (gammaproteobacteria, see Chapter five) and *Ca.* Profftella armatura (betaproteobacteria). *Profftella* has a genome of 470 kb including a plasmid, and seems to metabolically complement *Carsonella* by providing riboflavin and biotin (Nakabachi et al. 2013). *Profftella* is inserted in the tree of Fig. 5A.

Mealybugs harbor *Ca.* Tremblaya princeps and the related symbiont, *Ca.* Tremblaya phenacola, which are listed as unclassified betaproteobacteria. Remarkably, *Ca.* Tremblaya princeps contains diverse gammaproteobacterial close to *Sodalis* within its cytoplasm, as reported earlier in Chapter five (Husnik and McCutcheon 2016). There is metabolic complementation between the diverse endosymbionts, but only the betaproteobacterial primary symbiont shows phylogenetic congruence with the host. This probably indicates that *T. princeps* and related betaproteobacteria had a single origin, while secondary symbionts of the gammaproteobacteria class have been replaced several times (Husnik and McCutcheon 2016). *Ca.* Tremblaya phenacola does not contain secondary symbionts, but its genomic sequencing has shown an unprecedented case of genome fusion with genomic elements of a gammaproteobacterium (Gil et al. 2018). It has been suggested that this genome fusion must have occurred after the gammaproteobacterial had entered into the betaproteobacterium cytoplasm (Gil et al. 2018). The fusion mechanisms are currently unknown.

Nitrogen-Fixing Symbiosis of Betaproteobacteria with Insects

Betaproteobacterial symbionts have also been found in other sap-feeding hemipterans, the scale insects. In a few cases, N_2-fixing betaproteobacterial symbionts have been found inside insects. A diversity study in carmine cochineals showed that a novel betaproteobacterium, *Dactylopiibacterium carminicum*, was present in all carmine cochineal insects and species tested, including wild and domesticated ones (Ramírez-Puebla et al. 2010). *Dactylopiibacterium* belongs to the Rhodocyclaceae family of betaproteobacteria. No *nif* genes other than those from *Dactylopiibacterium* were found in the carmine cochineals. *Dactylopiibacterium* "is well adapted to a low oxygen life style that is compatible with nitrogen fixation, which is oxygen sensitive" (Vera Ponce de León et al. 2017). Like other nitrogen-fixing betaproteobacteria, *Dactylopiibacterium* has a large genome. The N_2 fixation products excreted to the insect are unknown.

Dactylopiibacterium probably evolved from plant endophytes and may still have a dual lifestyle inside insects and inside plants (Vera Ponce de León et al. 2017). This has been inferred from genomic analysis showing the presence of enzymes that would degrade salicylic acid, polygalacturonic acid and pectin, and of proteins for the transport of auxin, a common plant hormone. Such functional traits could be useful if the bacteria reside within plants as endophytes. Alternatively, *Dactylopiibacterium* may use such plant-associated traits in the insect's gut to degrade sap and plant-derived components. Betaproteobacterial endophytes may be cultured in the laboratory. However, *Azoarcus* attains a viable non-culturable state when fixing nitrogen in association with plants (Hurek et al. 2002). This may well happen with the related nitrogen-fixing *Dactylopiibacterium* inside insects, since we have been unable to grow *Dactylopiibacterium* in the laboratory up till now.

In sum, betaproteobacteria symbionts of fungi, plants and insects have a diversity of micro niches and functions. As within deltaproteobacteria (Chapter four), nitrogen fixation is not uncommon in betaproteobacteria, presumably allowing their respective hosts to survive under low nitrogen conditions. Functions other than nitrogen fixation are found among betaproteobacterial symbionts which, for example, may produce or degrade diverse metabolites, including insecticides, and enlarge the host habitat range. Similarities clearly exist with the endosymbionts of insects belonging to the gammaproteobacterial class that have been presented in Chapter five. However, we still have an incomplete picture of many interesting aspects of the association of betaproteobacteria with plants, fungi and animals.

Acknowledgements

Research in Paola Bonfante lab was funded by Università di Torino (60% projects). Dr A. Desirò is thanked for providing Fig 1. In Mexico, this work was sponsored by grants CONACyT Basic Science 253116 and PAPIIT (UNAM) IN207718 to E.M-R.

References

Agnolucci, M., F. Battini, C. Cristani and M. Giovannetti. 2015. Diverse bacterial communities are recruited on spores of different arbuscular mycorrhizal fungal isolates. Biology and Fertility of Soils 51: 379–389.

Alves, J.M., M.G. Serrano, F. Maia da Silva, L.J. Voegtly, A.V. Matveyev, M.M. Teixeira et al. 2013. Genome evolution and phylogenomic analysis of Candidatus Kinetoplastibacterium, the betaproteobacterial endosymbionts of Strigomonas and Angomonas. Genome Biol. Evol. 5: 338–350. doi:10.1093/gbe/evt012.

Ambrosio, R., J.C. Ortiz-Marquez and L. Curatti. 2017. Metabolic engineering of a diazotrophic bacterium improves ammonium release and biofertilization of plants and microalgae. Metab. Eng. 40: 59–68. doi:10.1016/j.ymben.2017.01.002.

Aoki, S., M. Ito and W. Iwasaki. 2013. From β- to α-proteobacteria: the origin and evolution of rhizobial nodulation genes nodIJ. Mol. Biol. Evol. 30: 2494–2508. doi:10.1093/molbev/mst153.

Bageshwar, U.K., M. Srivastava, P. Pardha-Saradhi, S. Paul, S. Gothandapani, R.S. Jaat et al. 2017. An environment friendly engineered *Azotobacter* can replace substantial amount of urea fertilizer and yet sustain same wheat yield. Appl. Environ. Microbiol. pii: AEM.00590-17. doi:10.1128/AEM.00590-17.

Balandreau, J., V. Viallard, B. Cournoyer, T. Coenye, S. Laevens and P. Vandamme. 2001. *Burkholderia cepacia* genomovar III Is a common plant-associated bacterium. Appl. Environ. Microbiol. 67: 982–985.

Baumann, P. 2005. Biology bacteriocyte-associated endosymbionts of plant sap-sucking insects. Ann. Rev. Microbiol. 59: 155–189.

Bennett, G.M. and N.A. Moran. 2013. Small, smaller, smallest: the origins and evolution of ancient dual symbioses in a Phloem feeding insect. Genome Biol. Evol. 5: 1675–1688. doi:10.1093/gbe/evt118.

Beukes, C.W., S.N. Venter, I.J. Law, F.L. Phalane and E.T. Steenkamp. 2013. South African papilionoid legumes are nodulated by diverse burkholderia with unique nodulation and nitrogen-fixation loci. PLoS One 8: e68406. doi:10.1371/journal.pone.0068406.

Bianciotto, V., C. Bandi, D. Minerdi, M. Sironi, H.V. Tichy and P. Bonfante. 1996. An obligately endosymbiotic mycorrhizal fungus itself harbors obligately intracellular bacteria. Applied Environ. Microbiol. 62: 3005–3010.

Boden, R., L.P. Hutt and A.W. Rae. 2017. Reclassification of Thiobacillus aquaesulis (Wood & Kelly 1995) as *Annwoodia aquaesulis* gen. nov., comb. nov., transfer of Thiobacillus (Beijerinck, 1904) from the Hydrogenophilales to the Nitrosomonadales, proposal of Hydrogenophilalia class. nov. within the 'Proteobacteria', and four new families within the orders Nitrosomonadales and Rhodocyclales. Int. J. Syst. Evol. Microbiol. 67: 1191–1205. doi:10.1099/ijsem.0.001927.

Bonfante, P. and A. Desirò. 2017. Who lives in a fungus? The diversity, origins and functions of fungal endobacteria living in Mucoromycota. The ISME Journal 11: 1727–1735.

Bonfante, P. and A. Genre. 2010. Mechanisms underlying beneficial plant-fungus interactions in mycorrhizal symbiosis. Nature Commun. 1: 48–58.

Bonfante, P. and I.A. Anca. 2009. Plants, mycorrhizal fungi, and bacteria: a network of interactions. Annu. Rev. Microbiol. 63: 363–383.

Bontemps, C., G.N. Elliott, M.F. Simon, F.B. Dos Reis Junior, E. Gross, R.C. Lawton et al. 2010. *Burkholderia* species are ancient symbionts of legumes. *In*: Molecular Ecology, Vol. 19, No. 1, 01.2010, pp. 44–52.

Bontemps, C., M.A. Rogel, A. Wiechmann, A. Mussabekova, S. Moody, M.F. Simon et al. 2016. Endemic *Mimosa* species from Mexico prefer alphaproteobacterial rhizobial symbionts. New Phytol. 209: 319–333. doi:10.1111/nph.13573.

Boucias, D.G., A. Garcia-Maruniak, R. Cherry, H. Lu, J.E. Maruniak and V.U. Lietze. 2012. Detection and characterization of bacterial symbionts in the Heteropteran, *Blissus insularis*. FEMS Microbiol. Ecol. 82: 629–641. doi:10.1111/j.1574-6941.2012.01433.x.

Caballero-Mellado, J., J. Onofre-Lemus, P. Estrada-de Los Santos and L. Martínez-Aguilar. 2007. The tomato rhizosphere, an environment rich in nitrogen-fixing *Burkholderia* species with capabilities of interest for agriculture and bioremediation. Appl. Environ. Microbiol. 73: 5308–5319.

Cardinale, M., M. Grube, J. Vieira de Castro, H. Muller and G. Berg. 2012. Bacterial taxa associated with the lung lichen *Lobaria pulmonaria* are differentially shaped by geography and habitat. FEMS Microbiol. Lett. 329: 111–115.

Chen, W.M., E.K. James, A.R. Prescott, M. Kierans and J.I. Sprent. 2003. Nodulation of *Mimosa* spp. by the beta-proteobacterium *Ralstonia taiwanensis*. Mol. Plant Microbe Interact. 16: 1051–1061.

Chen, W.M., C.H. Wu, E.K. James and J.S. Chang. 2008. Metal biosorption capability of *Cupriavidus taiwanensis* and its effects on heavy metal removal by nodulated *Mimosa pudica*. J. Hazard Mater. 151: 364–371.

Chistoserdova, L., M.G. Kalyuzhnaya and M.E. Lidstrom. 2009. The expanding world of methylotrophic metabolism. Annu. Rev. Microbiol. 63: 477–499. doi:10.1146/annurev.micro.091208.073600.

De León-Martínez, J.A., G. Yañez-Ocampo and A. Wong-Villarreal. 2017. *Burkholderia* species associated with legumes of Chiapas, Mexico, exhibit stress tolerance and growth in aromatic compounds. Rev. Argent. Microbiol. 49: 394–401. doi:10.1016/j.ram.2017.04.009.

Degli Esposti, M. 2014. Bioenergetic evolution in proteobacteria and mitochondria. Genome Biol.

Evol. 6: 3238–3251. doi:10.1093/gbe/evu257.

Degli Esposti, M. and E. Martinez-Romero. 2017. The functional microbiome of arthropods. PLoS One 12: e0176573. doi:10.1371/journal.pone.0176573.

Demont, N., F. Debellé, H. Aurelle, J. Dénarié and J.C. Promé. 1993. Role of the *Rhizobium meliloti nodF* and *nodE* genes in the biosynthesis of lipo-oligosaccharidic nodulation factors. J. Biol. Chem. 268: 20134–20142.

Desirò, A., A. Salvioli, E.L. Ngonkeu, S.J. Mondo, S. Epis, A. Faccio et al. 2014. Detection of a novel intracellular microbiome hosted in arbuscular mycorrhizal fungi. ISME J. 8: 257–270.

Deveau, A., G. Bonito, J. Uehling, M. Paoletti, M. Becker, S. Bindschedler et al. 2018. Bacterial - Fungal Interactions: ecology, mechanisms and challenges. FEMS Microbiol. Rev. 42: 335–352, fuy008,https://doi.org/10.1093/femsre/fuy008.

Dione. N., J. Rathored, E. Tomei, J.C. Lagier, S. Khelaifia, C. Robert et al. 2017. Dakarella massiliensis gen. nov., sp. nov., strain ND3(T): a new bacterial genus isolated from the female genital tract. New Microbes New Infect. 18: 38–46. doi:10.1016/j.nmni.2017.05.003.

Dobritsa, A.P. and M. Samadpour. 2016. Transfer of eleven species of the genus *Burkholderia* to the genus *Paraburkholderia* and proposal of *Caballeronia gen. nov.* to accommodate twelve species of the genera *Burkholderia* and *Paraburkholderia*. Int. J. Syst. Evol. Microbiol. 66: 2836–2846. doi:10.1099/ijsem.0.001065.

Dolatabadi, S., K. Scherlach, M. Figge, C. Hertweck, J. Dijksterhuis, S.B. Menken et al. 2016. Food preparation with mucoralean fungi: A potential biosafety issue? Fungal Biol. 120: 393–401.

Dörr, J., T. Hurek and B. Reinhold-Hurek. 1998. Type IV pili are involved in plant-microbe and fungus-microbe interactions. Mol. Microbiol. 30: 7–17.

Eberl, L. and P. Vandamme. 2016. Members of the genus *Burkholderia*: good and bad guys. F1000Res. 5. pii: F1000 Faculty Rev-1007. doi:10.12688/f1000research.8221.1.

Estrada, G.A., V.L.D. Baldani, D.M. de Oliveira, S. Urquiaga and J.I. Baldani. 2013. Selection of phosphate-solubilizing diazotrophic *Herbaspirillum* and *Burkholderia* strains and their effect on rice crop yield and nutrient uptake. Plant Soil 369: 115. https://doi.org/10.1007/s11104-012-1550-1557.

Estrada-De Los Santos, P., R. Bustillos-Cristales and J. Caballero-Mellado. 2001. *Burkholderia*, a genus rich in plant-associated nitrogen fixers with wide environmental and geographic distribution. Appl. Environ. Microbiol. 67: 2790–2798.

Fujimura, R., A. Nishimura, S. Ohshima, Y. Sato, T. Nishizawa, K. Oshima et al. 2014. Draft genome sequence of the Betaproteobacterial endosymbiont associated with the fungus *Mortierella elongata* FMR23-6. Genome Announ. 2: e01272-14.

Futahashi, R., K. Tanaka, M. Tanahashi, N. Nikoh, Y. Kikuchi, B.L. Lee et al. 2013. Gene expression in gut symbiotic organ of stinkbug affected by extracellular bacterial symbiont. PLoS One 8: e64557. doi:10.1371/journal.pone.0064557.

Garbaye, J. 1994. Interactions between mycorrhizal fungi and other soil organisms. Plant Soil 159: 123–132.

Garcia, J.R., A.M. Laughton, Z. Malik, B.J. Parker, C.S.L. Trincot, S. Chiang et al. 2014. Partner associations across sympatric broad-headed bug species and their environmentally acquired bacterial symbionts. Mol. Ecol. 23: 1333–1347. doi:10.1111/mec.12655.

Geddes, B.A., M.H. Ryu, F. Mus, A. Garcia Costas, J.W. Peters, C.A. Voigt et al. 2015. Use of plant colonizing bacteria as chassis for transfer of N-fixation to cereals. Curr. Opin. Biotechnol. 32: 216–222. doi:10.1016/j.copbio.2015.01.004.

Ghignone, S., A. Salvioli, I. Anca, E. Lumini, G. Ortu, L. Petiti et al. 2012. The genome of the obligate endobacterium of an AM fungus reveals an interphylum network of nutritional interactions. ISME J. 6: 136–145.

Gil, R., C. Vargas-Chavez, S. López-Madrigal, D. Santos-García, A. Latorre and A. Moya. 2018. *Tremblaya phenacola* PPER: an evolutionary beta-gammaproteobacterium collage. ISME J. 12: 124–135. doi:10.1038/ismej.2017.144.

Gonella, E., I. Negri, M. Marzorati, M. Mandrioli, L. Sacchi, M. Pajoro et al. 2011. Bacterial endosymbiont localization in *Hyalesthes obsoletus*, the insect vector of Bois noir in *Vitis vinifera*. Appl. Environ. Microbiol. 77: 1423–1435.

Gui, H., W.M. Purahong, K.D. Hyde, J. Xu and P.E. Mortimer. 2017. The arbuscular mycorrhizal fungus Funneliformis mosseae alters bacterial communities in subtropical forest soils during litter decomposition. Front. Microbiol. 8: 1120.

Harrison, P.W., R.P. Lower, N.K. Kim and J.P. Young. 2010. Introducing the bacterial 'chromid': not a chromosome, not a plasmid. Trends Microbiol. 18: 141–148. doi:10.1016/j.tim.2009.12.010.

Hesselmann, R.P., C. Werlen, D. Hahn, J.R. van der Meer and A.J. Zehnder. 1999. Enrichment, phylogenetic analysis and detection of a bacterium that performs enhanced biological phosphate removal in activated sludge. Syst. Appl. Microbiol. 22(3): 454–65.

Huggett, M.J., D.H. Hayakawa and M.S. Rappé. 2012. Genome sequence of strain HIMB624, a cultured representative from the OM43 clade of marine Betaproteobacteria. Stand. Genomic Sci. 6: 11-20.

Hurek, T., L.L. Handley, B. Reinhold-Hurek and Y. Piché. 2002. *Azoarcus* grass endophytes contribute fixed nitrogen to the plant in an unculturable state. Mol. Plant Microbe Interact. 15: 233–242.

Hurek, T. and B. Reinhold-Hurek. 2003. *Azoarcus* sp. strain BH72 as a model for nitrogen-fixing grass endophytes. J. Biotechnol. 106: 169–178.

Husnik, F. and J.P. McCutcheon. 2016. Repeated replacement of an intrabacterial symbiont in the tripartite nested mealybug symbiosis. Proc. Natl. Acad. Sci. USA 113: E5416–E5424. doi:10.1073/ pnas.1603910113.

Iffis, B., M. St-Arnaud and M. Hijri. 2014. Bacteria associated with arbuscular mycorrhizal fungi within roots of plants growing in a soil highly contaminated with aliphatic and aromatic petroleum hydrocarbons. FEMS Microbiol. Lett. 358: 44–54.

Itoh, H., M. Aita, A. Nagayama, X.Y. Meng, Y. Kamagata, R. Navarro et al. 2014. Evidence of environmental and vertical transmission of *Burkholderia* symbionts in the oriental chinch bug, *Cavelerius saccharivorus* (Heteroptera: Blissidae). Appl. Environ. Microbiol. 80: 5974–5983. doi:10.1128/AEM.01087-14.

Jeanbille, M., M. Buée, C. Bach, A. Cébron, P. Frey-Klett, M.P. Turpault et al. 2016. Soil parameters drive the structure, diversity and metabolic potentials of the bacterial communities across temperate beech forest soil sequences. Microbial. Ecol. 71: 482–493.

Jiang, Y., W. Wang, Q. Xie, N. Liu, L. Liu, D. Wang et al. 2017. Plants transfer lipids to sustain colonization by mutualistic mycorrhizal and parasitic fungi. Science 356: 1172–1175.

Joo, G.J., S.M. Kang, M. Hamayun, S.K. Kim, C.I. Na, D.H. Shin et al. 2009. *Burkholderia* sp. KCTC 11096BP as a newly isolated gibberellin producing bacterium. J. Microbiol. 47: 167–171. doi:10.1007/s12275-008-0273-1.

Keymer, A., P. Pimprikar, V. Wewer, C. Huber, M. Brands, S.L. Bucerius et al. 2017. Lipid transfer from plants to arbuscular mycorrhiza fungi. eLife 6: e29107.

Kikuchi, Y., X.Y. Meng and T. Fukatsu. 2005. Gut symbiotic bacteria of the genus *Burkholderia* in the broad-headed bugs *Riptortus clavatus* and *Leptocorisa chinensis* (Heteroptera: Alydidae). Appl. Environ. Microbiol. 71: 4035–4043. doi:10.1128/AEM.71.7.4035-4043.2005.

Kikuchi, Y., T. Hosokawa and T. Fukatsu. 2007. Insect-microbe mutualism without vertical transmission: a stinkbug acquires a beneficial gut symbiont from the environment every generation. Appl. Environ. Microbiol. 73: 4308–4316.

Kikuchi, Y., T. Hosokawa and T. Fukatsu. 2008. "Chapter II: Diversity of bacterial symbiosis in stinkbugs," in Microbial Ecology Research Trends, ed Van Dijk, T. (Nova Science Publishers, Inc.), 39–63.

Kikuchi, Y., T. Hosokawa and T. Fukatsu. 2011. An ancient but promiscuous host–symbiont association between *Burkholderia* gut symbionts and their heteropteran hosts. ISME J. 5: 446–460. doi:10.1038/ismej.2010.150.

Kikuchi, Y., M. Hayatsu, T. Hosokawa, A. Nagayama, K. Tago and T. Fukatsu. 2012. Symbiont-mediated insecticide resistance. Proc. Natl. Acad. Sci. USA 109: 8618–8622. doi:10.1073/pnas.1200231109.

Kim, J.K., N.H. Kim, H.A. Jang, Y. Kikuchi, C.H. Kim, T. Fukatsu et al. 2013. Specific midgut region controlling the symbiont population in an insect-microbe gut symbiotic association. Appl. Environ. Microbiol. 79: 7229–7233. doi:10.1128/AEM.02152-13.

Kim, J.K., H.A. Jang, Y.J. Won, Y. Kikuchi, S.H. Han, C.H. Kim et al. 2014. Purine biosynthesis-deficient *Burkholderia* mutants are incapable of symbiotic accommodation in the stinkbug. ISME J. 8: 552–563. doi:10.1038/ismej.2013.168.

Kim, J.K. and BB.L. Lee. 2015. Symbiotic factors in Burkholderia essential for establishing an association with the bean bug, Riptortus pedestris. Arch. Insect Biochem. Physiol. 88: 4–17. doi: 10.1002/arch.21218.

Koga, R., G.M. Bennett, J.R. Cryan and N.A. Moran. 2013. Evolutionary replacement of obligate symbionts in an ancient and diverse insect lineage. Environ. Microbiol. 15: 2073–2081.

Kuechler, S.M., Y. Matsuura, K. Dettner and Y. Kikuchi. 2016. Phylogenetically diverse *Burkholderia* associated with midgut crypts of spurge bugs, *Dicranocephalus* spp. (Heteroptera: Stenocephalidae). Microbes Environ. 31: 145–153. doi:10.1264/jsme2.ME16042.

Lackner, G., N. Moebius, L.P. Partida-Martinez, S. Boland and C. Hertweck. 2011. Evolution of an endofungal lifestyle: deductions from the Burkholderia rhizoxinica genome. BMC Genomics 12: 210.

Lackner, L.L. and J.M. Nunnari. 2009. The molecular mechanism and cellular functions of mitochondrial division. Biochim. Biophys. Acta 1792: 1138–1144.

Lastovetsky, O.A., M.L. Gaspar, S.J. Mondo, K.M. LaButti, L. Sandor, I.V. Grigoriev et al. 2016. Lipid metabolic changes in an early divergent fungus govern the establishment of a mutualistic symbiosis with endobacteria. Proc. Natl. Acad. Sci. USA 113: 15102–15107.

Lee, Y.M., E.H. Kim and H.K. Lee. 2014. Biodiversity and physiological characteristics of Antarctic and Arctic lichens-associated bacteria. World J. Microbiol. Biotechnol. 30: 2711–2721.

Leveau, J.H.J. and G.M. Preston. 2008. Bacterial mycophagy: Definition and diagnosis of a unique bacterial-fungal interaction. New Phytologist 177: 859–876.

Li, Y., H. Xue, S.Q. Sang, C.L. Lin and X.Z. Wang. 2017a. Phylogenetic analysis of family Neisseriaceae based on genome sequences and description of *Populibacter corticis* gen. nov., sp. nov., a member of the family Neisseriaceae, isolated from symptomatic bark of *Populus × euramericana* canker. PLoS One 12: e0174506. doi:10.1371/journal.pone.0174506.

Li, Z., Q. Yao, S.P. Dearth, M.R. Entler, H.F. Castro Gonzalez, J.K. Uehling et al. 2017b. Integrated proteomics and metabolomics suggests symbiotic metabolism and multimodal regulation in a fungal-endobacterial system. Environ. Microbiol. 19: 1041–1053.

Long, L., Q. Lin, Q. Yao and H. Zhu. 2017. Population and function analysis of cultivable bacteria associated with spores of arbuscular mycorrhizal fungus Gigaspora margarita. 3 Biotech 7: 8.

López-Guerrero, M.G., E. Ormeño-Orrillo, J.L. Acosta, A. Mendoza-Vargas, M.A. Rogel, M.A. Ramírez et al. 2012. Rhizobial extrachromosomal replicon variability, stability and expression in natural niches. Plasmid. 68: 149–158. doi:10.1016/j.plasmid.2012.07.002.

Luginbuehl, L.H., G.N. Menard, S. Kurup, H. Van Erp, G.V. Radhakrishnan, A. Breakspear et al. 2017. Fatty acids in arbuscular mycorrhizal fungi are synthesized by the host plant. Science 356: 1175–1178.

Lumini, E., V. Bianciotto, P. Jargeat, M. Novero, A. Salvioli, A. Faccio et al. 2007. Presymbiotic growth and sporal morphology are affected in the arbuscular mycorrhizal fungus Gigaspora margarita cured of its endobacteria. Cell. Microbiol. 9: 1716–1729.

Martin, F., A. Kohler, C. Murat, R. Balestrini, P.M. Coutinho, O. Jaillon et al. 2010. Périgord black truffle genome uncovers evolutionary origins and mechanisms of symbiosis. Nature 464: 1033–1038.

Martínez-Romero, E. 2012. How do microbes enhance the carrying capacity of their habitats? Expert Opin. Environ. Biol 1: 1 http://dx.doi.org/10.4172/2325-9655.1000e103.

Martinez-Romero, J.C., E. Ormeno-Orrillo, M.A. Rogel, A. Lopez-Lopez and E. Martinez-Romero. 2010. Trends in rhizobial evolution and some taxonomic remarks. pp. 301–316. *In*: Pontarotti, P. (ed.). Evolutionary Biology Concepts, Molecular and Morphological Evolution. Springer, Berlin, Germany.

McCutcheon, J.P. and N.A. Moran. 2010. Functional convergence in reduced genomes of bacterial symbionts spanning 200 My of evolution. Genome Biol. Evol. 2: 708–718. doi:10.1093/gbe/evq055.

McCutcheon, J.P. and P.J. Keeling. 2014. Endosymbiosis: protein targeting further erodes the organelle/symbiont distinction. Curr Biol. 24: R654–R655.

Moawad, M. and E.L. Schmidt. 1987. Occurrence and nature of mixed infections in nodules of field-grown soybeans (*Glycine max*). Biol. Fertil. Soils 5: 112–114.

Moebius, N., Z. Üzüm, J. Dijksterhuis, G. Lackner and C. Hertweck. 2014. Active invasion of bacteria into living fungal cells. eLife 3: e03007.

Mondo, S.J., K.H. Toomer, J.B. Morton, Y. Lekberg and T.E. Pawlowska. 2012. Evolutionary stability in a 400-million-year-old heritable facultative mutualism. Evolution 66: 2564–2574.

Moran, N.A., P. Tran and N.M. Gerardo. 2005. Symbiosis and insect diversification: an ancient symbiont of sap-feeding insects from the bacterial phylum Bacteroidetes. Appl. Environ. Microbiol. 71: 8802–8810.

Morris, J.J., R.E. Lenski and E.R. Zinser. 2012. The Black Queen Hypothesis: evolution of dependencies through adaptive gene loss. MBio. 3(2). pii: e00036-12. doi:10.1128/mBio.00036-12.

Moulin, L., A. Munive, B. Dreyfus and C. Boivin-Masson. 2001. Nodulation of legumes by members of the beta-subclass of Proteobacteria. Nature 411: 948–950.

Nagai, F., M. Morotomi, H. Sakon and R. Tanaka. 2009. *Parasutterella excrementihominis* gen. nov., sp. nov., a member of the family Alcaligenaceae isolated from human faeces. Int. J. Syst. Evol. Microbiol. 59: 1793–1797. doi:10.1099/ijs.0.002519-0.

Nakabachi, A., R. Ueoka, K. Oshima, R. Teta, A. Mangoni, M. Gurgui et al. 2013. Defensive bacteriome symbiont with a drastically reduced genome. Curr. Biol. 23: 1478–1484.

Noda, H., K. Watanabe, S. Kawai, F. Yukuhiro, T. Miyoshi, M. Tomizawa et al. 2012. Bacteriome-associated endosymbionts of the green rice leafhopper *Nephotettix cincticeps* (Hemiptera: Cicadellidae). Appl. Entomol. Zool. 47: 217–225.

Ohbayashi, T., K. Takeshita, W. Kitagawa, N. Nikoh, R. Koga, X.Y. Meng et al. 2015. Insect's intestinal organ for symbiont sorting. Proc. Natl. Acad. Sci. USA 112: E5179–E5188. doi:10.1073/pnas.1511454112.

Olsson, S., P. Bonfante and T.E. Pawlowska. 2017. Ecology and evolution of fungal-bacterial interactions. pp. 563–583. *In*: Dighton, J. and P. Oudemans (eds.). The Fungal Community: Its Organization and Role in the Ecosystem, 4th edn. CRC Press, Taylor & Francis: Boca Raton, FL, USA.

Onofre-Lemus, J., I. Hernández-Lucas, L. Girard and J. Caballero-Mellado. 2009. ACC (1-aminocyclopropane-1-carboxylate) deaminase activity, a widespread trait in *Burkholderia* species, and its growth-promoting effect on tomato plants. Appl. Environ. Microbiol. 75: 6581–6590. doi:10.1128/AEM.01240-09.

Ormeño-Orrillo, E., M.A. Rogel, L.M. Chueire, J.M. Tiedje, E. Martínez-Romero and M. Hungria. 2012. Genome sequences of *Burkholderia* sp. strains CCGE1002 and H160, isolated from legume nodules in Mexico and Brazil. J. Bacteriol. 194: 6927. doi:10.1128/JB.01756-12.

Ormeño-Orrillo, E., M. Hungria and E. Martínez-Romero. 2013. Dinitrogen fixing prokaryotes. *In*: Rosenberg, E. et al. (eds.). The Prokaryotes–Prokaryotic Physiology and Biochemistry. Springer-Verlag Berlin Heidelberg 2013.doi:10.1007/978-3-642-30141-4_72.

Oyserman, B.O., F. Moya, C.E. Lawson, A.L. Garcia, M. Vogt, M. Heffernen et al. 2016. Ancestral genome reconstruction identifies the evolutionary basis for trait acquisition in polyphosphate accumulating bacteria. ISME J. 10: 2931–2945. doi:10.1038/ismej.2016.67.

Parker, M.A. 2015. A single sym plasmid type predominates across diverse chromosomal lineages of *Cupriavidus* nodule symbionts. Syst. Appl. Microbiol. 38: 417–423. doi:10.1016/j. syapm.2015.06.003.

Partida-Martinez, L.P. and C. Hertweck. 2005. Pathogenic fungus harbours endosymbiotic bacteria for toxin production. Nature 437: 884–888.

Partida-Martinez, L.P., I. Groth, I. Schmitt, W. Richter, M. Roth and C. Hertweck. 2007a. *Burkholderia rhizoxinica* sp. nov. and *Burkholderia endofungorum* sp. nov., bacterial endosymbionts of the plant-pathogenic fungus Rhizopus microsporus. Int. J. Syst. Evol. Microbiol. 57: 2583–2590.

Partida-Martinez, L.P., S. Monajembashi, K.O. Greulich and C. Hertweck. 2007b. Endosymbiont-dependent host reproduction maintains bacterial-fungal mutualism. Curr. Biol. 17: 773–777.

Partida-Martinez, L.P., C.F. de Looss, K. Ishida, M. Ishida, M. Roth, K. Buder et al. 2007c. Rhizonin, the first mycotoxin isolated from the zygomycota, is not a fungal metabolite but is produced by bacterial endosymbionts. Applied Environ. Microbiol. 73: 793–797.

Partida-Martinez, L.P. 2013. A model for bacterial-fungal interactions. LAP Lambert Academic Publishing, GmbH & Co. KG, Saarbrücken, Germany.

Partida-Martinez, L.P. 2017. The fungal holobiont: Evidence from early diverging fungi. Environ. Microbiol. 19: 2919–2923.

Peay, K.G., P.G. Kennedy and J.M. Talbot. 2016. Dimensions of biodiversity in the Earth mycobiome. Nature Rev. Microbiol. 14: 434–447.

Peix, A., P.F. Mateos, C. Rodríguez-Barrueco, E. Martínez-Molina and E. Velázquez. 2001. Growth promotion of common bean (*Phaseolus vulgaris* L.) by a strain of *Burkholderia cepacia* under growth chamber conditions. Soil Biol. Biochem. 33: 1927–1935.

Pent, M., K. Põldmaa and M. Bahram. 2017. Bacterial communities in boreal forest mushrooms are shaped both by soil parameters and host identity. Front. Microbiol. 8: 836.

Platero, R., E.K. James, C. Rios, A. Iriarte, L. Sandes, M. Zabaleta et al. 2016. Novel *Cupriavidus* strains isolated from root nodules of native Uruguayan *Mimosa* species. Appl. Environ. Microbiol. 82: 3150–3164. doi:10.1128/AEM.04142-15.

Poddar, A., R.T. Lepcha and S.K. Das. 2014. Taxonomic study of the genus *Tepidiphilus*: transfer of *Petrobacter succinatimandens* to the genus *Tepidiphilus* as *Tepidiphilus succinatimandens* comb. nov., emended description of the genus *Tepidiphilus* and description of *Tepidiphilus thermophilus* sp. nov., isolated from a terrestrial hot spring. Int. J. Syst. Evol. Microbiol. 64: 228–235. doi:10.1099/ijs.0.056424-0.

Ramírez-Puebla, S.T., M. Rosenblueth, C.K. Chávez-Moreno, M.C. Catanho Pereira De Lyra, A. Tecante and E. Martínez-Romero. 2010. Molecular phylogeny of the genus *Dactylopius* (Hemiptera: Dactylopiidae) and identification of the symbiotic bacteria. Environ. Entomol. 39: 1178–1183.

Rangel-Castro, J.I., J.J. Levenfors and E. Danell. 2002. Physiological and genetic characterization of fluorescent Pseudomonas associated with *Cantharellus cibarius*. Can. J. Microbiol. 48: 739–748.

Reis, V.M., P. Estrada-de los Santos, S. Tenorio-Salgado, J. Vogel, M. Stoffels, S. Guyon et al. 2004. *Burkholderia tropica* sp. nov., a novel nitrogen-fixing, plant-associated bacterium. Int. J. Syst. Evol. Microbiol. 54: 2155–2162.

Roncato-Maccari, L.D., H.J. Ramos, F.O. Pedrosa, Y. Alquini, L.S. Chubatsu, M.G. Yates et al. 2003. Endophytic *Herbaspirillum seropedicae* expresses *nif* genes in gramineous plants. FEMS Microbiol. Ecol. 45: 39–47. doi:10.1016/S0168-6496(03)00108-9.

Rosenblueth, M. and E. Martínez-Romero. 2006. Bacterial endophytes and their interactions with hosts. Mol. Plant Microbe Interact. 19: 827–837.

Rosenblueth, M., A. López-López, J. Martínez, M.A. Rogel, I. Toledo and E. Martínez-Romero. 2012. Seed bacterial endophytes: common genera, seed-to-seed variability and their possible role in plants. Proc. XXVIIIth IHC – IS on Envtl., Edaphic & Gen. Factors Affecting Plants, Seeds and Turfgrass. pp. 39–48. Eds.: G.E. Welbaum et al. Acta Hort. 938, ISHS.

Russell, J.A., C.S. Moreau, B. Goldman-Huertas, M. Fujiwara, D.J. Lohman and N.E. Pierce. 2009. Bacterial gut symbionts are tightly linked with the evolution of herbivory in ants. Proc. Natl. Acad. Sci. USA 106: 21236–21241. doi:10.1073/pnas.0907926106.

Salvioli, A., S. Ghignone, M. Novero, L. Navazio, F. Venice, P. Bagnaresi and P. Bonfante. 2016. Symbiosis with an endobacterium increases the fitness of a mycorrhizal fungus, raising its bioenergetic potential. ISME J. 10: 130–144.

Salvioli, A., J. Lipuma, F. Venice, L. Dupont and P. Bonfante. 2017. The endobacterium of an arbuscular mycorrhizal fungus modulates the expression of its toxin–antitoxin systems during the life cycle of its host. ISME J. 11: 2394–2398.

Sato, Y., K. Narisawa, K. Tsuruta, M. Umezu, T. Nishizawa, K. Tanaka et al. 2010. Detection of betaproteobacteria inside the mycelium of the fungus *Mortierella elongata*. Microbes Environ. 25: 321–324.

Sawana, A., M. Adeolu and R.S. Gupta. 2014. Molecular signatures and phylogenomic analysis of the genus *Burkholderia*: Proposal for division of this genus into the emended genus *Burkholderia* containing pathogenic organisms and a new genus *Paraburkholderia* gen. nov. harboring environmental species. Front. Genet. 5: 429. doi:10.3389/fgene.2014.00429.

Scheublin, T.R., I.R. Sanders, C. Keel and J.R. van der Meer. 2010. Characterisation of microbial communities colonising the hyphal surfaces of arbuscular mycorrhizal fungi. ISME J. 4: 752–763.

Setten, L., G. Soto, M. Mozzicafreddo, A.R. Fox, C. Lisi, M. Cuccioloni et al. 2013. Engineering *Pseudomonas protegens* Pf-5 for nitrogen fixation and its application to improve plant growth under nitrogen-deficient conditions. PLoS One 8(5): e63666. doi:10.1371/journal.pone.0063666.

Spatafora, J.W., Y. Chang, G.L. Benny, K. Lazarus, M.E. Smith, M.L. Berbee et al. 2016. A phylum-level phylogenetic classification of zygomycete fungi based on genomescale data. Mycologia 108: 1028–1046.

Splivallo, R., A. Deveau, N. Valdez, N. Kirchhoff, P. Frey-Klett and P. Karlovsky. 2015. Bacteria associated with truffle-fruiting bodies contribute to truffle aroma. Environ. Microbiol. 17(8): 2647–60.

Stephen, J. and M.S. Jisha. 2011. Gluconic acid production as the principal mechanism of mineral phosphate solubilization by *Burkholderia* sp. (MTCC 8369). J. Tropical Agricult. 49: 99–103.

Swain, D.M., S.K. Yadav, I. Tyagi, R. Kumar, R. Kumar, S. Ghosh et al. 2017. A prophage tail-like protein is deployed by Burkholderia bacteria to feed on fungi. Nature Commun. 8: 404.

Tedersoo, L. 2017. Biogeography of Mycorrhizal Symbiosis. Springer Verlag, Switzerland.

Tran Van, V., O. Berge, S. Ngo ke, J. Balandreu and T. Heulin. 2000. Repeated beneficial effect of rice inoculation with a strain of *Burkholderia vietnamiensis* on early and late yield components in low fertility sulphate acid soils of Vietnam. Plant Soil 218: 273–284.

Uehling, J., A. Gryganski, K. Hameed, T. Tschaplinski, P.K. Misztal, S. Wu et al. 2017. Comparative genomics of Mortierella elongata and its bacterial endosymbiont *Mycoavidus cysteinexigens*. Environ. Microbiol. 19: 2964–2983.

Uzum, Z., A. Silipo, G. Lackner, A. De Felice, A. Molinaro and C. Hertweck. 2015. Structure, genetics and function of an exopolysaccharide produced by a bacterium living within fungal hyphae. Chembiochem. 16: 387–392.

Vannini, C., A. Carpentieri, A. Salvioli, M. Novero, M. Marsoni, L. Testa et al. 2016. An interdomain network: the endobacterium of a mycorrhizal fungus promotes antioxidative responses in both fungal and plant hosts. New Phytologist 211: 265–275.

Venice, F., M.C. de Pinto, M. Novero, S. Ghignone, A. Salvioli and P. Bonfante. 2017. Gigaspora margarita with and without its endobacterium shows adaptive responses to oxidative stress. Mycorrhiza 27: 747–759.

Vera-Ponce de León, A., E. Ormeño-Orrillo, S.T. Ramírez-Puebla, P. González-Román, M. Rosenblueth, M. Degli Esposti et al. 2017. *Candidatus* Dactylopiibacterium carminicum, a nitrogen-fixing symbiont of the cochineal insect *Dactylopius coccus* (Hemiptera: Coccoidea: Dactylopiidae). Genome Biol. Evol. 9: 2237–2250. https://doi.org/10.1093/gbe/evx156.

Vial, L., M.C. Groleau, V. Dekimpe and E. Déziel. 2007. *Burkholderia* diversity and versatility: an inventory of the extracellular products. J. Microbiol. Biotechnol. 17: 1407–1429.

Vial, L., A. Chapalain, M.C. Groleau and E. Déziel. 2011. The various lifestyles of the *Burkholderia cepacia* complex species: a tribute to adaptation. Environ. Microbiol. 13: 1–12. doi:10.1111/j.1462-2920.2010.02343.x.

Wewer, V., M. Brands and P. Dörmann. 2014. Fatty acid synthesis and lipid metabolism in the obligate biotrophic fungus Rhizophagus irregularis during mycorrhization of *Lotus japonicus*. Plant J. 79: 398–412. doi:10.1111/tpj.12566.

Wexler, H.M., D. Reeves, P.H. Summanen, E. Molitoris, M. McTeague, J. Duncan et al. 1996. *Sutterella wadsworthensis* gen. nov., sp. nov., bile-resistant microaerophilic *Campylobacter gracilis*-like clinical isolates. Int. J. Syst. Bacteriol. 46: 252–28.

Woese, C.R., W.G. Weisburg, B.J. Paster, C.M. Hahn, R.S. Tanner, N.R. Krieg et al. 1984. The phylogeny of purple bacteria: the beta subdivision. Syst. Appl. Microbiol. 5: 327–336.

Wu, L., K. Ma and Y. Lu. 2009. Prevalence of betaproteobacterial sequences in *nifH* gene pools associated with roots of modern rice cultivars. Microb. Ecol. 57: 58–68. doi:10.1007/s00248-008-9403-x.

Xu, Y., Buss, E.A. and D.G. Boucias. 2016a. Environmental transmission of the gut symbiont *Burkholderia* to phloem-feeding *Blissus insularis*. PLoS One. 11(8): e0161699. doi:10.1371/journal.pone.0161699.

Xu, Y., Buss, E.A. and D.G. Boucias. 2016b. Culturing and characterization of gut symbiont *Burkholderia* spp. from the Southern chinch bug, *Blissus insularis* (Hemiptera: Blissidae). Appl. Environ. Microbiol. 82: 3319–3330. doi:10.1128/AEM.00367-16.

Zampieri, E., M. Chiapello, S. Daghino, P. Bonfante and A. Mello. 2016. Soil metaproteomics reveals an inter-kingdom stress response to the presence of black truffles. Sci. Rep. 6: 25773.

Zhang, T., Y. Yan, S. He, S. Ping, K.M. Alam, Y. Han et al. 2012. Involvement of the ammonium transporter AmtB in nitrogenase regulation and ammonium excretion in *Pseudomonas stutzeri* A1501. Res. Microbiol. 163: 332–339. doi:10.1016/j.resmic.2012.05.002.

Zheng, J.Z., R. Wang, R.R. Liu, J.J. Chen, Q. Wei, X.Y. Wu et al. 2017. The structure and evolution of beta-rhizobial symbiotic genes deduced from their complete genomes. Immunome Res. 13: 131. doi: 10.4172/17457580.1000131.

7

From Alphaproteobacteria to Proto-Mitochondria

Mauro Degli Esposti, *Julio Martínez* and
Esperanza Martínez-Romero

Introduction

The class of alphaproteobacteria is the second largest among proteobacteria after that of gammaproteobacteria, according to the number of genomes and proteins present in current NCBI databases and also in terms of taxonomic richness (Schulz et al. 2017), but it is the most diverse in both phylogenetic breadth and functional properties (Garrity et al. 2005, Gupta and Mok 2007, Williams et al. 2007, Ferla et al. 2013, Degli Esposti and Martinez-Romero 2017). Alphaproteobacteria include some of the most widespread and economically important prokaryotes, from agriculture to biotechnology and human health. Indeed, entire groups of this class are noxious pathogens of humans and animals with some, in particular the Rickettsiales, living solely as obligate endocellular parasites of eukaryotes, from protists to humans. Alphaproteobacteria share the propensity to intimately associate with plants and animals with other bacteria, but this propensity is so diffuse among their various phylogenetic groups that it has sustained the concept that proto-mitochondria originated within this bacterial class (Yang et al. 1985, Esser et al. 2004, Wu et al. 2004, Fitzpatrick et al. 2006, Williams et al. 2007, Atteia et al. 2009, Gray et al. 1999, Gray 2012, Ferla et al. 2013).

Initially, the alpha subdivision was dominated by purple non sulfur bacteria, such as *Rhodobacter* and *Rhodospirillum*, which have played a major role in the history of microbiology (Truper and Pfennig 1981, Woese et al. 1984). Contrary

Center for Genomic Sciences, UNAM Campus de Cuernavaca, Cuernavaca, 62130 Morelos, Mexico.
*Corresponding author: mauro1italia@gmail.com

to the initial observations (Woese et al. 1984, Woese 1987), the great majority of the two thousand taxa now classified within the class (Chapter four) are not photosynthetic (Louca et al. 2016, Degli Esposti and Martinez Romero 2017). This underlies the emerging concept that photosynthesis has been acquired and lost multiple times via Lateral Gene Transfer (LGT), in proteobacteria as in other bacterial *phyla* (Martin et al. 2018).

Alphaproteobacteria are subdivided in an increasing number of orders and unclassified taxa. The latest PATRIC web repository https://www.patricbrc.org/view/Taxonomy/28211#view_tab=taxontree (accessed on 9 Dec 2017) lists a total of 16 orders of alphaproteobacteria. The orders with the largest number of taxa remain, in decreasing number of available genomes: Rhizobiales, Rhodobacterales, Sphingomonadales, Rhodospirillales and Rickettsiales, followed by Caulobacterales and Pelagibacterales. These are the same orders considered 10 years ago by Williams et al. (2007), following almost exactly their phylogenetic sequence from the latest divergent to the most basal. The basic phylogenetic tree of alphaproteobacteria was established in previous works (Lee et al. 2005, Gupta 2005), which had not considered *Pelagibacter* and its relatives. The phylogenetic position of these common marine bacteria has remained controversial since the first tree of reference reported by Williams et al. (2007), in part due to the diversity of molecular approaches used (Ferla et al. 2013, Luo 2015). However, a major factor influencing the relative phylogenetic position of *Pelagibacter* and its relatives, now forming the order of Pelagibacterales, is the taxonomic sampling of the alphaproteobacterial organisms considered, as exemplified by phylogenetic trees of the ubiquitous cytochrome *b* protein (Fig. 1).

The taxonomic span or breadth of alphaproteobacteria has increased considerably in the last few years, owing to programs designed to reduce the human bias in current genomic databases and the dramatic increase in metagenomic information on previously unclassified organisms (see Table 1 in Chapter four, and also later). It now appears established that *Magnetococcus* and its relatives (grouped in the order of Magnetococcales; Bazylinski et al. 2013a) form the basal branch of the class (Schübbe et al. 2009, Ferla et al. 2013, Degli Esposti et al. 2014, Wang and Wu 2015). When these and various unclassified alphaproteobacteria are considered, the phylogenetic trees contain Pelagibacterales in an intermediate branch, as shown in Fig. 1B—in agreement with some articles (Viklund et al. 2012, 2013, Luo 2015)—rather than in a deep branch as reported in other papers (Williams et al. 2007, Georgiades et al. 2011 Smith et al. 2012, Ferla et al. 2013) and shown in Fig. 1A, where Magnetococcales were not picked by the initial blast search.

The issue of the taxonomic position of Pelagibacterales is of particular importance for the origin of proto-mitochondria, since the works that consider these organisms as basal to the alpha class also imply that Pelagibacterales subtended the phylogenetic origin of proto-mitochondria (Thrash et al. 2011, Georgiades et al. 2011, Ferla et al. 2013). This topic will be elaborated further in the final part of the chapter because of its relevance to the the origin of proto-mitochondria.

(A) *Pelagibacter* without unclassified alpha

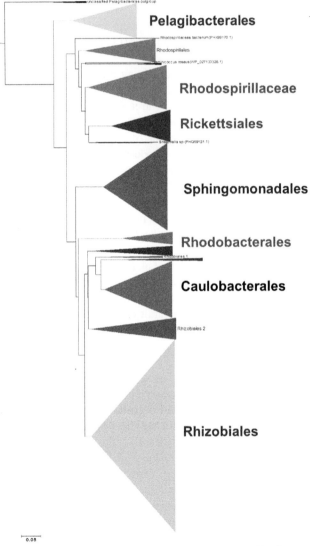

0.05

Fig. 1 cont. ...

... Fig. 1 cont.

(B) *Magnetococcus* with unclassified alpha

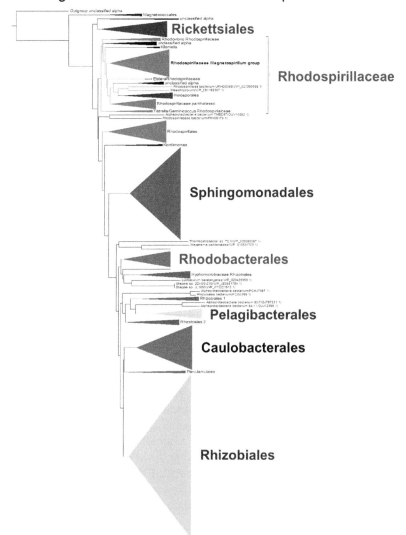

Fig. 1. Phylogenetic tree of alphaproteobacteria using cytochrome *b* as the universal marker. (A) The NJ tree was obtained by a wide DeltaBLAST (Boratyn et al. 2012) using as a query the cytochrome *b* of *Pelagibacter ubique* over all major taxa of alphaproteobacteria excuding Magnetococcales and unclassified alphaproteobacteria. (B) The NJ tree was obtained by a wider DeltaBLAST than in A, using as a query the cytochrome *b* of *Magnetococcus marinus* including Magnetococcales and unclassified alphaproteobacteria.

An Iron Wire in the Evolution of Alphaproteobacteria

The controversy surrounding the origin of proto-mitochondria and the taxonomic position of Pelagibacterales mentioned above has overshadowed the significance of Magnetococcales in the evolution of alphaproteobacteria, especially in regard to their functional properties. Some of these properties have already been discussed in Chapter five in comparison with deep-branching gammaproteobacteria. The type organism of Magnetococcales was previously characterized as microaerophilic magnetic coccus, strain MC-1 (Frankel et al. 1997). Renamed as *Magnetococcus marinus*, it entered the phylogeny of alphaproteobacteria the same year (Esser et al. 2007) in which the controversial issue regarding Pelagibacterales vs. proto-mitochondria started (Williams et al. 2007). Although earlier reported to be at the basis of the alphaproteobacterial lineage (Schübbe et al. 2009), *Magnetococcus* has been considered in major phylogenetic works only since 2013 (Bazylinski et al. 2013a, Ferla et al. 2013, Degli Esposti et al. 2014, Ji et al. 2017). Some initial uncertainties regarding the affiliation to the alpha class (Esser et al. 2007, Schübbe et al. 2009, Thiergart et al. 2012) have been set aside by the consistent finding of its basal position among alphaproteobacteria in phylogenetic trees obtained with different markers (Bazylinski et al. 2013a, Ferla et al. 2013, Degli Esposti et al. 2014, Morillo et al. 2014, Wang and Wu 2015, Ji et al. 2017).

Genomic analysis of *Magnetococcus* has shown a high degree of chimaerism, since the affiliation of all coded proteins produce a mosaic picture in regard to their closest homologs, with only about one third of proteins showing close affiliation to alphaproteobacteria (Esser et al. 2007, Schübbe et al. 2009, Ji et al. 2017). This result has been recently confirmed by using a simplified version of all reciprocal blast approach focused on proteins for energy metabolism (Degli Esposti 2017). When not arising from a secondary evolutionary event as in *Ca.* Tremblaya (see Chapter six), genome chimaerism may be considered a telltale of antiquity for a bacterium, as in the case of Acidithiobacillia (Chapter five). On one side, chimaerism reflects taxonomically deep evolutionary relationships with ancestral lineages, while on the other side it derives from affiliations with more modern taxa of the same class. Intriguingly, the same applies to magnetotaxis, which is the distinctive trait of *Magnetococcus* (Schübbe et al. 2009, Ji et al. 2017). Of note is that *Magnetospirillum magneticum* has been reported to display the second largest degree of genomic chimaerism among alphaproteobacteria after *Magnetococcus* (Esser et al. 2007).

Magnetococcus and other Magnetococcales share magnetotaxis with three genera of alphaproteobacteria belonging to the family Rhodospirillaceae - notably *Magnetospirillum*, where this peculiar trait was first described, as well as *Magnetovibrio* and *Magnetospira* (Bazylinski et al. 2013a, b, Lefèvre and Bazylinski 2013, Lin et al. 2017a). Among the whole *phylum* of proteobacteria, magnetotaxis is additionally present in some deltaproteobacteria such as *Geobacter magneticus* (Lefèvre et al. 2012, Lin et al. 2017a) and a few gammaproteobacteria (Lefèvre and Bazylinski 2013, Leão et al. 2016). However, it is relatively common in organisms of the *phylum* Nitrospirae (Jogler et al. 2011, Lin et al. 2017a,b, Wang and Chen 2017, Lin et al. 2017c—cf. Chapter three), from which it is supposed to have been vertically transmitted to proteobacteria (Lefèvre et al. 2013, Zeytuni et al. 2015, Lin et al. 2017a).

Magnetococcus and Magnetospirilli, as well as magnetotactic Nitrospirae, characteristically contain an array of intracellular vesicles enriched in Fe compounds such as magnetite, the magnetosomes (Lefèvre et al. 2013, Bazylinski et al. 2013b, Li et al. 2014, Dziuba et al. 2016, Uebe and Schüler 2016, Kolinko et al. 2016, Lin et al. 2017a). Fundamentally, magnetosomes serve to orient the bacteria towards water columns with oxygen gradient, so as to locate areas containing micromolar concentration of oxygen which are optimal for their growth (Lefèvre et al. 2013, Bazylinski et al. 1988, 2013b, Li et al. 2014, Uebe and Schüler 2016, Lin et al. 2017a). Hence, these microoxia-seeking bacteria are facultatively anaerobes in the most poignant significance of the term (cf. Chapter two). Besides sulfur oxidation, a trait already discussed in Chapter five, *Magnetococcus* shares with Magnetospirilli another metabolic trait that is typical of anaerobes, strictly or facultatively, namely the ionmotive Rnf complex (Biegel et al. 2011). The Rnf complex was originally discovered in *Rhodobacter* and hence named Rhodobacter nitrogen fixation (Biegel et al. 2011), since it physiologically reduces the ferredoxin that feeds electrons into the nitrogenase reaction. Reduction of ferredoxin is the reverse reaction of the thermodynamically more favourable reduction of NADH and is driven by the ionmotive function of the enzyme complex, which can pump either Na^+ or H^+ across the membrane (Biegel et al. 2011). It is likely that the Rnf complex predominantly functions in the same way in magnetotactic alphaproteobacteria, for both *Magnetococcus* and Magnetospirilli have the nitrogenase complex and therefore are capable of fixing nitrogen as free-living diazotrophs (Schübbe et al. 2009, Lefèvre et al. 2013, Bazylinski et al. 2013b, Dziuba et al. 2016). See also Chapter three, four and five for Rnf distribution among other facultatively anaerobic bacteria.

From Molecular Phylogeny to Functional Evolution of Alphaproteobacteria

A survey of the distribution of the Rnf complex in currently available genomes (January 2018) has revealed its presence in 27 alphaproteobacterial taxa, 21 more than eight years ago (Biegel et al. 2011). The majority of Rnf-containing taxa belong to the family Rhodospirillaceae, especially Magnetospirilli, but also four metagenomic *Azospirillum* strains that have been found in human gut and are strictly anaerobes (Fig. 2). These organisms are *Azospirillum* sp. CAG:239 and sp. CAG:260 (Nielsen et al. 2014, Degli Esposti et al. 2016), which are closely related to *Azospirillum* sp. 51_20 and *Azospirillum* sp. 47_25, respectively (Brown et al. 2016). Remarkably, they neither have membrane quinones nor cytochromes and thus rely only on the Rnf complex for pumping ions across the membrane to drive ATP production via protonmotive ATP synthase, as presented in Fig. 2A of

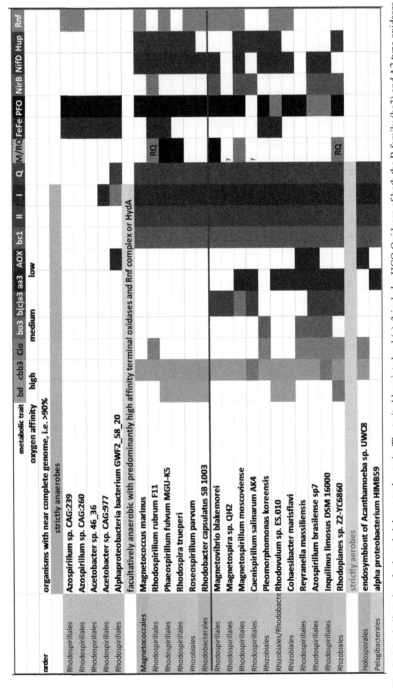

Fig. 2. Metabolic traits of selected alphaproteobacteria. The trait abbreviated as b(c)a3 includes HCO Oxidases of both the B family (ba3) and A2 type oxidases of the A family (aa3 oxidases) that often have a cytochrome *c* fused with subunit 2 as in *Bacillus* caa3 oxidases. Other traits are abbreviated as in chapter five.

Chapter two. Such a simplified protonmotive circuit is typical of strictly anaerobes of the Firmicutes *phylum*, in which ferredoxin is reduced by catabolic enzymes or hydrogenases (Biegel et al. 2011). In such metagenomic *Azospirillum* strains, ferredoxin is reduced primarily by Pyruvate-Ferredoxin Oxidoreductase (PFO; Fig. 2, cf. Chapter four) and can also be re-oxidized by one or more forms of [FeFe]-hydrogenases that are coded in their genomes and appear to be related to those present in anaerobic eukaryotes such as *Entamoeba* (Degli Esposti et al. 2016). This metabolic path is shared with strictly anaerobic prokaryotes such as Clostridiales and also aerotolerant sulfate-reducing deltaproteobacteria (see Chapter four), as well as strict anaerobes of the betaproteobacteria class such as *Sutterella* (see Chapter six). Therefore, it constitutes a basal trait inherited by proteobacteria from ancestral lineages of anaerobic prokaryotes, surviving as a metabolic adaptation to an anaerobic niche like the animal and human gut, which in several aspects appears to be a nutrient-rich environment favouring the survival of strictly anaerobes (Nielsen et al. 2014, Degli Esposti and Martinez-Romero 2017).

The scattered taxonomic distribution of the anaerobic traits that are concentrated in the above mentioned *Azospirillum* strains (Fig. 2) suggests important considerations regarding the functional evolution of alphaproteobacteria: (1) the progenitors of extant Rhodospirillaceae might have been strictly anaerobes and thus likely to be the most ancestral of the class—possibly even 'older' than the Great Oxygenation Event (GOE, Chapter two); (2) ancestral alphaproteobacteria adapted to the microoxic conditions that followed the GOE by acquiring or inheriting high affinity terminal oxidases such as the cbb3 oxidase present in *Magnetococcus* (Fig. 2); (3) subsequent adaptation to increased levels of oxygen entailed the additional acquisition of low affinity terminal oxidases and then a progressive loss of anaerobic traits. In this rationale, extant alphaproteobacteria that combine anaerobic traits such as PFO, [FeFe]-hydrogenases and the Rnf complex with high affinity terminal oxidases could be considered closest to the ancestors of the class. Genomic analysis indicates that such a combination is currently present in a handful of contemporary bacteria besides *Magnetococcus* (Fig. 2): photosynthetic Rhodospirillaceae closely related to *Rhodospirillum* like *Phaeospirillum* (Imhoff 2005, Degli Esposti and Martinez-Romero 2017); *Roseospirillum parvum*, a photosynthetic member of the Rhizobiales order that is classified within the family Rhodobiaceae (Glaeser and Overmann 1999, Permentier et al. 2000); and *Rhodobacter capsulatus*, the type species of Rhodobacterales (Gupta 2005, Williams et al. 2007). At difference with *Magnetococcus*, these organisms additionally contain an ubiquinol oxidase of bd-I type (Fig. 2), which has an oxygen affinity comparable to that of cbb3 oxidases (see Chapter two).

While the absence of low affinity terminal oxidases of the A family is typical for *Rhodospirillum* and related organisms (Imhoff 2005, Degli Esposti et al. 2014), it is exceptional in the case of *Rhodobacter capsulatus*, since its closely related *R. sphaeroides*, as most other members of the parent Rhodobacteraceae family, have one or more oxidases of the A family (Pereira et al. 2001, Degli Esposti et al. 2014, Degli Esposti and Martinez-Romero 2017). Indeed, the first crystal structure of a family A oxidase has been obtained in *Paracoccus denitrificans*, a non-photosynthetic member of the same family (Iwata et al. 1995, Pereira et al. 2001). Therefore, it is quite possible that *R. capsulatus* has lost its original low affinity oxidase as a result of secondary evolutionary events, as in the case of other alphaproteobacteria (Fig. 2, cf. Degli Esposti and Martinez-Romero 2017). The same concept applies to *Roseospirillum parvum*, since all its relatives of the Rhodobiaceae family have at least one low affinity terminal oxidase of the A family (Degli Esposti and Martinez-Romero 2017). Following these considerations, then *Rhodospirillum/Phaeospirillum* (Imhoff et al. 1998) remain the alphaproteobacterial taxa that constitutively possesses the rare combination of anaerobic traits with high affinity terminal oxidases only, as in *Magnetococcus*. Hence, *Phaeospirillum* and *Rhodospirillum* may be viewed, functionally, as the most ancestral members of the alphaproteobacteria class, besides Magnetococcales.

Similar to sulfur-oxidizing taxa among the gammaproteobacteria (Chapter five), the genomic distribution of functional traits strongly indicates that the family Rhodospirillaceae is among the most basal in alphaproteobacteria, for it contains taxa possessing all the anaerobic and aerobic traits that are present in the whole class (Fig. 2). This conclusion is supported only in part by phylogenetic trees of ubiquitous proteins such as cytochrome *b* (Fig. 1), but could be in accord with the evidence that members of the family Rhodospirillaceae do not follow the tree topology of the rest of the Rhodospirillales order (see Ferla et al. 2013, and reference therein). However, it contrasts many reported trees of alphaproteobacteria in which members of the Rhodospirillaceae diverge later than either Rickettsiales or Pelagibacterales (Gupta and Mok 2007, Williams et al. 2007, Ferla et al. 2013, Viklund et al. 2013, Wang and Wu 2015). Those trees are often based upon rRNA phylogenies and consequently suffer from the dichotomy between ribosomal genes and the phenotypic traits of bacterial organisms, a problem that has been discussed previously in Chapters four and five of the book. The streamlined nature of the genome of Pelagibacterales, as well as the eroded and reduced genome of intracellular parasites of the Rickettsiales and Holosporales order, do not leave much space for metabolic traits that are no more used after diversification from ancestral alphaproteobacteria (Emelyanov 2003, Wu et al. 2004, Georgiades et al. 2011, Wang and Wu 2015). Indeed, the majority of the taxa classified under these orders currently have an aerobic metabolism that is based upon low affinity terminal oxidases, as mitochondria (Kurland and Andersson 2000, Emelyanov 2003, Giovannoni et al. 2005b, Boussau et al. 2004, Georgiades et al. 2011, Morris and Schmidt 2013). This functional evidence is in stark contrast with the basal position assigned to Rickettsiales or Pelagibacterales in traditional phylogenetic trees of alphaproteobacteria (Fig. 1, cf. Williams et al. 2007), a problem which will be discussed further in regard to the origin of proto-mitochondria at the end of this chapter.

The problems just discussed raise a fundamental question: how can we recognise secondary loss of aerobic traits or their possible acquisition by LGT rather than vertical inheritance from ancestral lineages? Evidently, phylogenetic trees cannot provide an adequate answer (Chapter one). Integrated approaches are thus required, as pioneered by Degli Esposti et al. (2014). Central to these approaches is the systematic classification of the diverse variants of COX and other terminal oxidases that are present in alphaproteobacteria and define the plasticity of their aerobic metabolism (Imhoff 2005, Morris and Schmidt 2013, Degli Esposti 2014, Wang and Wu 2015). The following section will provide an updated view of the currently known variety of heme *a*-containing terminal oxidases.

The Variety of Heme a Oxidases in Alphaproteobacteria

The variety of aerobic metabolism present in alphaproteobacteria has been recognised at the very beginning of their classification on the basis of 16S rRNA in the 80s. In his classical review on bacterial evolution, Woese (1987) remarked: 'Aerobic metabolism also appears to have arisen a number of times in the alpha subdivision alone', citing Woese et al. (1984). Compared with the delta subdivision, alphaproteobacteria obviously emerged as a fundamentally aerobic lineage of proteobacteria, even if its initial members were predominantly facultatively anaerobes and photosynthetic (Schultz and Weaver 1982, Woese et al.1984, Woese 1987, Boussau et al. 2004, Imhoff 2005, Garrity et al. 2005). After the sequencing of the first *Rickettia* genome (Andersson et al. 1998), strictly aerobes became a fundamental part of the alphaproteobacterial lineage (Müller and Martin 1999, Kurland and Andersson 2000, Boussau et al. 2004, Gupta 2005). However, strictly aerobes constitute a minority of today's alphaproteobacteria with sequenced genomes, as previously found for other bacteria (Morris and Schmidt 2013). The great majority of alphaproteobacteria, therefore, are facultatively anaerobes or microaerobic (Morris and Schmidt 2013, Martin 2017). These definitions of aerobic metabolism, presented in Chapter two of the book, do not clarify how such a wide set of organisms live in so many different environments, adapting to oxygen levels that vary from basically zero—as in parts of animal guts, underground niches and sulfidic ocean zones—to fully aerated, as in the photic zone of the oceans (Louca et al. 2016). Such adaptations primarily depend upon the genomic endowment and functional expression of terminal oxidases that differ in structure, regulation and fundamental biochemical properties, in particular the affinity for oxygen (Morris and Schmidt 2013, Degli Esposti et al. 2014, Degli Esposti and Martinez-Romero 2017). The different types of bacterial terminal oxidases have been introduced in Chapter two. Here the subtypes and variants of heme *a* oxidases are discussed in detail, since their presence and combination with other terminal oxidases determine the environmental adaptation of alphaproteobacteria, as previously mentioned.

Terminal oxidases containing heme *a* belong to family A and B according to the classification of Pereira et al. (2001), now widely accepted (Sousa et al. 2012, Gao et al. 2012, Sharma and Wikström 2014, Degli Esposti 2014). The classification is based upon conserved signatures of the proton channels that are present in the structure of the largest catalytic subunit COX1 (Iwata et al. 1995), which is shared by all Heme Copper Oxygen reductases, HCO (Pereira et al. 2001). Family B oxidases lack one of these channels (Pereira et al. 2001, Radzi Noor and Soulimane 2012) and consequently have a reduced capacity of proton pumping (Han et al. 2011, Sharma and Wikström 2014). However, this structural distinction is becoming blurred, following the accumulation of COX1 sequences in the last few years, especially from metagenomic studies (see following section of this chapter). It thus appears that there is a continuum in the variations of COX1 sequences associated with proton channels and their function, from the cbb3 oxidases of the C family to the most ancestral forms of aa3 oxidases of the A family (Ducluzeau et al. 2014). What is becoming particularly interesting in the context of the evolution of aerobic metabolism is that COX1 proteins now document the gradual transition from family B to family A oxidases in current genomes of alphaproteobacteria. Species of this class contain all the variants of both family A and B that have been recognized so far, including a novel operon for type A2 oxidases that has been found in a few alphaproteobacteria and other bacterial taxa (Fig. 3A, *vedi infra*). This operon is currently named CyoCAB from the sequence of the Conserved Domain (CD—Marchler-Bauer et al. 2015) definition of its subunits: CyoC for COX3, CyoA for COX2 and CyoB for COX1, with a Cu-assembly protein of the SCO (Synthesis of Cytochrome *c* Oxidase) family intermixed between CyoC and CyoA. The CD of CyoA, CyoB and CyoC are usually associated with ubiquinol oxidases belonging to type A1 (Matsutani et al. 2014), as shown in Fig. 3B, but they are assigned also to catalytic subunits of cytochrome *c* oxidases of the same A family.

The real peculiarity of the CyoCAB operon resides in the sequence of its genes, since the great majority of gene clusters for heme *a*-containing oxidases start with COX2 (Degli Esposti 2014) and rarely contain SCO proteins, for example in aerotolerant *Desulfovibrio* (Fig. 3B and Chapter four). The operon was initially discovered in the genome of two marine magnetotactic Rhodospirillaceae, *Magnetovibrio* (Bazyliski et al. 2013b, Trubitsyn et al. 2016) and *Magnetospira* (Ji et al. 2014), which do not have the low affinity cytochrome *c* oxidases of A1 type that is present in freshwater Magnetospirilli (Figs. 2 and 3B - M.D.E., M. Mentel, W.F. Martin and F. Sousa, manuscript in preparation). *Magnetovibrio* has a more pronounced anaerobic physiology than Magnetospirilli (Bazyliski et al. 2013b); accordingly, its genome contains the complete set of enzymes for the biosynthesis of menaquinone as in *Phaeospirillum* genome (Fig. 2, cf. Duquesne et al. 2012, Degli Esposti 2017). Notably, menaquinone is typical of anaerobic respiratory chains (see Chapters two and five). The evidence just mentioned suggests that CyoCAB operon oxidases may have higher affinity for oxygen than type A1 oxidases of Magnetospirilli, or have a specialized function related to Fe metabolism. At present, these possibilities remain speculative, since no biochemical data on the enzyme coded by the CyoCAB operon is available yet.

However, some analogies could be drawn with cytochrome *c* oxidase I of *Aquifex*, which appears to have the same gene sequence as the CyoCAB operon of *Magnetovibrio* (Fig. 3A). This oxidase, which has been classified as a typical A2 type by Pereira et al. (2001), has a gene cluster concatenated with that of cytochrome *c* oxidase II or cox2, which is classified instead among family B

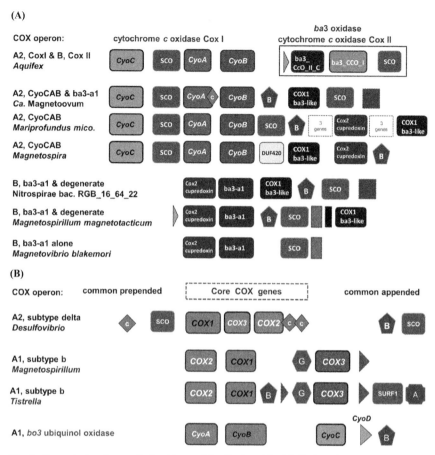

Fig. 3. Gene clusters for terminal oxidases of family A and B. (A) Variants in heme *a*-containing terminal oxidases of various phyla. (B) Subtypes of family A terminal oxidases, from *Desulfovibrio* to alphaproteobacteria. Modified from Degli Esposti (2014). Symbols with A and B identify ctaA and ctaB, respectively.

oxidases (Pereira et al. 2001, Prunetti et al. 2011, Gao et al. 2012). The latter has been characterized biochemically after purification and displays the dual function of oxidizing either ubiquinol or cytochrome *c* (Gao et al. 2012), a hybrid property that may be present in other family B oxidases. Intriguingly, the CyoCAB operon of *Magnetovibrio* and *Magnetospira* is associated with a deranged form of what appears a B family oxidase that is structurally different from cox2 of *Aquifex*, since its subunit 1 does not have the conserved Y280 and also the subsequent doublet of histidine ligands for CuB (M.D.E., unpublished results). A similarly deranged oxidase is associated with the gene cluster of another subtype of B family oxidases that is present in Magnetospirilli (Fig. 3A, cf. Degli Esposti et al. 2014), which retains the doublet of histidine ligands but not the conserved Y280. The absence of this conserved residue has been recognized earlier in the COX1 subunit of some B family oxidases (Sousa et al. 2012), including that present in *Magnetospirillum* (Ducluzeau et al. 2014), and considered to indicate the loss of cytochrome *c* oxidase function. The *Magnetospirillum* B family oxidase, without conserved Y280, corresponds to the so-called cytochrome *a*1-like that is expressed in *Magnetospirillum magnetotacticum* under microooxic conditions (Tanimura and Fukumori 2000). Because this *Magnetospirillum* oxidase had been previously labelled cytochrome *a*1-like (Tanimura and Fukumori 2000), it will, henceforth, be called subtype **ba3-a1** (Fig. 3). Its homologs are generally present in taxa having the CyoCAB operon, thereby indicating possible evolutionary connections between the two different enzymes. To uncover these connections, a detailed analysis of their distribution, genetic neighborhood and protein sequence has ben undertaken, obtaining the following information.

CyoCAB operon oxidases are present in two other alphaproteobacteria besides *Magnetovibrio* and *Magnetospira: Terasakiella* sp. *PR1*, taxonomically classified among the Rhizobiales (Han et al. 2106) but containing Mam proteins typical of magnetotactic organisms (Kolinko et al. 2012, 2016), and the unclassified alphaproteobacterium RIFOXYD12_FULL_60_8 (Anantharaman et al. 2016, Fig. 4). In these four species, the operon is associated with one or more subunits of the above mentioned deranged B family oxidase. This association is not present in the genome of other proteobacterial organisms that have the CyoCAB operon: strains of the ancestral betaproteobacterium *Ca.* Accumulibacter (see Chapter six), a couple of gammaproteobacteria such as *Sulfuritalea hydrogenivorans*, several metagenomic deltaproteobacteria such as deltaproteobacteria bacterium GWA2_55_82 (Anantharaman et al. 2016) and, remarkably, the same zetaproteobacteria that also have heme *a* synthase (Fig. 4, cf. Table 2 in Chapter four). In the latter group, however, *Mariprofundus micogutta* has both subunits of a deranged B family oxidase intermixed with other genes downstream the CyoCAB operon (Fig. 3A), suggesting genomic erosion of the latter oxidase. In the genome of *Candidatus* Magnetoovum, a magnetotactic Nitrospirae (Kolinko et al. 2016), the gene for subunit 1 of the same deranged oxidase (labelled ba3-like in Fig. 3) follows that of the ctaB gene at the end of the CyoCAB operon (Fig. 3A, top). This gene sequence is unusual, since other Nitrospirae have either a ba3-a1 oxidase concatenation with deranged oxidases, or CyoCAB operons alone as in deltaproteobacteria. However, all Nitrospirae do not have the *ctaA* gene for heme *a* synthase. Consequently, CyoCAB operon and also ba3-a1 oxidases present in this *phylum* cannot have heme *a* prosthetic groups. Future biochemical analysis will verify whether these oxidases actually have only *b* hemes, similar to the cbb3 oxidases, or rather resemble the bo3 oxidases, since the majority of Nitrospirae do have the *ctaB* gene for heme *o* biosynthesis, often in association with the gene clusters of the oxidases (Fig. 3A).

Overall, the association of the above mentioned oxidases does not follow a recognizable pattern, even if it may reflect some ancient event in the evolution of these and other terminal oxidases. Conversely, the occurrence of concatenated clusters of HCO appears to be relatively frequent in diverse bacterial groups,

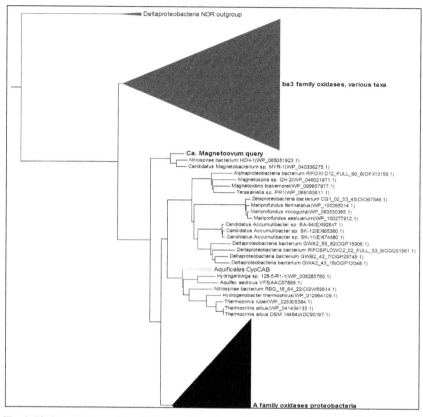

Fig. 4. Phylogenetic tree of CyoB oxidases. The NJ tree derived from a phylogenetically broad blast of the COX1 subunit of the CyoCAB operon of family A cytochrome c oxidases. The tree was rooted by the outgroup of NOR oxidases at the top. Note that the Ca. Magnetoovum (cf. Fig. 3) protein is the most ancestral in CyoCAB clade.

as previously documented for the concatenated family A oxidases present in alphaproteobacteria such as *Methylocella* (Degli Esposti et al. 2014). However, the concentration of these unusual oxidases in Fe-metabolizing organisms— magnetotactic Nitrospirae, magnetotactic alphaproteobacteria and Fe^{II}-oxidizing zetaproteobacteria—suggests an ancient connection with Fe chemolithotrophy. Of note is that the uptake of Fe and its subsequent incorporation in magnetosomes require oxidoreduction reactions involving terminal oxidases (Lefèvre et al. 2013, Li et al. 2014, Kolinko et al. 2016, Lin et al. 2017a), which may also be contributed by CyoCAB operon or B family oxidases. These enzymes may have an intermediate affinity for oxygen compared with the high affinity of cbb3 oxidases that are present in the majority of the above taxa (cf. Li et al. 2014), thereby enabling the removal of oxygen over a wide range of ambient concentrations so as to limit spontaneous Fe^{II} auto-oxidation with respect to its uptake and metabolism in bacterial cells

(Fullerton et al. 2017, Chiu et al. 2017—see also the section on zetaproteobacteria in Chapter four of the book).

Ultimately, the CyoCAB operon may represent an ancient form of cytochrome *c* oxidase that originally evolved in response to increasing levels of oxygen in primordial seas and then survived subsequent evolutionary changes in species living in niches preserving the original conditions. This possibility has implications for the evolution of alphaproteobacteria and, in a broader perspective, of the aerobic metabolism in prokaryotes.

Examples of Common and Well Known Alphaproteobacteria

After having considered the phylogeny and functional evolution of the alphaproteobacteria lineage, a brief review of some of the most common and representative members of the class is presented. Within alphaproteobacteria, there are some of the most widely spread bacteria in Nature, namely *Pelagibacter* and *Wolbachia*. The former thrives in oceans, the latter inside insects and nematodes. Methylotrophic *Methylobacterium* that consumes the plant wall biosynthesis byproduct methanol and most of nitrogen-fixing rhizobia are also successful and widespread members of alphaproteobacteria.

Pelagibacter

Pelagibacter was first reported as the SAR11 cluster organisms having a novel 16S rRNA gene sequence within the alphaproteobacteria, originally obtained from the Sargasso Sea by a culture-independent approach (Giovannoni et al. 1990). SAR11 clade bacteria are the most abundant group of planktonic cells in marine systems, accounting for a third of cells present in surface oceanic waters; in some regions they are up to 50% of the total surface microbial community (Morris et al. 2002). Hence, they were considered as the first candidate for the most successful organism on Earth (Morris et al. 2002, Giovannoni et al. 2005b, Giovannoni and Vergin 2012). *Pelagibacter* constitutes a separate phylogenetic group within the alphaproteobacteria (Williams et al. 2007, Ferla et al. 2013) but, as mentioned earlier in this chapter, its phylogenetic position has been variable (Ferla et al. 2013). Although *Pelagibacter* isolates are genetically diverse, analysis of their genome showed that they share a large core and extensive synteny of various gene clusters, perhaps due to strong genomic streamlining (Grote et al. 2012, Giovannoni et al. 2005b) since they do not have pseudogenes and are small, with an average size of 1.3 Mb. *Pelagibacter* genomes are AT-rich as other bacterial genomes that are reduced, for example those from insect endosymbionts presented in Chapter five.

Pelagibacter cells are small with seemingly efficient surface/volume ratios (Steindler et al. 2011). Their small cell volume may be related to efficient adaptation to low nutrient marine environments, with high affinity transporters and a very efficient metabolism to use organic matter in oligotrophic conditions. Notably, *Pelagibacter* has a proteorhodopsin that pumps protons upon absorption of light and increases the supply of ATP during periods of carbon limitation (Giovannoni et al. 2005a), similar to some marine CFB (see Chapter three). SAR11 strains have genes for glycolysis and C1 metabolism, require reduced sulfur compounds for growth (Tripp et al. 2008) and show a conditional auxotrophy for glycine. Originally, *Pelagibacter* was uncultured, but now it may be cultured (Carini et al. 2013), thus allowing transcriptomic and proteomic analysis that have helped to clarify its metabolism (Smith et al. 2013). *Pelagibacter* has phages, including novel viruses just recently described, which apparently do not have effects on diminishing the *Pelagibacter* populations in the ocean (Mizuno et al. 2016).

Wolbachia

Among the various obligate parasites of the Rickettsiales order, *Wolbachia* organisms are the master manipulators of sex in insects (Werren et al. 2008) and responsible for insect speciation in some cases (Telschow et al. 2005). Estimates on the degree of insect infection by *Wolbachia* range from 16 to 76% (Dobson 2003). Since insects are the dominant animals on Earth, then their associated *Wolbachia* are numerous and widely distributed. The origin of wolbachias in nematodes is unclear. *Wolbachia* symbionts of nematodes provide vitamins and ATP to their hosts and are essential for host development (Darby et al. 2012). Remarkably, part of the eye damage inflicted in humans by the filarial nematode of river blindness is due to *Wolbachia* parasites (Tamarozzi et al. 2011, Gillette-Ferguson et al. 2004). Hence, antibiotics against *Wolbachia* are being used in the fight of river blindness (Hoerauf et al. 2003). The dissemination of *Wolbachia* in mosquitos has been studied (de Oliveira et al. 2017) with increasing interest because the bacteria exert effects on the immune response of the insects (Dieme et al. 2017, Pan et al. 2018) and affect their transmission of virus. For example, mosquitoes infected with *Wolbachia* were less susceptible to infections by dengue (Moreira et al. 2009, Mousson et al. 2012, Geoghegan et al. 2017, Joanne et al. 2017, Terradas et al. 2017), but unfortunately they became more susceptible to West Nile virus infections (Dodson et al. 2014). Thus, the use of *Wolbachia* to prevent viral transmission to humans has been questioned (Popovici et al. 2010).

It is estimated that *Wolbachia* organisms evolved around 500 million years ago, before the divergence of the slow growing *Bradyrhizobium* and the fast growing rhizobia. Only a low number of genes are conserved between rhizobia and wolbachias (E.M-R., unpublished data). None of the conserved genes were located in the *Rhizobium* extrachromosomal replicons such as plasmids, except for one common gene that was located in a rhizobial chromid. This suggests that *Rhizobium* plasmids were genomic innovations that occurred after the split of *Wolbachia* and rhizobia along the evolution of today's alphaprotobacteria. In contrast to the large genomes of rhizobia, normally over six Mb, *Wolbachia* genomes are reduced to small sizes ranging from 0.86 to 1.5 Mb (Ramírez-Puebla et al. 2015). Although maternal inheritance of *Wolbachia* is present in various insects, horizontal transfer of *Wolbachia* transfer between different insects, and maybe also to plants, has been documented. For instance, parasitic wasps may transfer *Wolbachia* between different insect species (Cook and Butcher 1999).

Wolbachia organisms have been taxonomically organized into supergroups, many of which include only one or few strains, with one species formally recognized, *W. pipentis*. Most *Wolbachia* strains remain uncultured and their effects exerted on the hosts are just beginning to be understood. For example, the mechanisms that underlie male killing and the masculinization inhibition in insects that are induced by *Wolbachia* infection have been explored (Fukui et al. 2015). Taxonomically, these intracellular bacteria are classified within the family of Anaplasmataceae, despite recent additions and re-organization to the Rickettsiales order (Szokoli et al. 2016).

Methylobacterium

Known *Methylobacterium* bacteria were originally designated as PPFMs, for pink pigmented facultative methylotrophs (Holland and Polacco 1994). When plant mutants in the urease gene were tested for their enzymatic activity, urease was still detected, indicating the presence of a bacterial protein of resident methylobacteria (Holland and Polacco 1992). Methylobacteria are common endophytes that use methanol as carbon and energy source; methanol is released as a byproduct of plant wall synthesis. They grow selectively in media with added methanol without any other carbon source needed—see Chapter two for more information on methylotrophy. The characteristic pink color of *Methylobacterium* is due to carotenoids, but the genes involved in their biosynthesis are not known (Van Dien et al. 2003). Proteomic studies have revealed rare proteins that are present in methylobacteria and may be involved in the production of carotenoids (Kumar et al. 2014).

Methylobacterium is the type genus of its own family of Methylobacteraceae, which currently contain four other genera, the most widespread of which is *Microvirga* forming root-nodules (Ardley et al. 2012). There are also taxa revealed from metagenomic studies that are associated with the family (Beck et al. 2015).

Methylobacteria produce the plant hormones cytokinins (Lidstrom and Chistoserdova 2002) and also auxins (Ivanova et al. 2001). They thus promote growth in inoculation assay in plants, as well as in mosses (Tani et al. 2012). No specificity in the bacteria-plant interaction has been revealed; however, not all the strains of *Methyobacterium* promote the growth of all plant host species (Tani et al. 2012). In one case of scotch pines, methylobacteria have been found as intracellular symbionts (Koskimaki et al. 2015); they also become intracellular when they harbor *nod* genes (similar to those from rhizobia) and form nitrogen fixing bacteroids in few legume nodules, for example *Methylobacterium nodulans* in different *Crotalaria* species and *Microvirga* in lupinus (Ardley et al. 2012). Intriguingly, methylobacteria that are capable of forming nodules are not pink pigmented (Sy et al. 2001). They probably arose by acquiring *nod* genes from rhizobia via LGT.

Rhizobia

Rhizobium was initially called *Bacterium radiobacter* by Beijerinck. Nowadays, rhizobia is a common term used to collectively define a set of genera that are capable of forming nitrogen-fixing nodules in legumes (Poole et al. 2018). Taxonomically, rhizobia belong to different families of the order Rhizobiales, including Rhizobiaceae (*Rhizobium*), Phyllobacteraceae (*Mesorhizobium*) and Bradyrhizobiaceae. Overall, there are fifteen genera that contain such species (as mentioned in Chapter six), but nodulation is not universally distributed among these species. Nodulation is not uniformly distributed even among strains from a single nodulating species, which may thus contain non-symbiotic isolates (Segovia et al. 1991). The mechanisms underlying nodulation have been studied for a long time and are known to depend on *nod* genes; these genes code for Nod factors, variants of which could explain, to some extent, the specificity observed in the symbiosis of rhizobia with legumes. Nevertheless, there are rhizobial species that have a broader host range (Pueppke and Broughton 1999, Ormeño-Orrillo et al. 2012) and legumes that may be nodulated by many different species (Martinez-Romero 2003). There is phylogenetic incongruence among rhizobial housekeeping genes and symbiosis genes that may be explained by LGT, especially of *nod* and *nif* genes among bacterial species, conferring novel specificities and metabolic capabilities (López-Guerrero et al. 2012, Rogel et al. 2011).

The concept of symbiovar refers to the bacterial specificity for host plants that may be encoded in symbiosis plasmids, chromids or in symbiosis islands, which are contained in the chromosomes, for example in bradyrhizobia (Rogel et al. 2011). Novel symbiovars have been further described for different rhizobial species (Martinez-Hidalgo et al. 2015, 2016, Cobo-Diaz et al. 2014). LGT of symbiosis islands has been evidenced in *Mesorhizobium* (Sullivan and Ronson 1998, Nandasena et al. 2006) and in *Bradyrhizobium* (Parker 2012). *Bradyrhizobium* taxa belong to the phylogenetically deep branching family of Bradyrhizobiaceae, which is metabolically much more versatile than the Rhizobiaceae (one of its genus is a specialized nitrifier while another has the Fe^{II}-oxidizing trait, see Chapter two). Generally, bradyrhizobia have a few, if any, plasmids (Cytryn et al. 2008). In contrast, lateral transfer of plasmids seems to play a key role in the evolution of *Rhizobium*, *Sinorhizobium* and *Agrobacterium*. These taxa, indeed, are the latest diverging not only within the Rhizobiales order, but among all alphaproteobacteria (Williams et al. 2007, Ferla et al. 2013). In some rhizobia, up to around 50% of the genome is contained in extrachromosomal elements (Lopez-Guerrero et al. 2012). Moreover, plasmids have been exchanged among symbiotic *Rhizobium* and pathogenic *Agrobacterium* converting the latter pathogen into a nitrogen-fixing legume symbiont that carries the plant specificity of the donor (Martinez et al. 1987).

Besides forming nodules, rhizobia are soil and root colonizers and they are also found on the surface or inside seeds (Perez-Ramirez et al. 1998, López-López et al. 2010, Mora et al. 2014). In seeds, non-symbiotic rhizobia (López-López et al. 2010) have been found in addition to different symbiotic rhizobial species (Mora et al. 2014). Bacteria in seeds may be tolerant to desiccation, a desired trait when producing industrial rhizobial inoculants for agricultural crops. Rhizobia may colonize non-legume roots in large numbers and even promote their growth in a non-nodule based symbiosis, thus acting as other PGPR bacteria. Some strains of *Rhizobium leguminosarum* have been found associated also with rice roots (Yanni et al. 1997). Conversely, some strains of *Rhizobium phaseoli* are natural endophytes of maize (Gutierrez-Zamora and Martinez Romero 2001, Rosenblueth and Martinez-Romero 2004). In both cases, the rhizobium-cereal symbiosis was seemingly promoted by the succession or association of legume and non-legume crops that occurs in traditional agricultural systems.

Outstandingly, genes for rhizobial nodulation are only expressed in the presence of the host plant, or of its exuded molecules such as root flavonoids. In turn, Nod factors synthesized from enzymes encoded in rhizobial *nod* genes are inducers of plant gene expression (Journet et al. 1994, Smit et al. 2005). Among the genes thus expressed are the early nodulins that participate in a signaling cascade in plants, which culminates with the formation of a novel root or stem structure called nodule (Oldroyd 2013). In a few legumes, nodulation of photosynthetic bradyrhizobia may proceed without the involvement of Nod factors (Giraud et al. 2007). In nodules, rhizobial bacteroids (the term bacteroids indicates differentiated nitrogen fixing symbionts) are fed dicarboxylic acids by the plant, mainly malate or succinate (Ronson et al. 1981, Poole et al. 2018). Nitrogen fixation would not occur outside of the nodules, with few peculiar exceptions. Nitrogen fixation increased when plants were exposed to high CO_2 concentrations (Fischinger et al. 2010) or high light intensity, suggesting that the phosynthate supply (Baysdorfer and Bassham 1985) may limit nitrogen fixation. The respiratory chain of rhizobia changes upon differentiation into bacteroids in the nodules, which contain very low but constant concentrations of oxygen. Concomitant with the over-expression of the *nif* genes for N_2 fixation, bacteroids up-regulate their cbb3 oxidase, originally found in rhizobia as mentioned in Chapter two, while down-regulating other terminal oxidases with lower affinity for O_2 (Degli Esposti and Martinez-Romero 2016). As an example, Fig. 5 shows the respiratory chain of a well-known nodulating rhizobium of peas, *Rhizobium leguminosarum*.

In some cases, especially in legumes grown in temperate climates such as peas or alfalfa, bacteroids achieve a terminal differentiation process and become no longer viable (Mergaert et al. 2006, Montiel et al. 2016). In those cases, plant peptides similar to antimicrobial peptides participate both in the bacteroid differentiation process and in their subsequent loss of viability (Alunni and Gourion 2016). Efficiency of nitrogen fixation is higher in symbiosis associated with terminal differentiation, which seems to be a late acquisition in plant evolution as tropical legumes do not show such a process (Mergaert et al. 2006, Montiel et al. 2016).

Rhizobia deliver fixed nitrogen to plants in the form of ammonium, which is then processed by plant nitrogen assimilation enzymes (Ohyama and Kumazawa

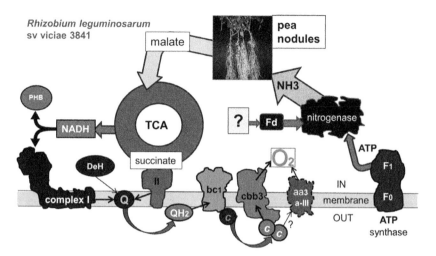

Fig. 5. Respiratory chain of a typical nodulating *Rhizobium*. The picture was modified from Degli Esposti and Martinez-Romero (2016).

1978, Poole et al. 2018). Amino acids have also been reported as rhizobial products that are delivered to plants, but these results turned out to be controversial (Waters et al. 1998, Li et al. 2002, Prell and Poole 2006, Dunn 2015). In some legumes, allantoin and allantoic acid are produced from fixed nitrogen precursors in non-infected legume cells and used for the transport of nitrogen compounds to aerial parts of the host plant (Herridge and Peoples 1990). However, this process occurs only in tropical legumes due to the low solubility of these ureides at low temperatures. Different types of rhizobial secretion systems have a role in the symbiotic interaction (Nelson and Sadowsky 2015). The diversity of nutrients from roots and soils seems to have been a driver to promote rhizobial differentiation and speciation in such a way to avoid bacterial nutrient competition (López-Guerrero et al. 2013). It is not uncommon that different strains from a single species have diverse phenotypic traits with regard to carbon and nitrogen metabolism. A consequence of this is that phenotypes are not good characteristic traits for a species description in taxonomy studies (Ormeño-Orrillo et al. 2013). Furthermore, most of the genes encoding nutritional capabilities in the rhizosphere are encoded in plasmids or genome islands; therefore, they may be easily lost or gained, supporting the concept that gene content constitutes the main variation among rhizobial genomes.

Rhizobia became one of the most studied symbiosis models in microbiology with their own developments in genetic strategies (Olson et al. 1985, Martínez-Romero et al. 1990, Hernandez-Lucas et al. 2002). Such strategies and further functional genomic and metabolic analysis (Degli Esposti and Martinez-Romero 2016) will continue to open wide and far reaching possibilities towards rhizobial biotechnological and commercial applications (Jiménez-Gómez et al. 2018).

Rhodospirillum

Rhodospirillum rubrum is the last example of a common, cultivated member of the alphaproteobacteria class that will be presented here. It is a photosynthetic bacterium of intense red colour living in diverse freshwater environments, including mud and sludge. *R. rubrum* was discovered over a century ago (Molisch 1907) for its characteristic spiral shape, hence *spirillum* from the Latin word for spiral (Imhoff 2005). The red colour derives from pigments associated with the photosynthetic apparatus, which is particularly developed when the bacterium grows under anoxic conditions (Imhoff 2005). *R. rubrum* has been used since the 60s as a model organism for studying anoxygenic photosynthesis, as well as other bioenergetic systems of electron transport. This is due to its extreme metabolic versatility and genomic plasticity in response to oxygen (Imhoff 2005, Munk et al. 2011). *Rhodospirillum* is also the type genus of the Rhodospirillaceae family and its parent Rhodospirillales order (Imhoff 2005), which have been previously mentioned along this chapter. Its best studied species is *R. rubrum*, which was identified early as a prototypic member of the alpha subdivision of proteobacteria (Woese et al. 1984, Woese 1987, Lee et al. 2005, Imhoff 2005). Recently, this organism has been exploited also for the biotechnological production of bio-plastics based upon its storage compounds (Karmann et al. 2016, Heinrich et al. 2016).

R. rubrum is also notable for its production of the rare amino analog of ubiquinone, rhodoquinone (Moore and Folkers 1965)—see also the end of chapter four. This quinone, usually abbreviated as RQ, is present together with ubiquinone in the membranes of *R. rubrum* and its closely related *Pararhodospirillum photometricum*, formerly belonging to the same genus (Imhoff 2005, Lakshmi et al. 2014), and in a few other alphaproteobacteria (Hiraishi and Hoshino 1984, Hiraishi et al. 1995, Okamura et al. 2009a,b, Srinivas et al. 2014). Rhodoquinone has a redox potential much lower than that of ubiquinone and therefore is very suitable for anaerobic electron transport, which can use multiple donors and acceptors in *R.rubrum* owing to the richness of redox enzymes coded in its genome (Imhoff 2005, Munk et al. 2011). However, this bacterium does not possess low affinity terminal oxidases (Fig. 2), thereby explaining its preference for a facultatively anaerobic lifestyle. Its relative *P. photometricum* actually cannot grow above micromolar concentrations of O_2 (Imhoff 2005). *R. rubrum* will be further mentioned in the final part of this chapter for it has been proposed as a possible relative of proto-mitochondria (Esser et al. 2004, Atteia et al. 2009, Thiergart et al. 2012, Abishek et al. 2013).

New Information from Unclassified Alphaproteobacteria of Metagenomes

There has always been a strong human bias in the taxonomic representation of bacterial genomes in NCBI databases due to medical and economic reasons (Mukherjee et al. 2017). This bias has been progressively reduced by ongoing efforts to fill the sequence space of under-represented groups of bacteria that have been discovered in environmental and microbiome studies (Rinke et al. 2013, Schulz et al. 2017, Mukherije et al. 2017, Spang et al. 2017). Current genome databases still maintain a disproportion of taxonomic richness, measured in number of taxa defined on the basis of thresholds for rRNA similarity (Schulz et al. 2017), vs. phylogenetic diversity of proteins and their function (Anton et al. 2014, Louca et al. 2016, Probst et al. 2017, Bergauer et al. 2017, Bowers et al. 2017). However, it appears that the increase in phylogenetic diversity contributed by the rapid expansion of metagenomic assembled genomes (MAGs) is approaching saturation for the major proteobacterial classes of gamma-, beta- and alphaproteobacteria (Schulz et al. 2017, Mukherije et al. 2017, Parks et al. 2017).

Although an actual representation of bacterial taxa in environmental communities may be obscured in metagenomic studies of soil and other solid samples, due to the differential lysis of distinct species, such problems can be minimized using methodological strategies (de Castro et al. 2011, Parks et al. 2017). In the case of alphaproteobacteria, significant gains in phylogenetic diversity have been obtained in metagenomic studies of seawater (Tully et al. 2017, Parks et al. 2017) and groundwater environments (Ananthamaran et al. 2016, Probst et al. 2017, Schulz et al. 2017, Parks et al. 2017). Unexpected aspects of what appear to be primordial features of alphaproteobacteria have recently emerged from the analysis of the increasing wealth of genomic information derived from metagenomic studies (Brindefalk et al. 2011, Degli Esposti et al. 2016, Anantharam et al. 2016, Probst et al. 2017, Tully et al. 2017, Bergauer et al. 2017, Eme et al. 2017). For example, MAG's of unclassified alphaproteobacteria occupy the deepest branches in the great majority of phylogenetic trees of proteins required for the biosynthesis of membrane quinones (Ravcheev and Thiele 2016, Degli Esposti 2017).

The recent work by Parks et al. (2017) has added several novel MAG's, called UBA for Uncultivated Bacteria and Archaea, to both classified and unclassified groups of alphaproteobacteria. Of note, UBA taxa are not equally distributed in the various genera of the family of Rhodospirillaceae which has been discussed previously in this chapter. While only one UBA is associated with the *Rhodospirillum* genus, which previousy had four recognized species (Imhoff 2005), 37 UBA have been associated with the genus of predatory *Micavibrio*, doubling the number of its strains identified before (Parks et al. 2017); this clearly indicates that the phylogenetic diversity of *Micavibrio* is much larger than previously estimated. Although *Micavibrio* is usually considered an unclassified alphaproteobacterium (Davidov et al. 2006), phylogenetic trees of bioenergetic proteins such as cytochrome *b* indicate its close association with some MAG's from groundwater metagenomes (Probst et al. 2017) and members of the Rhodospirillaceae family such as *Nitrospirillum* (see later Chapter eight). A potential association of *Micavibrio*-like environmental rRNA's to Nitrospirae (Dolinšek et al. 2013) may thus be discarded. The discovery of several new *Micavibrio* taxa living in diverse environments (Parks et al. 2017) is of particular interest to the origin of mitochondria, since this predatory alphaproteobacterium has been suggested to be a relative of proto-mitochondria

(Davidov et al. 2006). The issue of the origin of proto-mitochondria is discussed below—see also Chapter eight.

Alphaproteobacteria and Proto-mitochondria

This section will summarize the approaches and findings obtained by studying the bacterial ancestry of mitochondria from the bacterial angle, i.e., the guest in the symbiogenic process leading to the eukaryotic cell (Margulis 1970). Subsequent Chapter eight will deal with the complementary approach of studying mitochondria and their origin from the eukaryotic angle, i.e., from the host side of the symbiosis, focusing on eukaryogenesis itself (Ettema 2016). The fundamental object of the current section is to provide evidence sustaining the concept that proto-mitochondria originated from alphaproteobacteria (Gray et al. 1999, Gray 2012), a concept that has been assumed along this and other chapters of the book. Previous evidence included the similarity in rRNA (Yang et al. 1985, Ferla et al. 2013) and the homology between alphaproteobacterial proteins and those coded in the mitochondrial DNA (mtDNA) of eukaryotes (Andersson et al. 1998, Gray et al. 1999, Emelyanov 2003, Willams et al. 2007, Brindefalk et al. 2011, Gray 2012, Wang and Wu 2015). Additional evidence has been obtained recently by genomic analysis of metabolic traits such as ubiquinone biosynthesis (Pelosi et al. 2016, Degli Esposti 2017), structure of the COX operon (Degli Esposti 2014) and regulatory subunits of ATP synthase that are functionally equivalent in the mitochondrial and alphaproteobacterial enzyme complex (García-Trejo et al. 2016, Mendoza-Hoffmann et al. 2018). This evidence is summarized in Table 1.

The items listed in Table 1 also include the presence of RQ, a membrane quinone initially discovered in *R. rubrum* (see the previous section dedicated to this

Table 1. Proteins and traits that are shared by mitochondria and alphaproteobacteria.

Protein and metabolic trait	Present in other proteobacteria	Reference
UbiL, ubiquinone biosynthesis	NO	Pelosi et al. 2016
CoQ9, ubiquinone biosynthesis	NO	Degli Esposti 2017
AOX, alternative ubiquinol oxidase, energy conservation	a few gammaproteobacteria	Atteia et al. 2004
Cox11/CtaG assembly of cytochrome c oxidase	NO	Degli Esposti et al. 2014, Pittis and Gabaldon 2016
Subunit zeta ATP synthase, energy conservation	NO	Garcia-Trejo et al. 2016, Mendoza-Hoffmann et al. 2018
Supernumerary subunit B17.2 13 kDa of complex 1	NO	Yip et al. 2011
Rhodoquinone, energy conservation	a few beta protecbacteria	Hiraishi and Hoshino 1984, Müller et al. 2012

bacterium) and then found in several metazoans adapted to anaerobiosis (Müller et al. 2012). Of note is that the presence of RQ in eukaryotes and its exclusive distribution in alphaproteobacteria, plus a few betaproteobacteria (which anyway do not share with mitochondria the proteins shown in Table 1), would in principle exclude the origin of proto-mitochondria from any other bacterial group. The fundamental question then becomes: from which lineage of alphaproteobacteria did proto-mitochondria come from?

This question remains open today since there is no consensus on the answer. Various studies (see Thiergart et al. 2012 for an earlier survey) have pinpointed to different orders of alphaproteobacteria that might include close relatives to proto-mitochondria, from Rhizobiales (Yang et al. 1985) to Pelagibacterales (Williams et al. 2007). An exhaustive study based on the phylogeny of the 16S rRNA gene has indicated that proto-mitochondria would be a sister group of contemporary Rickettsiales, both having Pelagibacterales as common ancestors (Ferla et al. 2013). This phylogeny would be consistent with the popular view initiated by Lynn Margulis (1970) that mitochondria derive from bacteria with strong aerobic metabolism (Andersson et al. 1998, Gabaldon and Huynen 2003, Boussau et al. 2004, Brindefalk et al. 2011, Wang and Wu 2015). However, several problems considerably weaken the possibility of a close relatedness between Pelagibacterales, Rickettsiales and proto-mitochondria.

Regarding the closeness between Rickettsiales and mitochondria, the first major problem is biological and is often overlooked, for example by Ferla et al. (2013). The genome of Rickettsiales is fast evolving and rapidly eroding because these bacteria are obligate intracellular parasites, a condition which prevents the evolutionary adaptation to changing environmental conditions that the genome of free-living bacteria must undergo for the species to survive and reproduce. Once established within the primordial eukaryotic cells, proto-mitochondria also became isolated from the outside environment and thus evolved a lifestyle equivalent to that of intracellular parasites (Ku et al. 2015). Indeed, genes encoded by mtDNA have a higher mutation rate than nuclear-encoded genes that end up in mitochondria (Gray et al. 1999, Wang and Wu 2015). Inevitably, this high mutation rate and the compositional bias in gene sequence due to a very high AT content—a genomic feature which is shared by mitochondria, Rickettsiales and Pelagibacterales (Andersson et al. 1998, Gray 2012, Grote et al. 2012, Rodríguez-Ezpeleta and Embley 2012), as well as gammaproteobacterial symbionts of insects (Chapter five)—produce artifacts in phylogenetic trees of either RNA or protein-coding genes (Gray 2012, Rodríguez-Ezpeleta and Embley 2012, Viklund et al. 2012, 2013). The artifacts consist of evident tree distortions technically called long branch attraction, which we have previously encountered in Chapter five in relation to sulfur-oxidizing symbionts. Sequences of either genes or proteins of fast evolving organisms are attracted to each other in any type of phylogenetic tree, producing erroneous sister clades (Philippe and Forterre 1999, Gray 2012, Wang and Wu 2015). A clear example of long-branch attraction between Rickettsiales and mitochondria is seen in the 16S rRNA trees reported by Ferla et al. (2013).

A similar situation applies to Pelagibacterales of the SAR11 clade and Rickettsiales, both of which have genome reduction and high AT content as discussed by Rodríguez-Ezpeleta and Embley (2012). Consequently, even if Pelagibacterales are free-living, and thus would not have the same accelerated mutation rate as endocellular parasites, their unusual genomic features, including very high AT content, likely derive from phenomena of convergent evolution, rather than reflecting evolutionary relatedness (Rodríguez-Ezpeleta and Embley 2012, Degli Esposti et al. 2014). The phylogenetic placement of the group of Pelagibacterales within the alphaproteobacteria has been debated extensively, as mentioned in the Introduction of this chapter (cf. Fig. 1). However, it appears that the concept of their affiliation to Rickettsiales (Williams et al. 2007, Grote et al. 2012, Ferla et al. 2013) has progressively given way to the more realistic possibility that they constitute a highly derived, but not very ancient lineage among alphaproteobacteria (Rodríguez-Ezpeleta and Embley 2012, Degli Esposti et al. 2014, Wang and Wu 2015). Although their taxonomic position remains essentially unresolved, it is becoming more central than originally thought by increasing the phylogenetic diversity of the alphaproteobacterial taxa with which they are compared, as shown by the tree in Fig. 1B. In any case, detailed phylogenetic analysis has progressively indicated that the quest for the closest relatives to mitochondria should point to a direction different from Pelagibacterales and related oceanic taxa, to paraphrase the conclusion by Rodríguez-Ezpeleta and Embley (2012).

This may well apply to a small group of oceanic metagenomic bacteria that Brindefalk et al. (2011) reported to display the closest bioenergetic proteins coded by mtDNA. These bacterial sequences were later identified to cluster with the taxon of alphaproteobacterium HIMB59 (Grote et al. 2012, Viklund et al. 2013), which is currently classified among Pelagibacterales, according to NCBI websites (https://www.ncbi.nlm.nih.gov/Taxonomy/Browser/wwwtax.cgi?id=744985 accessed 9 March 2018).

The study of Brindefalk et al. (2011) represents a paradigm for investigating the closest living relatives of proto-mitochondria by analyzing the sequence of major bioenergetic proteins encoded in mtDNA. This approach was pioneered by the same group of Andersson (Andersson et al. 1998) and later expanded by also including mitochondrial proteins that are now encoded in the nuclear DNA of eukaryotes (Esser et al. 2004, Thiergart et al. 2012, Wang and Wu 2015, Degli Esposti 2016). However, extensive experimentation has clarified that not all mitochondrial proteins have the same capacity of resolving deep phylogenetic relationships, while some are much better than others. These include cytochrome *b*, COX1 and ND5 among mtDNA-encoded proteins (Zardoya and Meyer 1996, Esser et al. 2004, Degli Esposti et al. 2014, Havird and Santos 2014), as well as the Nad7/NuoD/49kDa subunit of complex I (Brindefalk et al. 2011, Degli Esposti 2016) and the cytochrome *c*1 subunit of the bc1 complex (Degli Esposti 2016). Such proteins have strong phylogenetic signal, a technical term which indicates a sequence length above 200 amino acids providing a combination of highly conserved residues required for function with hyper-variable stretches corresponding to external or transmembrane regions, which can undergo large sequence changes without altering the structure-function of the protein. Concatenation of these proteins with other mitochondrial sequences generally does not add significant phylogenetic resolution (Esser et al. 2004), while taking away information from those regions that do not have detailed 3D structures of reference to guide their proper alignment (see Chapter one). Indeed, trees based upon concatenated sequences of proteins often fail to resolve the order of deep branches (Thiergart et al. 2014), a problem of particular severity for the ancestry of mitochondria, which undoubtedly evolved from an early branching group of alphaproteobacteria.

Using concatenations of hybrid sets of mtDNA-coded and nuclear coded mitochondrial proteins with diverse phylogenetic signal, Wang and Wu (2015) have placed proto-mitochondria within the Rickettsiales order. They claimed to have reduced the almost intractable problem of long branch attraction by increasing the phylogenetic depth of their analysis with newly sequenced genomes of Rhodospirillales and Rickettsiales (Wang and Wu 2015). However, the overall set of taxa examined was limited and biased, thereby undermining the conclusion of Wang and Wu (2015). Indeed, if a similar analysis were to be repeated today with all the available genomes of unclassified proteobacteria, the results and consequent conclusion would be different (cf., Fig. 1).

The truth of the matter is that the branching order of mitochondrial proteins vs. their alphaproteobacterial homologs depends upon the phylogenetic signal of the same proteins, the breadth of the taxonomic sampling of both the bacterial and eukaryotic species analyzed and the depth of the rooting taxa. Even small changes in any of these major variables can produce largely different trees, with diverse topologies of the closest nodes to the mitochondrial clade (Rochette et al. 2014, Burki 2014, Derelle et al. 2015). Ultimately, proto-mitochondria arose within the alphaproteobacterial lineage, most likely a few hundred million years after the separation from other proteobacteria and just before the branching of gamma and beta lineages (Degli Esposti 2016). Very few extant taxa can be related to the ancestral alphaproteobacteria that lived just before or at the same time of proto-mitochondria. It thus looks unlikely that their contemporary relatives can be clearly found by even the most sophisticated phylogenetic analysis based upon trees (Müller et al. 2012). The examination of the distribution of metabolic traits among extant species (Fig. 2) provides an alternative source of information, which clearly points to contemporary members of the Rhodospirillaceae family as likely relatives of those ancestral alphaproteobacteria from which proto-mitochondria then arose. Next chapter will verify this possibility further from the angle of the host in the eukaryogenesis process.

To conclude this chapter, the only certainty we currently have regarding the origin of mitochondria from the standpoint of the bacterial symbionts is that proto-mitochondria originated from an early lineage of ancestral alphaproteobacteria. Consequently, the phylogenesis of alphaproteobacteria is intimately linked to the origin of mitochondria. However, phylogenetic trees can provide limited information due to their intrinsic limits and the problems discussed in the final part of the chapter. Alternatively, functional analysis extended to the ever increasing wealth of genomic data does provide valuable information, which so far pinpoints to facultatively anaerobes of the Rhodospirillaceae family as the most likely relatives of proto-mitochondria. This possibility is consistent with the results of early studies (Esser et al. 2004, Thiergart et al. 2012) and in part (regarding the Rhodospirillaceae *Tistrella* and the origin of ubiquinone, cf. Degli Esposti 2017) with those previously reported (Degli Esposti 2014). However, it definitively contrasts with several other studies that suggested a mitochondrial ancestry around or within the Rickettsiales order of intracellular parasites, a proposition that would never be able to answer the logical question: which organism were *Rickettsia*-like proto-mitochondria parasites of?

Acknowledgements

This work was sponsored by grants CONACyT Basic Science 253116 and PAPIIT (UNAM) IN207718 to E.M-R. We thank Francisco Mendoza-Hoffmann and José Garcia-Trejo for enthusiastic discussion on aspects of mitochondrial and bacterial evolution.

References

Abhishek, A., A. Bavishi, A. Bavishi and M. Choudhary. 2011. Bacterial genome chimaerism and the origin of mitochondria. Can. J. Microbiol. 57: 49–61. doi:10.1139/w10-099.

Alunni, B. and B. Gourion. 2016. Terminal bacteroid differentiation in the legume-*Rhizobium* symbiosis: nodule-specific cysteine-rich peptides and beyond. New Phytol. 211: 411–417. doi:10.1111/nph.14025.

Andersson, S.G., A. Zomorodipour, J.O. Andersson, T. Sicheritz-Pontén, U.C. Alsmark, R.M. Podowski et al. 1998. The genome sequence of *Rickettsia prowazekii* and the origin of mitochondria. Nature 396: 133–140.

Anton, B.P., S. Kasif, R.J. Roberts and M. Steffen. 2014. Objective: biochemical function. Front. Genet. 5: 210. doi:10.3389/fgene.2014.00210.

Ardley, J.K., S.E. Parker, R.D. De Meyer, R.D. Trengove, G.W. O'Hara, W.G. Reeve et al. 2012. *Microvirga lupini* sp. nov., *Microvirga lotononidis* sp. nov. and *Microvirga zambiensis* sp. nov. are alphaproteobacterial root-nodule bacteria that specifically nodulate and fix nitrogen with geographically and taxonomically separate legume hosts. Int. J. Syst. Evol. Microbiol. 62: 2579–2588. doi:10.1099/ijs.0.035097-0.

Atteia, A., A. Adrait, S. Brugière, M. Tardif, R. van Lis, O. Deusch et al. 2009. A proteomic survey of *Chlamydomonas reinhardtii* mitochondria sheds new light on the metabolic plasticity of the organelle and on the nature of the alpha-proteobacterial mitochondrial ancestor. Mol. Biol. Evol. 26: 1533–1548. doi:10.1093/molbev/msp068.

Baysdorfer, C. and J.A. Bassham. 1985. Photosynthate supply and utilization in alfalfa: a developmental shift from a source to a sink limitation of photosynthesis. Plant Physiol. 77: 313–317.

Bazylinski, D.A., D.B. Frankel and H.W. Jannasch. 1988. Anaerobic magnetite production by a marine, magnetotactic bacterium. Nature 334: 518–519. doi:10.1038/334518a0.

Bazylinski, D.A., T.J. Williams, C.T. Lefèvre, R.J. Berg, C.L. Zhang, S.S. Bowser et al. 2013a. *Magnetococcus marinus gen. nov., sp. nov.*, a marine, magnetotactic bacterium that represents a novel lineage (*Magnetococcaceae* fam. nov., *Magnetococcales* ord. nov.) at the base of the Alphaproteobacteria. Int. J. Syst. Evol. Microbiol. 63: 801–808. doi:10.1099/ijs.0.038927-0.

Bazylinski, D.A., T.J. Williams, C.T. Lefèvre, D. Trubitsyn, J. Fang, T.J. Beveridge et al. 2013b. *Magnetovibrio blakemorei gen. nov., sp. nov.*, a magnetotactic bacterium (Alphaproteobacteria:

Rhodospirillaceae) isolated from a salt marsh. Int. J. Syst. Evol. Microbiol. 63: 1824–1833. doi:10.1099/ijs.0.044453-0.

Beck, D.A., T.L. McTaggart, U. Setboonsarng, A. Vorobev, L. Goodwin, N. Shapiro et al. 2015. Multiphyletic origins of methylotrophy in Alphaproteobacteria, exemplified by comparative genomics of Lake Washington isolates. Environ. Microbiol. 17: 547–554. doi:10.1111/1462-2920.12736.

Bergauer, K., A. Fernandez-Guerra, J.A.L. Garcia, R.R. Sprenger, R. Stepanauskas, M.G. Pachiadaki et al. 2017. Organic matter processing by microbial communities throughout the Atlantic water column as revealed by metaproteomics. Proc. Natl. Acad. Sci. USA. pii: 201708779. doi:10.1073/pnas.1708779115.

Biegel, E., S. Schmidt, J.M. González and V. Müller. 2011. Biochemistry, evolution and physiological function of the Rnf complex, a novel ion-motive electron transport complex in prokaryotes. Cell Mol. Life Sci. 68: 613–634. doi:10.1007/s00018-010-0555-8.

Boratyn, G.M., A.A. Schäffer, R. Agarwala, S.F. Altschul, D.J. Lipman and T.L. Madden. 2012. Domain enhanced lookup time accelerated BLAST. Biol. Direct. 7: 12.

Boussau, B., E.O. Karlberg, A.C. Frank, B.A. Legault and S.G. Andersson. 2004. Computational inference of scenarios for alpha-proteobacterial genome evolution. Proc. Natl. Acad. Sci. USA 101: 9722–9727.

Bowers, R.M., N.C. Kyrpides, R. Stepanauskas, M. Harmon-Smith, D. Doud, T.B.K. Reddy et al. 2017. Minimum information about a single amplified genome (MISAG) and a metagenome-assembled genome (MIMAG) of bacteria and archaea. Nat. Biotechnol. 35: 725–731. doi:10.1038/nbt.3893.

Brindefalk, B., T.J. Ettema, J. Viklund, M. Thollesson and S.G. Andersson. 2011. A phylometagenomic exploration of oceanic alphaproteobacteria reveals mitochondrial relatives unrelated to the SAR11 clade. PLoS One 6(9): e24457. doi:10.1371/journal.pone.0024457.

Brown, C.T., M.R. Olm, B.C. Thomas and J.F. Banfield. 2016. Measurement of bacterial replication rates in microbial communities. Nat. Biotechnol. 34: 1256–1263. doi:10.1038/nbt.3704.

Burki, F. 2014. The eukaryotic tree of life from a global phylogenomic perspective. Cold Spring Harb Perspect Biol. 6: a016147.

Carini, P., L. Steindler, S. Beszteri and S.J. Giovannoni. 2013. Nutrient requirements for growth of the extreme oligotroph 'Candidatus Pelagibacter ubique' HTCC1062 on a defined medium. ISME J. 7: 592–602. doi:10.1038/ismej.2012.122.

Chiu, B.K., S. Kato, S.M. McAllister, E.K. Field and C.S. 2017. Novel pelagic iron-oxidizing zetaproteobacteria from the chesapeake bay oxic-anoxic transition zone. Front. Microbiol. 8: 1280. doi:10.3389/fmicb.2017.01280.

Cobo-Díaz, J.F., P. Martínez-Hidalgo, A.J. Fernández-González, E. Martínez-Molina, N. Toro, E. Velázquez et al. 2014. The endemic *Genista versicolor* from Sierra Nevada National Park in Spain is nodulated by putative new *Bradyrhizobium* species and a novel symbiovar (sierranevadense). Syst. Appl. Microbiol. 37: 177–185. doi:10.1016/j.syapm.2013.09.008.

Cook, J.M. and R.D.J. Butcher. 1999. The transmission and effects of *Wolbachia* bacteria in parasitoids. Researches Population Ecol. 41: 15–28.

Cytryn, E.J., S. Jitacksorn, E. Giraud and M.J. Sadowsky. 2008. Insights learned from pBTAi1, a 229-kb accessory plasmid from *Bradyrhizobium* sp. strain BTAi1 and prevalence of accessory plasmids in other *Bradyrhizobium* sp. strains. ISME J. 2: 158–170. doi:10.1038/ismej.2007.105.

Darby, A.C., S.D. Armstrong, G.S. Bah, G. Kaur, M.A. Hughes, S.M. Kay et al. 2012. Analysis of gene expression from the *Wolbachia* genome of a filarial nematode supports both metabolic and defensive roles within the symbiosis. Genome Res. 22: 2467–2477. doi:10.1101/gr.138420.112.

Davidov, Y., D. Huchon, S.F. Koval and E. Jurkevitch. 2006. A new alpha-proteobacterial clade of Bdellovibrio-like predators: Implications for the mitochondrial endosymbiotic theory. Environ. Microbiol. 8: 2179–2188.

de Castro, A.P., B.F. Quirino, H. Allen, L.L. Williamson, J. Handelsman and R.H. Krüger. 2011. Construction and validation of two metagenomic DNA libraries from Cerrado soil with high clay content. Biotechnol. Lett. 33: 2169–2175. doi:10.1007/s10529-011-0693-6.

Degli Esposti, M. 2014. Bioenergetic evolution in proteobacteria and mitochondria. Genome Biol. Evol. 6: 3238–3251.

Degli Esposti, M., B. Chouaia, F. Comandatore, E. Crotti, D. Sassera, P.M. Lievens et al. 2014. Evolution of mitochondria reconstructed from the energy metabolism of living bacteria. PLoS One. 9: e96566. doi: 10.1371/journal.pone.0096566.

Degli Esposti, M. 2016. Late mitochondrial acquisition, really? Genome Biol. Evol. 8: 2031–2035. doi:10.1093/gbe/evw130.

Degli Esposti, M., D. Cortez, L. Lozano, S. Rasmussen, H.B. Nielsen and E. Martinez Romero. 2016. Alpha proteobacterial ancestry of the [Fe-Fe]-hydrogenases in anaerobic eukaryotes. Biol. Direct. 11: 34. doi:10.1186/s13062-016-0136-3.

Degli Esposti, M. 2017. A journey across genomes uncovers the origin of ubiquinone in cyanobacteria. Genome Biol. Evol. 9: 3039–3053. doi: 10.1093/gbe/evx225.

Degli Esposti, M. and E. Martinez-Romero. 2017. The functional microbiome of arthropods. PLoS One. 12: e0176573. doi: 10.1371/journal.pone.0176573.

de Oliveira, S., D.A.M. Villela, F.B.S. Dias, L.A. Moreira and R. Maciel de Freitas. 2017. How does competition among wild type mosquitoes influence the performance of *Aedes aegypti* and dissemination of *Wolbachia pipientis*? PLoS Negl. Trop. Dis. 11(10): e0005947. doi:10.1371/journal.pntd.0005947.

Derelle, R., G. Torruella, V. Klimeš, H. Brinkmann, E. Kim, Č. Vlček et al. 2015. Bacterial proteins pinpoint a single eukaryotic root. Proc. Natl. Acad. Sci. USA 112: E693–E699. doi:10.1073/pnas.1420657112.

Dieme, C., B. Rotureau and C. Mitri. 2017. Microbial pre-exposure and vectorial competence of *Anopheles* mosquitoes. Front. Cell. Infect. Microbiol. doi.org/10.3389/fcimb.2017.00508.

Dobson, S.L. 2003. *Wolbachia pipientis*: Impotent by association. *In*: Bourtzis, K. and T.A. Miller (eds.). Insect Symbiosis. Series: Contemporary Topics in Entomology. Boca Raton, USA. CRC Press.

Dodson, B.L., G.L. Hughes, O. Paul, A.C. Matacchiero, L.D. Kramer and J.L. Rasgon. 2014. *Wolbachia* enhances West Nile virus (WNV) infection in the mosquito *Culex tarsalis*. PLoS Negl. Trop. Dis. 8(7): e2965. doi:10.1371/journal.pntd.0002965.

Dolinšek, J., I. Lagkouvardos, W. Wanek, M. Wagner and H. Daims. 2013. Interactions of nitrifying bacteria and heterotrophs: identification of a *Micavibrio*-like putative predator of *Nitrospira* spp. Appl. Environ. Microbiol. 79: 2027–2037. doi:10.1128/AEM.03408-12.

Dunn, M.F. 2015. Key roles of microsymbiont amino acid metabolism in rhizobia-legume interactions. Crit. Rev. Microbiol. 41: 411–51. doi:10.3109/1040841X.2013.856854.

Duquesne, K., V. Prima, B. Ji, Z. Rouy, C. Médigue, E. Talla et al. 2012. Draft genome sequence of the purple photosynthetic bacterium *Phaeospirillum molischianum* DSM120, a particularly versatile bacterium. J. Bacteriol. 194: 3559–3560. doi:10.1128/JB.00605-12.

Dziuba, M., V. Koziaeva, D. Grouzdev, E. Burganskaya, R. Baslerov, T. Kolganova et al. 2016. *Magnetospirillum caucaseum* sp. nov., *Magnetospirillum marisnigri* sp. nov. and *Magnetospirillum moscoviense* sp. nov., freshwater magnetotactic bacteria isolated from three distinct geographical locations in European Russia. Int. J. Syst. Evol. Microbiol. 66: 2069–2077. doi:10.1099/ijsem.0.000994.

Eme, L., A. Spang, J. Lombard, C.W. Stairs and T.J.G. Ettema. 2017. Archaea and the origin of eukaryotes. Nat. Rev. Microbiol. 15: 711–723. doi:10.1038/nrmicro.2017.133.

Emelyanov, V.V. 2003. Common evolutionary origin of mitochondrial and rickettsial respiratory chains. Arch. Biochem. Biophys. 420: 130–141.

Esser, C., N. Ahmadinejad, C. Wiegand, C. Rotte, F. Sebastiani, G. Gelius-Dietrich et al. 2004. A genome phylogeny for mitochondria among alpha-proteobacteria and a predominantly eubacterial ancestry of yeast nuclear genes. Mol. Biol. Evol. 21: 1643–1660.

Esser, C., W. Martin and T. Dagan. 2007. The origin of mitochondria in light of a fluid prokaryotic chromosome model. Biol. Lett. 3: 180–184.

Ettema, T.J. 2016. Evolution: Mitochondria in the second act. Nature 531: 39–40.

Ferla, M.P., J.C. Thrash, S.J. Giovannoni and W.M. Patrick. 2013. New rRNA gene-based phylogenies of the Alphaproteobacteria provide perspective on major groups, mitochondrial ancestry and phylogenetic instability. PLoS One 8: e83383. doi:10.1371/journal.pone.0083383.

Fischinger, S.A., M. Hristozkova, Z.A. Mainassara and J. Schulze. 2010. Elevated CO_2 concentration around alfalfa nodules increases N_2 fixation. J. Exp. Bot. 61: 121–130. doi:10.1093/jxb/erp287.

Fitzpatrick, D.A., C.J. Creevey and J.O. McInerney. 2006. Genome phylogenies indicate a meaningful alpha-proteobacterial phylogeny and support a grouping of the mitochondria with the Rickettsiales. Mol. Biol. Evol. 23: 74–85.

Frankel, R.B., D.A. Bazylinski, M.S. Johnson and B.L. Taylor. 1997. Magneto-aerotaxis in marine coccoid bacteria. Biophys. J. 73: 994–1000.

Fukui, T., M. Kawamoto, K. Shoji, T. Kiuchi, S. Sugano, T. Shimada et al. 2015. The endosymbiotic bacterium *Wolbachia* selectively kills male hosts by targeting the masculinizing gene. PLoS Pathog. 11(7): e1005048. doi:10.1371/journal.ppat.1005048.

Fullerton, H., K.W. Hager, S.M. McAllister and C.L. Moyer. 2017. Hidden diversity revealed by genome-resolved metagenomics of iron-oxidizing microbial mats from Lō'ihi Seamount, Hawai'i. ISME J. 11: 1900–1914. doi:10.1038/ismej.2017.40.

Gao, Y., B. Meyer, L. Sokolova, K. Zwicker, M. Karas, B. Brutschy et al. 2012. Heme-copper terminal oxidase using both cytochrome c and ubiquinol as electron donors. Proc. Natl. Acad. Sci. USA 109: 3275–3280. doi:10.1073/pnas.1121040109.

Gabaldón, T. and M.A. Huynen. 2003. Reconstruction of the proto-mitochondrial metabolism. Science 301: 609.

García-Trejo, J.J., M. Zarco-Zavala, F. Mendoza-Hoffmann, E. Hernández-Luna, R. Ortega and G. Mendoza-Hernández. 2016. The Inhibitory mechanism of the ζ subunit of the F1FO-ATPase nanomotor of paracoccus denitrificans and related α-proteobacteria. J. Biol. Chem. 291: 538–546. doi:10.1074/jbc.M115.688143.

Garrity, G.M., J.A. Bell and T. Lilburn. 2005. Class I. Alphaproteobacteria *class. nov.* Bergey's Manual® of Systematic Bacteriology. Springer US 1–574.

Geoghegan, V., K. Stainton, S.M. Rainey, T.H. Ant, A.A. Dowle, T. Larson et al. 2017. Perturbed cholesterol and vesicular trafficking associated with dengue blocking in *Wolbachia*-infected *Aedes* aegypti cells. Nat. Commun. 8(1): 526. doi:10.1038/s41467-017-00610-8.

Georgiades, K., M.A. Madoui, P. Le, C. Robert and D. Raoult. 2011. Phylogenomic analysis of *Odyssella thessalonicensis* fortifies the common origin of Rickettsiales, *Pelagibacter ubique* and *Reclimonas americana* mitochondrion. PLoS One 6(9): e24857. doi:10.1371/journal. pone.0024857.

Gillette-Ferguson, I., A.G. Hise, H.F. McGarry, J. Turner, A. Esposito, Y. Sun et al. 2004. *Wolbachia*-induced neutrophil activation in a mouse model of ocular onchocerciasis (river blindness). Infect. Immun. 72: 5687–5692.

Giovannoni, S.J., T.B. Britschgi, C.L. Moyer and K.G. Field. 1990. Genetic diversity in Sargasso Sea bacterioplankton. Nature 345: 60–63.

Giovannoni, S.J., L. Bibbs, J.C. Cho, M.D. Stapels, R. Desiderio, K.L. Vergin et al. 2005a. Proteorhodopsin in the ubiquitous marine bacterium SAR11. Nature 438: 82–85.

Giovannoni, S.J., H.J. Tripp, S. Givan, M. Podar, K.L. Vergin, D. Baptista et al. 2005b. Genome streamlining in a cosmopolitan oceanic bacterium. Science 309: 1242–1245.

Giovannoni, S.J. and K.L. Vergin. 2012. Seasonality in ocean microbial communities. Science 335: 671–676. doi:10.1126/science.1198078.

Giraud, E., L. Moulin, D. Vallenet, V. Barbe, E. Cytryn, J.C. Avarre et al. 2007. Legumes symbioses: absence of Nod genes in photosynthetic bradyrhizobia. Science 316: 1307–1312.

Glaeser, J. and J. Overmann. 1999. Selective enrichment and characterization of *Roseospirillum parvum*, gen. nov. and sp. nov., a new purple nonsulfur bacterium with unusual light absorption properties. Arch. Microbiol. 171: 405–416.

Gray, M.W., G. Burger and B.F. Lang. 1999. Mitochondrial evolution. Science 283: 1476–1481.

Gray, M.W. 2012. Mitochondrial evolution. Cold Spring Harb. Perspect. Biol. 4: a011403. doi:10.1101/cshperspect.a011403.

Gray, M.W. 2015. Mosaic nature of the mitochondrial proteome: Implications for the origin and evolution of mitochondria. Proc. Natl. Acad. Sci. USA 112: 10133–10138.

Grote, J., J.C. Thrash, M.J. Huggett, Z.C. Landry, P. Carini, S.J. Giovannoni et al. 2012. Streamlining and core genome conservation among highly divergent members of the SAR11 clade. MBio. 3. pii: e00252-12. doi:10.1128/mBio.00252-12.

Gupta, R.S. 2005. Protein signatures distinctive of alpha proteobacteria and its subgroups and a model for alpha-proteobacterial evolution. Crit. Rev. Microbiol. 31: 101–135.

Gupta, R.S. and A. Mok. 2007. Phylogenomics and signature proteins for the Alphaproteobacteria and its main groups. BMC Microbiol. 7: 106.

Gutiérrez-Zamora, M.L. and E. Martínez-Romero. 2001. Natural endophytic association between *Rhizobium etli* and maize (*Zea mays* L.). J. Biotechnol. 91: 117–126.

Han, H., J. Hemp, L.A. Pace, H. Ouyang, K. Ganesan, J.H. Roh et al. 2011. Adaptation of aerobic respiration to low O_2 environments. Proc. Natl. Acad. Sci. USA 108: 14109–14114. doi:10.1073/pnas.1018958108.

Han, S.B., Y. Su, J. Hu, R.J. Wang, C. Sun, D. Wu et al. 2016. *Terasakiella brassicae sp. nov.*, isolated from the wastewater of a pickle-processing factory, and emended descriptions of *Terasakiella pusilla* and the genus *Terasakiella*. Int. J. Syst. Evol. Microbiol. 66: 1807–1812. doi:10.1099/ijsem.0.000946.

Havird, J.C. and S.R. Santos. 2014. Performance of single and concatenated sets of mitochondrial genes at inferring metazoan relationships relative to full mitogenome data. PLoS One 9: e84080. doi:10.1371/journal.pone.0084080.

Heinrich, D., M. Raberg, P. Fricke, S.T. Kenny, L. Morales-Gamez, R.P. Babu et al. 2016. Synthesis gas (Syngas)-derived medium-chain-length polyhydroxyalkanoate synthesis in engineered *Rhodospirillum rubrum*. Appl. Environ. Microbiol. 82: 6132–6140.

Hernández-Lucas, I., P. Mavingui, T. Finan, P. Chain and E. Martínez-Romero. 2002. *In vivo* cloning strategy for *Rhizobium* plasmids. Biotechniques 33: 782–788.

Herridge, D.F. and M.B. Peoples. 1990. Ureide assay for measuring nitrogen fixation by nodulated soybean calibrated by N methods. Plant Physiol. 93: 495–503.

Hiraishi, A. and Y. Hoshino. 1984. Distribution of rhodoquinone in Rhodospirillaceae and its taxonomic implication. J. Gen. Appl. Microbiol. 30: 435–448.

Hiraishi, A., K. Urata and T. Satoh. 1995. A new genus of marine budding phototrophic bacteria, *Rhodobium* gen. nov., which includes *Rhodobium orientis sp. nov.* and *Rhodobium marinum* comb. nov. Int. J. Syst. Bacteriol. 45: 226–234.

Hoerauf, A., S. Mand, L. Volkmann, M. Büttner, Y. Marfo-Debrekyei, M. Taylor et al. 2003. Doxycycline in the treatment of human onchocerciasis: Kinetics of *Wolbachia* endobacteria reduction and of inhibition of embryogenesis in female *Onchocerca* worms. Microbes Infect. 5: 261–273.

Holland, M.A. and J.C. Polacco. 1992. Urease-null and hydrogenase-null phenotypes of a phylloplane bacterium reveal altered nickel metabolism in two soybean mutants. Plant Physiol. 98: 942–948.

Holland, M.A. and J.C. Polacco. 1994. PPFMs and other covert contaminants: Is there more to plant physiology than just plant? Ann. Rev. Plant Physiol. Plant Mol. Biol. 45: 197–209.

Imhoff, J.F. 2005. Genus I. *Rhodospirillum* Molisch 1907, 24AL emend. Imhoff, Petri and Süling 1998, 796. pp. 1–6. *In*: Brenner, D.J., N.R. Krieg, J.T. Staley and G.M. Garrity (eds.). Bergey's Mannual of Systematic Bacteriology, 2nd edn, Vol. 2, Part C, New York: Springer.

Imhoff, J.F., R. Petri and J. Süling. 1998. Reclassification of species of the spiral-shaped phototrophic purple non-sulfur bacteria of the alpha-Proteobacteria: Description of the new genera *Phaeospirillum* gen. nov., *Rhodovibrio* gen. nov., *Rhodothalassium* gen. nov. and *Roseospira* gen. nov. as well as transfer of *Rhodospirillum fulvum* to *Phaeospirillum fulvum* comb. nov., of *Rhodospirillum molischianum* to *Phaeospirillum molischianum* comb. nov., of *Rhodospirillum salinarum* to *Rhodovibrio salexigens*. Int. J. Syst. Bacteriol. 48: 793–798.

Ivanova, E.G., N.V. Doronina and Y.A. Trotsenko. 2001. Aerobic methylobacteria are capable of synthesizing auxins. Microbiology 70: 392–397.

Iwata, S., C. Ostermeier, B. Ludwig and H. Michel. 1995. Structure at 2.8 A resolution of cytochrome c oxidase from *Paracoccus denitrificans*. Nature 376: 660–669.

Ji, B., S.D. Zhang, P. Arnoux, Z. Rouy, F. Alberto, N. Philippe et al. 2014. Comparative genomic analysis provides insights into the evolution and niche adaptation of marine *Magnetospira* sp. QH-2 strain. Environ. Microbiol. 16: 525–544. doi:10.1111/1462-2920.12180.

Ji, B., S.D. Zhang, W.J. Zhang, Z. Rouy, F. Alberto, C.L. Santini et al. 2017. The chimeric nature of the genomes of marine magnetotactic coccoid-ovoid bacteria defines a novel group of Proteobacteria. Environ. Microbiol. 19: 1103–1119. doi:10.1111/1462-2920.13637.

Jiménez-Gómez, A., J.D. Flores-Félix, P. García-Fraile, P.F. Mateos, E. Menéndez, E. Velázquez et al. 2018. Probiotic activities of *Rhizobium laguerreae* on growth and quality of spinach. Sci. Rep. 8: 295. doi:10.1038/s41598-017-18632-z.

Joanne, S., I. Vythilingam, B.T. Teoh, C.S. Leong, K.K. Tan, M.L. Wong et al. 2017. Vector competence of Malaysian *Aedes albopictus* with and without *Wolbachia* to four dengue virus serotypes. Trop. Med. Int. Health. 22: 1154–1165. doi:10.1111/tmi.12918.

Jogler, C., Wanner, G., Kolinko, S., Niebler, M., Amann, R., Petersen, N., et al. 2011. Conservation of proteobacterial magnetosome genes and structures in an uncultivated member of the deep-branching *Nitrospira phylum*. Proc. Natl. Acad. Sci. USA 108: 1134–1139. doi:10.1073/pnas.1012694108.

Journet, E.P., M. Pichon, A. Dedieu, F. de Billy, G. Truchet and D.G. Barker. 1994. *Rhizobium meliloti* Nod factors elicit cell-specific transcription of the ENOD12 gene in transgenic alfalfa. Plant J. 6: 241–249.

Karmann, S., S. Follonier, M. Bassas-Galia, S. Panke and M. Zinn. 2016. Robust at-line quantification of poly(3-hydroxyalkanoate) biosynthesis by flow cytometry using a BODIPY 493/503-SYTO 62 double-staining. J. Microbiol. Methods 131: 166–171. doi:10.1016/j.mimet.2016.10.003.

Kolinko, S., C. Jogler, E. Katzmann, G. Wanner, J. Peplies and D. Schüler. 2012. Single-cell analysis reveals a novel uncultivated magnetotactic bacterium within the candidate division OP3. Environ. Microbiol. 14: 1709–1721. doi:10.1111/j.1462-2920.2011.02609.

Kolinko, S., M. Richter, F.O. Glöckner, A. Brachmann and D. Schüler. 2016. Single-cell genomics of uncultivated deep-branching magnetotactic bacteria reveals a conserved set of magnetosome genes. Environ. Microbiol. 18: 21–37. doi:10.1111/1462-2920.12907.

Koskimäki, J.J., A.M. Pirttilä, E.L. Ihantola, O. Halonen and A.C. Frank. 2015. The intracellular Scots pine shoot symbiont *Methylobacterium extorquens* DSM13060 aggregates around the host nucleus and encodes eukaryote-like proteins. MBio. 6. pii: e00039-15. doi:10.1128/mBio.00039-15.

Ku, C., S. Nelson-Sathi, M. Roettger, F.L. Sousa, P.J. Lockhart, D. Bryant et al. 2015. Endosymbiotic origin and differential loss of eukaryotic genes. Nature 524: 427–432. doi:10.1038/nature14963.

Kumar, D., A.K. Mondal, A.K. Yadav and D. Dash. 2014. Discovery of rare protein-coding genes in model methylotroph *Methylobacterium extorquens* AM1. Proteomics 14: 2790–2794. doi:10.1002/pmic.201400153. Epub 2014 Oct 2.

Kurland, C.G. and S.G. Andersson. 2000. Origin and evolution of the mitochondrial proteome. Microbiol. Mol. Biol. Rev. 64: 786–820.

Lakshmi, K.V., B. Divyasree, E.V. Ramprasad, Ch. Sasikala and Ch.V. Ramana. 2008. Reclassification of *Rhodospirillum photometricum* Molisch 1907, *Rhodospirillum sulfurexigens* Anil Kumar et al. 2008 and *Rhodospirillum oryzae* Lakshmi et al. 2013 in a new genus, *Pararhodospirillum* gen. nov., as *Pararhodospirillum photometricum* comb. nov., *Pararhodospirillum sulfurexigens* comb. nov. and *Pararhodospirillum oryzae* comb. nov., respectively, and emended description of the genus *Rhodospirillum*. Int. J. Syst. Evol. Microbiol. 64: 1154–1159. doi:10.1099/ijs.0.059147-0.

Leão, P., L.C. Teixeira, J. Cypriano, M. Farina, F. Abreu, D.A. Bazylinski et al. 2016. North-seeking magnetotactic Gammaproteobacteria in the southern hemisphere. Appl. Environ. Microbiol. 82: 5595–5602. doi:10.1128/AEM.01545-16.

Lee, K.B., C.T. Liu, Y. Anzai, H. Kim, T. Aono and H. Oyaizu. 2005. The hierarchical system of the 'Alphaproteobacteria': description of *Hyphomonadaceae* fam. nov., *Xanthobacteraceae* fam. nov. and *Erythrobacteraceae* fam. nov. Int. J. Syst. Evol. Microbiol. 55: 1907–1919.

Lefèvre, C.T., F. Abreu, M.L. Schmidt, U. Lins, R.B. Frankel, B.P. Hedlund et al. 2010. Moderately thermophilic magnetotactic bacteria from hot springs in Nevada. Appl. Environ. Microbiol. 76: 3740–3743. doi:10.1128/AEM.03018-09.

Lefèvre, C.T., N. Viloria, M.L. Schmidt, M. Pósfai, R.B. Frankel and D.A. Bazylinski. 2012. Novel magnetite-producing magnetotactic bacteria belonging to the Gammaproteobacteria. ISME J. 6: 440–450. doi:10.1038/ismej.2011.97.

Lefèvre, C.T. and D.A. Bazylinski. 2013. Ecology, diversity, and evolution of magnetotactic bacteria. Microbiol. Mol. Biol. Rev. 77: 497–526. doi:10.1128/MMBR.00021-13.

Lefèvre, C.T., D. Trubitsyn, F. Abreu, S. Kolinko, L.G. de Almeida, A.T. de Vasconcelos et al. 2013. Monophyletic origin of magnetotaxis and the first magnetosomes. Environ. Microbiol. 15: 2267–2274. doi:10.1111/1462-2920.12097.

Li, Y., R. Parsons, D.A. Day and F.J. Bergersen. 2002. Reassessment of major products of N_2 fixation by bacteroids from soybean root nodules. Microbiology 148: 1959–1966.

Li, Y., O. Raschdorf, K.T. Silva and D. Schüler. 2014. The terminal oxidase cbb3 functions in redox control of magnetite biomineralization in *Magnetospirillum gryphiswaldense*. J. Bacteriol. 196: 2552–2562. doi:10.1128/JB.01652-14.

Lidstrom, M.E. and L. Chistoserdova. 2002. Plants in the pink: Cytokinin production by *Methylobacterium*. J. Bacteriol. 184: 1818.

Lin, W., G.A. Paterson, Q. Zhu, Y. Wang, E. Kopylova, Y. Li et al. 2017a. Origin of microbial biomineralization and magnetotaxis during the Archean. Proc. Natl. Acad. Sci. USA 114: 2171–2176. doi:10.1073/pnas.1614654114.

Lin, W., Y. Pan and D.A. Bazylinski. 2017b. Diversity and ecology of and biomineralization by magnetotactic bacteria. Environ. Microbiol. Rep. 9: 345–356. doi:10.1111/1758-2229.12550.

Lin, W., G.A. Paterson, Q. Zhu, Y. Wang, E. Kopylova, Y. Li et al. 2017c. Reply to Wang and Chen: An ancient origin of magnetotactic bacteria. Proc. Natl. Acad. Sci. USA 114: E5019–E5020. doi:10.1073/pnas.1707301114.

López-Guerrero, M.G., E. Ormeño-Orrillo, J.L. Acosta, A. Mendoza-Vargas, M.A. Rogel, M.A. Ramírez et al. 2012. Rhizobial extrachromosomal replicon variability, stability and expression in natural niches. Plasmid. 68: 149–158. doi:10.1016/j.plasmid.2012.07.002.

López-Guerrero, M.G., E. Ormeño-Orrillo, M. Rosenblueth, J. Martinez-Romero and E. Martïnez-Romero. 2013. Buffet hypothesis for microbial nutrition at the rhizosphere. Front. Plant Sci.4: 188. doi:10.3389/fpls.2013.00188.

López-López, A., M.A. Rogel, E. Ormeño-Orrillo, J. Martínez-Romero and E. Martínez-Romero. 2010. *Phaseolus vulgaris* seed-borne endophytic community with novel bacterial species such as *Rhizobium endophyticum* sp. nov. Syst Appl. Microbiol. 33: 322–327. doi:10.1016/j.syapm.2010.07.005.

Louca, S., L.W. Parfrey and M. Doebeli. 2016. Decoupling function and taxonomy in the global ocean microbiome. Science 353: 1272–1277. doi:10.1126/science.aaf4507.

Luo, H. 2015. Evolutionary origin of a streamlined marine bacterioplankton lineage. ISME J. 9: 1423–1433. doi:10.1038/ismej.2014.227.

Marchler-Bauer, A., M.K. Derbyshire, N.R. Gonzales, S. Lu, F. Chitsaz, L.Y. Geer et al. 2015. CDD: NCBI's conserved domain database. Nucleic Acids Res. 43(Database issue): D222–D226. doi:10.1093/nar/gku1221.

Margulis, L. 1970. Origin of Eukaryotic Cells. Yale University Press, New Haven, CT.

Martin, W.F. and M. Müller. 1998. The hydrogen hypothesis for the first eukaryote. Nature 392: 37–41.

Martin, W.F. 2017. Physiology, anaerobes, and the origin of mitosing cells 50 years on. J. Theor. Biol. pii: S0022-5193(17)30004-8. doi:10.1016/j.jtbi.2017.01.004.

Martin, W.F., D.A. Bryant and J.T. Beatty. 2018. A physiological Perspective on the Origin and Evolution of Photosynthesis 42: 205–231. doi:10.1093/femsre/fux056.

Martínez, E., R. Palacios and F. Sánchez. 1987. Nitrogen-fixing nodules induced by *Agrobacterium tumefaciens* harboring *Rhizobium* phaseoli plasmids. J. Bacteriol. 169: 2828–2834.

Martínez-Hidalgo, P., J.D. Flores-Félix, E. Menéndez, R. Rivas, L. Carro, P.F. Mateos et al. 2015. *Cicer canariense*, an endemic legume to the Canary Islands, is nodulated in mainland Spain by fast-growing strains from symbiovar trifolii phylogenetically related to *Rhizobium leguminosarum*. Syst. Appl. Microbiol. 38: 346–350. doi:10.1016/j.syapm.2015.03.011.

Martínez-Hidalgo, P., J. Pérez-Yépez, E. Velázquez, R. Pérez-Galdona, E. Martínez-Molina and M. León-Barrios. 2016. Symbiovar loti genes are widely spread among *Cicer canariense* mesorhizobia, resulting in symbiotically effective strains. Plant Soil 398: 25–33. https://doi.org/10.1007/s11104-015-2614-2.

Martinez-Romero, E. 2003. Diversity of *Rhizobium-Phaseolus vulgaris* symbiosis: Overview and perspectives. Plant Soil. 252: 11–23.

Martínez-Romero, E., D. Romero and R. Palacios. 1990. The *Rhizobium* Genome. Crit. Rev. Plant Sci. 9: 59–93.

Mathuriya, A.S. 2016. Magnetotactic bacteria: nanodrivers of the future. Crit. Rev. Biotechnol. 36: 788–802. doi:10.3109/07388551.2015.

Matsutani, M., K. Fukushima, C. Kayama, M. Arimitsu, H. Hirakawa, H. Toyama et al. 2014. Replacement of a terminal cytochrome c oxidase by ubiquinol oxidase during the evolution of acetic acid bacteria. Biochim. Biophys. Acta. 1837: 1810–1820. doi:10.1016/j.bbabio.2014.05.355.

Mendoza-Hoffmann, F., Á. Pérez-Oseguera, M.Á. Cevallosn, M. Zarco-Zavala, R. Ortega, C. Peña-Segura et al. 2018. The Biological Role of the ζ Subunit as Unidirectional Inhibitor of the F(1)F(O)-ATPase of Paracoccus denitrificans. Cell Rep. 22: 1067–1078. doi:10.1016/j.celrep.2017.12.106.

Mergaert, P., T. Uchiumi, B. Alunni, G. Evanno, A. Cheron, O. Catrice et al. 2006. Eukaryotic control on bacterial cell cycle and differentiation in the *Rhizobium*-legume symbiosis. Proc. Natl. Acad. Sci. USA 103: 5230–5235.

Mizuno, C.M., R. Ghai, A. Saghaï, P. López-García and F. Rodriguez-Valera. 2016. Genomes of abundant and widespread viruses from the deep ocean. MBio. 7(4). pii: e00805-16. doi:10.1128/mBio.00805-16.

Molisch, H. 1907. Die Purpurbakterien Nach Neuen Untersuchungen. Jena: G. Fischer.

Montiel, J., A. Szűcs, I.Z. Boboescu, V.D. Gherman, E. Kondorosi and A. Kereszt. 2016. Terminal bacteroid differentiation is associated with variable morphological changes in legume species belonging to the inverted repeat-lacking clade. Mol. Plant-Microbe Interact 29: 210–219. https://doi.org/10.1094/MPMI-09-15-0213-R.

Moore, H.W. and K. Folkers. 1965. Coenzyme Q. LXII. structure and synthesis of rhodoquinone, a natural aminoquinone of the coenzyme Q group. J. Am. Chem. Soc. 87: 1409–1410.

Mora, Y., R. Díaz, C. Vargas-Lagunas, H. Peralta, G. Guerrero, A. Aguilar et al. 2014. Nitrogen-fixing rhizobial strains isolated from common bean seeds: phylogeny, physiology, and genome analysis. Appl. Environ. Microbiol. 80: 5644–5654. doi:10.1128/AEM.01491-14.

Moreira, L.A., I. Iturbe-Ormaetxe, J.A. Jeffery, G. Lu, A.T. Pyke, L.M. Hedges et al. 2009. A *Wolbachia* symbiont in *Aedes aegypti* limits infection with dengue, Chikungunya, and Plasmodium. Cell. 139: 1268–1278. doi:10.1016/j.cell.2009.11.042.

Morillo, V., F. Abreu, A.C. Araujo, L.G. de Almeida, A. Enrich-Prast, M. Farina et al. 2014. Isolation, cultivation and genomic analysis of magnetosome biomineralization genes of a new genus of South- seeking magnetotactic cocci within the Alphaproteobacteria. Front. Microbiol. 5: 72. doi:10.3389/fmicb.2014.00072.

Morris, R.L. and T.M. Schmidt. 2013. Shallow breathing: Bacterial life at low O(2). Nat. Rev. Microbiol. 11: 205–212. doi:10.1038/nrmicro2970.

Morris, R.M., M.S. Rappé, S.A. Connon, K.L. Vergin, W.A. Siebold, C.A., Carlson et al. 2002. SAR11 clade dominates ocean surface bacterioplankton communities. Nature 420: 806–810.

Mousson, L., K. Zouache, C. Arias-Goeta, V. Raquin, P. Mavingui and A.B. Failloux. 2012. The native *Wolbachia* symbionts limit transmission of dengue virus in *Aedes albopictus*. PLoS Negl. Trop. Dis. 6(12): e1989. doi:10.1371/journal.pntd.0001989.

Mukherjee, S., R. Seshadri, N.J. Varghese, E.A. Eloe-Fadrosh, J.P. Meier-Kolthoff, M. Göker et al. 2017. 1,003 reference genomes of bacterial and archaeal isolates expand coverage of the tree of life. Nat. Biotechnol. 35: 676–683. doi:10.1038/nbt.3886.

Müller, M. and W. Martin. 1999. The genome of Rickettsia prowazekii and some thoughts on the origin of mitochondria and hydrogenosomes. Bioessays 21: 377–381.

Müller, M., M. Mentel, J.J. van Hellemond, K. Henze, C. Woehle, S.B. Gould et al. 2012. Biochemistry and evolution of anaerobic energy metabolism in eukaryotes. Microbiol. Mol. Biol. Rev. 76: 444–495.

Munk, A.C., A. Copeland, S. Lucas, A. Lapidus, T.G. Del Rio, K. Barry et al. 2011. Complete genome sequence of *Rhodospirillum rubrum* type strain (S1). Stand. Genomic Sci. 4: 293–302. doi:10.4056/sigs.1804360.

Nandasena, K.G., G.W. O'hara, R.P. Tiwari and J.G. Howieson. 2006. Rapid in situ evolution of nodulating strains for *Biserrula pelecinus* L. through lateral transfer of a symbiosis island from the original mesorhizobial inoculant. Appl. Environ. Microbiol. 72: 7365–7367.

Nelson, M.S. and M.J. Sadowsky. 2015. Secretion systems and signal exchange between nitrogen-fixing rhizobia and legumes. Front. Plant Sci. 6: 491. doi:10.3389/fpls.2015.00491.

Nielsen, H.B., M. Almeida, A.S. Juncker, S. Rasmussen, J. Li, S. Sunagawa et al. 2014. Identification and assembly of genomes and genetic elements in complex metagenomic samples without using reference genomes. Nat. Biotechnol. 32: 822–828. doi:10.1038/nbt.2939.

Ohyama, T. and K. Kumazawa. 1978. Incorporation of ^{15}N into various nitrogenous compounds in intact soybean nodules after exposure to ^{15}N$_2$ gas. Soil Sci. Plant Nutrition. 24: 525–533. doi. org/10.1080/00380768.1978.10433132.

Okamura, K., T. Hisada, T. Kanbe and A. Hiraishi. 2009a. *Rhodovastum atsumiense gen.* nov., sp. nov., a phototrophic alphaproteobacterium isolated from paddy soil. J. Gen. Appl. Microbiol. 55: 43–50.

Okamura, K., T. Kanbe and A. Hiraishi. 2009b. *Rhodoplanes serenus* sp. nov., a purple non-sulfur bacterium isolated from pond water. Int. J. Syst. Evol. Microbiol. 59: 531–535. doi:10.1099/ ijs.0.000174-0.

Oldroyd, G.E. 2013. Speak, friend, and enter: Signalling systems that promote beneficial symbiotic associations in plants. Nat. Rev. Microbiol. 11: 252–263. doi:10.1038/nrmicro2990.

Olson, E.R., M.J. Sadowsky and D.P.S. Verma. 1985. Identification of genes involved in the *Rhizobium*-legume symbiosis by Mu-dI (Kan, lac)-generated transcription fusions. Nature Biotechnol. 3: 143–149. doi:10.1038/nbt0285-143.

Ormeño-Orrillo, E., P. Menna, L.G. Almeida, F.J. Ollero, M.F. Nicolás, E. Pains Rodrigues et al. 2012. Genomic basis of broad host range and environmental adaptability of *Rhizobium tropici* CIAT 899 and *Rhizobium* sp. PRF 81 which are used in inoculants for common bean (*Phaseolus vulgaris* L.). BMC Genomics 13: 735. doi:10.1186/1471-2164-13-735.

Ormeño-Orrillo, E. and E. Martínez-Romero. 2013. Phenotypic tests in *Rhizobium* species description: an opinion and (a sympatric speciation) hypothesis. Syst. Appl. Microbiol. 36: 145–147. doi:10.1016/j.syapm.2012.11.009.

Pan, X., A. Pike, D. Joshi, G. Bian, M.J. McFadden, P. Lu et al. 2018. The bacterium *Wolbachia* exploits host innate immunity to establish a symbiotic relationship with the dengue vector mosquito *Aedes aegypti*. ISME J. 12: 277–288. doi:10.1038/ismej.2017.174.

Parker, M.A. 2012. Legumes select symbiosis island sequence variants in *Bradyrhizobium*. Mol. Ecol. 21: 1769–1778. doi:10.1111/j.1365-294X.2012.05497.x.

Parks, D.H., C. Rinke, M. Chuvochina, P.A. Chaumeil, B.J. Woodcroft, P.N. Evans et al. 2017. Recovery of nearly 8,000 metagenome-assembled genomes substantially expands the tree of life. Nat. Microbiol. doi:10.1038/s41564-017-0012-7.

Pelosi, L., A.L. Ducluzeau, L. Loiseau, F. Barras, D. Schneider, I. Junier et al. 2016. Evolution of Ubiquinone Biosynthesis: Multiple Proteobacterial Enzymes with Various Regioselectivities To Catalyze Three Contiguous Aromatic Hydroxylation Reactions. mSystems 1. pii: e00091-16.

Pérez-Ramírez, N.O., M.A. Rogel, E. Wang, J.Z. Castellanos and E. Martínez-Romero. 1998. Seeds of *Phaseolus vulgaris* bean carry *Rhizobium etli*. FEMS Microbiol. Ecol. 26: 289–296.

Permentier, H.P., S. Neerken, K.A. Schmidt, J. Overmann and J. Amesz. 2000. Energy transfer and charge separation in the purple non-sulfur bacterium *Roseospirillum parvum*. Biochim. Biophys. Acta. 1460: 338–345.

Philippe, H. and P. Forterre. 1999. The rooting of the universal tree of life is not reliable. J. Mol. Evol. 49: 509–523.

Pittis, A.A. and T. Gabaldón. 2016. Late acquisition of mitochondria by a host with chimaeric prokaryotic ancestry. Nature 531: 101–104.

Poole, P., V. Ramachandran and J. Terpolilli. 2018. Rhizobia: From saprophytes to endosymbionts. Nature Rev. Microbiol. doi:10.1038/nrmicro.2017.171.

Popovici, J., L.A. Moreira, A. Poinsignon, I. Iturbe-Ormaetxe, D. McNaughton and S.L. O'Neill. 2010. Assessing key safety concerns of a *Wolbachia*-based strategy to control dengue transmission by *Aedes mosquitoes*. Mem. Inst. Oswaldo Cruz. 105: 957–964.

Prell, J. and P. Poole. 2006. Metabolic changes of rhizobia in legume nodules. Trends Microbiol. 14: 161–168.

Probst, A.J., C.J. Castelle, A. Singh, C.T. Brown, K. Anantharaman, I. Sharon et al. 2017. Genomic resolution of a cold subsurface aquifer community provides metabolic insights for novel microbes adapted to high CO(2) concentrations. Environ. Microbiol. 19: 459–474. doi:10.1111/1462-2920.13362.

Pueppke, S.G. and W.J. Broughton. 1999. *Rhizobium* sp. strain NGR234 and *R. fredii* USDA257 share exceptionally broad, nested host ranges. Mol. Plant Microbe Interact. 12: 293–318.

Radzi Noor, M. and T. Soulimane. 2012. Bioenergetics at extreme temperature: *Thermus thermophilus* ba(3)- and caa(3)-type cytochrome c oxidases. Biochim. Biophys. Acta. 1817: 638–649. doi:10.1016/j.bbabio.2011.08.004.

Ramírez-Puebla, S.T., L.E. Servín-Garcidueñas, E. Ormeño-Orrillo, A. Vera-Ponce de León, M. Rosenblueth, L. Delaye et al. 2015. Species in *Wolbachia*? Proposal for the designation of '*Candidatus* Wolbachia bourtzisii', '*Candidatus* Wolbachia onchocercicola', '*Candidatus* Wolbachia blaxteri', '*Candidatus* Wolbachia brugii', '*Candidatus* Wolbachia taylori', '*Candidatus* Wolbachia collembolicola' and '*Candidatus* Wolbachia multihospitum' for the different species within *Wolbachia* supergroups. Syst. Appl. Microbiol. 38: 390–399. doi:10.1016/j.syapm.2015.05.005.

Ricci, J.N., A.J. Michel and D.K. Newman. 2015. Phylogenetic analysis of HpnP reveals the origin of 2-methylhopanoid production in Alphaproteobacteria. Geobiology 13: 267–277. doi:10.1111/gbi.12129.

Rinke, C., P. Schwientek, A. Sczyrba, N.N. Ivanova, I.J. Anderson, J.F. Cheng et al. 2013. Insights into the phylogeny and coding potential of microbial dark matter. Nature 499: 431–437. doi:10.1038/nature12352.

Rochette, N.C., C. Brochier-Armanet and M. Gouy. 2014. Phylogenomic test of the hypotheses for the evolutionary origin of eukaryotes. Mol. Biol. Evol. 31: 832–845.

Rodríguez-Ezpeleta, N. and T.M. Embley. 2012. The SAR11 group of alpha-proteobacteria is not related to the origin of mitochondria. PLoS One 7(1): e30520. doi:10.1371/journal.pone.0030520.

Rogel, M.A., E. Ormeño-Orrillo and E. Martinez Romero. 2011. Symbiovars in rhizobia reflect bacterial adaptation to legumes. Syst. Appl. Microbiol. 34: 96–104. doi:10.1016/j.syapm.2010.11.015.

Ronson, C.W., P. Lyttleton and J.G. Robertson. 1981. C(4)-dicarboxylate transport mutants of *Rhizobium trifolii* form ineffective nodules on *Trifolium repens*. Proc. Natl. Acad. Sci. USA 78: 4284–4288.

Rosenblueth, M. and E. Martínez-Romero. 2004. *Rhizobium etli* maize populations and their competitiveness for root colonization. Arch. Microbiol. 181: 337–344.

Schübbe, S., T.J. Williams, G. Xie, H.E. Kiss, T.S. Brettin, D. Martinez et al. 2009. Complete genome sequence of the chemolithoautotrophic marine magnetotactic *Coccus* strain MC-1. Appl. Environ. Microbiol. 75: 4835–4852. doi:10.1128/AEM.02874-08.

Schultz, J.E. and P.F. Weaver. 1982. Fermentation and anaerobic respiration by *Rhodospirillum rubrum* and *Rhodopseudomonas capsulata*. J. Bacteriol. 149: 181–190.

Schulz, F., E.A. Eloe-Fadrosh, R.M. Bowers, J. Jarett, T. Nielsen, N.N. Ivanova et al. 2017. Towards a balanced view of the bacterial tree of life. Microbiome 5: 140. doi:10.1186/s40168-017-0360-9.

Segovia, L., D. Piñero, R. Palacios and E. Martínez-Romero. 1991. Genetic structure of a soil population of nonsymbiotic *Rhizobium leguminosarum*. Appl. Environ. Microbiol. 57: 426–433.

Smit, P., J. Raedts, V. Portyanko, F. Debellé, C. Gough, T. Bisseling et al. 2005. NSP1 of the GRAS protein family is essential for rhizobial Nod factor-induced transcription. Science 308: 1789–1791.

Smith, D.P., J.C. Thrash, C.D. Nicora, M.S. Lipton, K.E. Burnum-Johnson, P. Carini et al. 2013. Proteomic and transcriptomic analyses of "*Candidatus* Pelagibacter ubique" describe the first PII-independent response to nitrogen limitation in a free-living Alphaproteobacterium. MBio. 4(6): e00133-12. doi:10.1128/mBio.00133-12.

Spang, A., E.F. Caceres and T.J.G. Ettema. 2017. Genomic exploration of the diversity, ecology, and evolution of the archaeal domain of life. Science 357. pii: eaaf3883. doi:10.1126/science.aaf3883.

Srinivas, A., Ch. Sasikala and Ch.V. Ramana. 2014. *Rhodoplanes oryzae* sp. nov., a phototrophic alphaproteobacterium isolated from the rhizosphere soil of paddy. Int. J. Syst. Evol. Microbiol. 64: 2198–2203. doi:10.1099/ijs.0.063347-0.

Steindler, L., M.S. Schwalbach, D.P. Smith, F. Chan and S.J. Giovannoni. 2011. Energy starved *Candidatus* Pelagibacter ubique substitutes light-mediated ATP production for endogenous carbon respiration. PLoS One 6(5): e19725. doi:10.1371/journal.pone.0019725.

Sullivan, J.T. and C.W. Ronson. 1998. Evolution of rhizobia by acquisition of a 500-kb symbiosis island that integrates into a phe-tRNA gene. Proc. Natl. Acad. Sci. USA 95: 5145–5149.

Sy, A., E. Giraud, P. Jourand, N. Garcia, A. Willems, P. de Lajudie et al. 2001. Methylotrophic *Methylobacterium* bacteria nodulate and fix nitrogen in symbiosis with legumes. J. Bacteriol. 183: 214–220.

Szokoli, F., M. Castelli, E. Sabaneyeva, M. Schrallhammer, S. Krenek, T.G. Doak et al. 2016. Disentangling the taxonomy of Rickettsiales and description of two novel symbionts ("*Candidatus* Bealeia paramacronuclearis" and "*Candidatus* Fokinia cryptica") sharing the cytoplasm of the ciliate protist *Paramecium biaurelia*. Appl. Environ. Microbiol. 82: 7236–7247.

Tamarozzi, F., A. Halliday, K. Gentil, A. Hoerauf, E. Pearlman and M.J. Taylor. 2011. Onchocerciasis: the role of *Wolbachia* bacterial endosymbionts in parasite biology, disease pathogenesis, and treatment. Clin. Microbiol. Rev. 24: 459–468. doi:10.1128/CMR.00057-10.

Tani, A., Y. Takai, I. Suzukawa, M. Akita, H. Murase and K. Kimbara. 2012. Practical application of methanol-mediated mutualistic symbiosis between *Methylobacterium* species and a roof greening moss, *Racomitrium japonicum*. PLoS One 7: e33800. doi:10.1371/journal.pone.0033800.

Tanimura, Y. and Y. Fukumori. 2000. Heme-copper oxidase family structure of *Magnetospirillum magnetotacticum* 'cytochrome a1'-like hemoprotein without cytochrome c oxidase activity. J. Inorg. Biochem. 82: 73–78.

Telschow, A., P. Hammerstein and J.H. Werren. 2005. The effect of *Wolbachia* versus genetic incompatibilities on reinforcement and speciation. Evolution. 59: 1607–1619.

Terradas, G., S.L. Allen, S.F. Chenoweth and E.A. McGraw. 2017. Family level variation in *Wolbachia*-mediated dengue virus blocking in *Aedes aegypti*. Parasit. Vectors 10: 622. doi:10.1186/s13071-017-2589-3.

Thiergart, T., G. Landan, M. Schenk, T. Dagan, and W.F. Martin. 2012. An evolutionary network of genes present in the eukaryote common ancestor polls genomes on eukaryotic and mitochondrial origin. Genome Biol. Evol. 4: 466–485.

Thiergart, T., G. Landan and W.F. Martin. 2014. Concatenated alignments and the case of the \ disappearing tree. BMC Evol. Biol. 14: 266. doi:10.1186/s12862-014-0266-0.

Thrash, J.C., A. Boyd, M.J. Huggett, J. Grote, P. Carini, R.J. Yoder et al. 2011. Phylogenomic evidence for a common ancestor of mitochondria and the SAR11 clade. Sci. Rep. 1: 13. doi:10.1038/srep00013.

Tripp, H.J., J.B. Kitner, M.S. Schwalbach, J.W. Dacey, L.J. Wilhelm and S.J. Giovannoni. 2008. SAR11 marine bacteria require exogenous reduced sulphur for growth. Nature 452: 741–744. doi:10.1038/nature06776.

Trubitsyn, D., F. Abreu, F.B. Ward, T. Taylor, M. Hattori, S. Kondo et al. 2016. Draft genome sequence of *Magnetovibrio blakemorei* Strain MV-1, a marine vibrioid magnetotactic bacterium. Genome Announc. 4(6). pii: e01330-16. doi:10.1128/genomeA.01330-16.

Trüper, H.G. and N. Pfennig. 1981. Characterization and identification of the anoxygenic phototrophic bacteria. pp. 299–312. *In*: Starr, M.P., H. Stolp, H.G. Trüper, A. Balows and H.G. Schlegel (eds.). The Prokaryotes. Berlin-Heidelberg–New York Springer-Verlag.

Tully, B.J., R. Sachdeva, E.D. Graham and J.F. Heidelberg. 2017. 290 metagenome-assembled genomes from the Mediterranean Sea: a resource for marine microbiology. Peer J. 5: e3558. doi:10.7717/peerj.3558.

Uebe, R. and D. Schüler. 2016. Magnetosome biogenesis in magnetotactic bacteria. Nat. Rev. Microbiol. 14: 621–637. doi:10.1038/nrmicro.2016.99. Review.

Van Dien, S.J., C.J. Marx, B.N. O'Brien and M.E. Lidstrom. 2003. Genetic characterization of the carotenoid biosynthetic pathway in *Methylobacterium extorquens* AM1 and isolation of a colorless mutant. Appl. Environ. Microbiol. 69: 7563–7566.

Viklund, J., T.J. Ettema and S.G. Andersson. 2012. Independent genome reduction and phylogenetic reclassification of the oceanic SAR11 clade. Mol. Biol. Evol. 29: 599–615. doi:10.1093/molbev/msr203.

Viklund, J., J. Martijn, T.J. Ettema and S.G. Andersson. 2013. Comparative and phylogenomic evidence that the alphaproteobacterium HIMB59 is not a member of the oceanic SAR11 clade. PLoS One 8(11): e78858. doi:10.1371/journal.pone.0078858.

Wang, S. and Y. Chen. 2017. Origin of magnetotaxis: Vertical inheritance or horizontal transfer? Proc. Natl. Acad. Sci. USA 114: E5016–E5018. doi:10.1073/pnas.1706937114.

Wang, Z. and M. Wu. 2015. An integrated phylogenomic approach toward pinpointing the origin of mitochondria. Sci. Rep. 5: 7949. doi:10.1038/srep07949.

Waters, J.K., B.L. Hughes, 2nd, L.C. Purcell, K.O. Gerhardt, T.P. Mawhinney and D.W. Emerich. 1998. Alanine, not ammonia, is excreted from N_2-fixing soybean nodule bacteroids. Proc. Natl. Acad. Sci. USA 95: 12038–12042.

Werren, J.H., L. Baldo and M.E. Clark. 2008. Wolbachia: Master manipulators of invertebrate biology. Nat. Rev. Microbiol. 6: 741–751. doi:10.1038/nrmicro1969.

Williams, K.P., B.W. Sobral and A.W. Dickerman. 2007. A robust species tree for the Alphaproteobacteria. J. Bacteriol. 189: 4578–4586.

Woese, C.R., E. Stackebrandt, W.G. Weisburg, B.J. Paster, M.T. Madigan, C.M.R. Fowler et al. 1984. The phylogeny of purple bacteria: The alpha subdivision. Syst. Appl. Microbiol. 5: 315–326.

Woese, C.R. 1987. Bacterial evolution. Microbiol. Rev. 51: 221–271.

Wu, M., L.V. Sun, J. Vamathevan, M. Riegler, R. Deboy, J.C. Brownlie et al. 2004. Phylogenomics of the reproductive parasite *Wolbachia pipientis* wMel: a streamlined genome overrun by mobile genetic elements. PLoS Biol. 2: E69.

Yang, D., Y. Oyaizu, H. Oyaizu, G.J. Olsen and C.R. Woese. 1985. Mitochondrial origins. Proc. Natl. Acad. Sci. USA 82: 4443–4447.

Yanni, Y.G., R.Y. Rizk, V. Corich, A. Squartini, K. Ninke, S. Philip-Hollingsworth et al. 1997. Natural endophytic association between *Rhizobium leguminosarum* bv. trifolii and rice roots and assessment of its potential to promote rice growth. Plant Soil. 194: 99–114.

Yip, C.Y., M.E. Harbour, K. Jayawardena, I.M. Fearnley and L.A. Sazanov. 2011. Evolution of respiratory complex I: "supernumerary" subunits are present in the alpha-proteobacterial enzyme. J. Biol. Chem. 286: 5023–5033. doi:10.1074/jbc.M110.194993.

Zardoya, R., A. Meyer. 1996. Phylogenetic performance of mitochondrial protein-coding genes in resolving relationships among vertebrates. Mol. Biol. Evol. 13(7): 933–42.

Zeytuni, N., S. Cronin, C.T. Lefèvre, P. Arnoux, D. Baran, Z. Shtein et al. 2015. MamA as a model protein for structure-based insight into the evolutionary origins of magnetotactic bacteria. PLoS One. 10(6): e0130394. doi:10.1371/journal.pone.0130394.

8

From Proto-Mitochondria to the Mitochondrial Organelles of Our Cells

Mauro Degli Esposti

~~~~~~

The Mitochondrial Organelles of Our Cells

This final chapter of the book presents mitochondria as we know them, namely as the bioenergy-producing organelles of our cells. How did proto-mitochondria from alphaproteobacteria became these organelles that are indispensable for the health of the majority of our cells is the central theme that will be developed along the chapter, mainly from the perspective of the host of the symbiotic event that led to the genesis of eukaryotic cells, usually called eukaryogenesis. After detailed presentation of the mitochondrial organelles (Fig. 1), the chapter will discuss the approach of studying the origin of mitochondria from the eukaryotic angle, namely from the host side of the symbiosis (Ettema 2016). The transformation of free-living alphaproteobacteria into proto-mitochondria and their evolution into permanent organelles within eukaryotic cells is one major transition in evolution and therefore a fundamental subject of research and conjectures in biology. Since Lynn Margulis re-introduced the concept of eukaryotic evolution by symbiosis with bacteria that evolved into mitochondria (Sagan 1967, Margulis 1970, Gray 2012, Martin et al. 2015), a large body of literature has accumulated, which has been further expanded last year on the occasion of the 50th anniversary of the seminal paper by Sagan (1967), previous maiden name of Margulis (Lazcano and Peretó 2016, Lane 2017, Martin 2017a).

Center for Genomic Sciences, UNAM Campus de Cuernavaca, Cuernavaca, 62130 Morelos, Mexico.
Email: mauro1italia@gmail.com

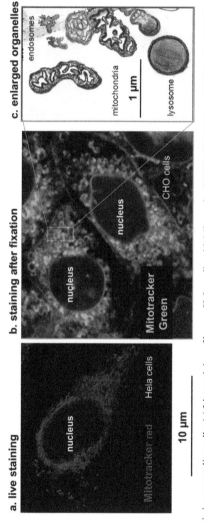

Fig. 1. Mitochondria in mammalian cells. (a) Live staining of human HeLa cells with MitotrackerRED©. (b) Staining of CHO cells with MitotrackerGREEN© after fixation onto coverslips with a standard protocol containing paraformaldehyde. Images were obtained with confocal microscopy with a x40 objective as in a. (c) Drawing of the membrane organelles concentrated in the perinuclear region of the cells in b.

The mitochondrial organelle defines eukaryotic cells (Lane 2006, Lane and Martin 2010) and thus it is at least as old as the first proto-eukaryotic organism, LECA - Last Eukaryotic Common Ancestor (Burki 2014, Roger et al. 2017). Among all cellular compartments in eukaryotic cells, only mitochondria have retained structural membrane components, RNA elements and a circular DNA chromosome that are directly related to bacteria. To provide an updated and informative presentation of mitochondria, their distinctive characters are grouped here in three major categories: cytology, including morphological and structural features of the organelles; physiology, encompassing the fundamental functional properties of the organelles; and phylogeny, which puts together the various ways the origin of mitochondria has been studied recently, predominantly from the eukaryotic angle (see Chapter seven for the complementary approach from the bacterial angle).

In 40 years of research on diverse aspects of mitochondria, the author has accumulated considerable experience and lots of information on the subject discussed in this chapter; therefore, several statements in what follows do not need citation of the work by other scientists. The opinions expressed here are also based on a deep knowledge of mitochondria that has been acquired in diverse fields, from bioenergetics to cell death and molecular phylogeny. Hence, the reader should not expect a neutral, dubitative position on hot issues that surround, in particular, the origin of mitochondria. This is going to be Mauro Degli Esposti's version of the origin of mitochondria.

Cytology—Morphological and Structural Features of Mitochondria

Mitochondria are easily recognized in cytological images by their shape, network-like distribution and double membrane structure, generally showing invaginations of the inner membrane that form the characteristic cristae. The structure, density and morphology of the cristae vary considerably in protists and algae with respect to the classical images of mitochondrial in mammalian cells. Figure 1 shows some such images obtained by the author using live staining with a red fluorescent dye that accumulates in mitochondria in response to their membrane potential (Fig. 1 left). The staining is concentrated around the nucleus, the peri-nuclear region in which several other membrane organelles cluster (Fig. 1, right). Most cytological studies are undertaken with cells or tissues slices that are fixed before mounting on the slides, a process that rapidly kills cells and consequently alters the fine morphology of mitochondria, because they have lost their membrane potential following treatment with membrane-destroying fixing agents. Under these conditions, mitochondria can be stained by specific dyes that do not respond to membrane potential such as Mitotracker Green®, as shown in the central image of Fig. 1. The morphological transition of mitochondria from the elongated shape in live cells to the donut shape in fixed cells is typically observed under conditions of stress and apoptosis (Mannella 2006, Youle and van der Bliek 2012). In healthy cells, such a transition (sometimes called 'spaghetti to orecchiette' by culinary analogy) can be reversible and is modulated by the nuclear-encoded system of fusion and fission of mitochondria (Mannella 2006, Youle and van der Bliek 2012).

We now know the intimate details of the dynamic process underlying the cytological features of mitochondrial organelles thanks to a wealth of studies in the field of cell death, which was revolutionized twenty years ago by the discovery that mitochondria play a central role in the pathways of programmed cell death. The intriguing question then arose: why are mitochondria targeted by the inborn programs of cell suicide in animal cells? One hypothesis put forward by the author and other scientists is based upon the evolutionary scenario of eukaryogenesis: the invading alphaproteobacterial guest was initially treated by the host as a parasite, to eliminate via molecular tools specifically attacking its membrane, which contains lipids such as cardiolipin that are now present only in mitochondria (Degli Esposti 2002, Cristea and Degli Esposti 2004).

Cardiolipin and its metabolites have been traditionally considered to play a role in shaping the membrane curvature and consequently the morphology of mitochondrial cristae. However, we now know major protein factors that are responsible for the formation and morphology of these cristae. Recent advances have clarified the molecular aspects of cristae formation and stabilization, which were previously considered morphological features exhibiting wide variability and limited correlations with functional aspects of the organelle. Studies in yeast led to the discovery of a multiprotein system that regulates not only the formation of the cristae, but also the contact sites between the inner and outer membrane of mitochondria (von der Malsburg et al. 2011). This system is called MICOS, which stands for mitochondrial contact site and cristae organizing system (von der Malsburg et al. 2011, Muñoz-Gómez et al. 2015, 2017, Wollweber et al. 2017). Phylogenetic analysis of the major MICOS proteins has shown that the system is present in all eukaryotic lineages which have aerobic mitochondria exhibiting cristae with different morphologies, while it is absent in eukaryotic organisms that have mitochondria-related organelles (MRO) such as mitosomes (see later the part on mitochondrial function), which are involved in anaerobic metabolism or have just residual functions (Muñoz-Gómez et al. 2015). Moreover, the largest and apparently most conserved proteins of eukaryotic MICOS contains a C-terminal domain which is also found in a variety of alphaproteobacteria, especially Rhizobiales, Rhodobacterales and Rhodospirillales (Muñoz-Gómez et al. 2015, 2017). Several organisms of these orders have been found earlier to have so-called intra-cytoplasmic membranes (ICM) due to invagination of the bacterial inner membrane, in some cases clearly resembling mitochondrial cristae (reviewed in Degli Esposti 2014).

Cristae-like ICM are particularly striking in methane and nitrite oxidizing bacteria, which are present in both alpha- and gammaproteobacteria (Degli Esposti 2014), thereby contradicting the claim that MICOS homologs and the membrane invaginations they produce would be specifically present in alphaproteobacteria (Muñoz-Gómez et al. 2015, 2017). Nevertheless, one intriguing piece of evidence emerging from the work of Muñoz-Gómez et al. (2015, 2017) is that both Rickettsiales and Pelagibacterales do not have MICOS 'homologs' and correspondingly do not exhibit ICM or other types of membrane invaginations. This may be correlated with a different structure of their ATP synthase complex, since these alphaproteobacteria also lack the zeta subunit that is present in the rest of the class (Mendoza-Hoffmann and Garcia-Trejo, unpublished data). Of note is that MRO, which do not show cristae, also lack the ATP synthase complex (Müller et al. 2012).

ATP synthase, a very ancient enzyme that has been introduced in Chapter one as rotor stator ATPase, has recently emerged as the key molecule determining the structure and shape of membrane cristae (Minauro-Sanmiguel et al. 2005, Mannella 2016, Dudkina et al. 2010, Davies et al. 2011). It is the dimer of the transmembrane enzyme complex that produces diverse structures of the cristae. Early biochemical evidence for a dimeric organization of the mitochondrial ATP synthase complex (Minauro-Sanmiguel et al. 2005) has been solidified by direct evidence obtained with cryo-electron tomography (cryo-ET), a sophisticated microscopic technique for 3D reconstruction of cellular compartments (Mannella et al. 1994, Mannella 2006, Dudkina et al. 2010, Davies et al. 2011). Cryo-ET images of isolated mitochondria from different eukaryotic organisms show oligomeric rows of ATP synthase dimers which shape the ridges of lamellar cristae or subtend the elongated form of tubular cristae, depending upon different oligomeric arrays that the dimers can produce in the three dimensions (Dudkina et al. 2010, Davies et al. 2011, Mühleip et al. 2016). The basis of these oligomeric arrays is the conical shape of the dimer of the mitochondrial ATP synthase complex (Minauro-Sanmiguel et al. 2005), which drives the curvature of the inner membrane at its edge, where the dimers are concentrated (Davies et al. 2011). The cryo-ET evidence has thus clarified that the shape and overall structure of mitochondrial cristae is determined by the oligomeric arrangement of the ATP synthase dimers, which in turn are regulated by the presence or absence of different subunits or sequence variations in key subunits that hold together the dimer interfaces (Minauro-Sanmiguel et al. 2005, Davies et al. 2011, Mühleip et al. 2016). It is plausible that the same applies to the ICM of some proteobacteria, but studies in this direction are just at their beginning (Mendoza-Hoffmann et al. 2018). So far, it seems, no biochemical characterization of the oligomeric state of the ATP synthase complex has been carried out in bacterial species that regularly show cristae-like structures, as reviewed by Degli Esposti (2014).

Physiology—Functional Properties of Mitochondria and Their Derived Organelles

Biochemically, mitochondria are the most active compartment of eukaryotic cells. They contain the hub of central metabolism, the TCA cycle, which is interconnected with fatty acid oxidation, the urea cycle of amino acid degradation and the biosynthetic pathways of several amino acids, bases of nucleic acids and cofactors of various kinds. Mitochondria are also involved in the biosynthesis of ubiquinone, glycerol-based phospholipids such as cardiolipin and ceramide-based sphingolipids, as well as in maintaining the redox balance of the cells with diverse anti-oxidants proteins. All these activities are carried out by a complement of proteins which generally includes no more than 1000 polypeptides genuinely associated with the organelles, the mitochondrial proteome (Gray 2012, 2015). Of these, very few are coded by mitochondrial DNA (mtDNA) and are thus directly related to the alphaproteobacterial lineage from which proto-mitochondria originated. The number and type of proteins that are encoded in mtDNA varies considerably in the diverse groups in which eukaryotes are classified today (Roger et al. 2017), with a minimal of two in *Chromera*, an Apicomplexan relative of the malaria pathogen *Plasmodium* (Roger et al. 2017), to a maximum of nearly 60 in some Jakobida flagellates (Gray 2015). There actually are organelles clearly derived from mitochondria that have completely lost their own DNA altogether and therefore contain only nuclear-encoded proteins. These organelles are particular types of MRO, the hydrogenosomes and mitosomes—see Müller et al. (2012) for a review.

MRO are present in most supergroups in which eukaryotes are currently classified (Burki 2014), as recently reviewed by Roger et al. (2017). Fully aerobic mitochondria containing all the metabolic systems mentioned above are actually restricted to the taxonomic supergroup from which fungi and animals have evolved. However, basal organisms of this supergroup also contain mitosomes (*Entamoeba*) or hybrid organelles between mitochondria and hydrogenosomes (*Acanthamoeba*), as well as classical hydrogenosomes as in pathogenic Chytrid fungi (Müller et al. 2012, Stairs et al. 2015, Roger et al. 2017). This evidence suggests that proto-mitochondria contained the metabolic traits of both aerobic and anaerobic metabolism, which have been differentially distributed or sequentially lost along the diverse lineages of eukaryotes depending upon the prevalent aerobic or anaerobic conditions in which these organisms have adapted. Hybrid situations also exist, as best represented by the mitochondria of Nematodes (Fig. 2A). In larval or juvenile stages, nematode worms have standard aerobic mitochondria resembling those of mammalian cells, but when they mature inside the guts of animals where they spend the rest of their life, the mitochondria become anaerobic. The part of the respiratory chain that oxidizes ubiquinol to oxygen is completely repressed, so that the mitochondria of adult worms such as *Ascaris* have no cytochrome (Fig. 2A). However, they are able to re-oxidize ubiquinol or, most commonly, rhodoquinol produced by complex I and other dehydrogenases by expressing fumarate reductase (Fdr), a paralogue of complex II which catalyzes the reverse reaction of the TCA cycle at the level of succinate, namely the reduction of fumarate to succinate (Fig. 2A). Excess succinate is then exported out of mitochondria and the cells. This short-circuit of both the TCA cycle and the respiratory chain is capable, with the support of a few other enzymes (Müller et al. 2012), to consume metabolites obtained from the host and produce sufficient ATP by exploiting the protonmotive capacity of complex I alone (Fig. 2). A similar respiratory chain is expressed in *E. coli* grown under anaerobic conditions and in several other bacteria, including

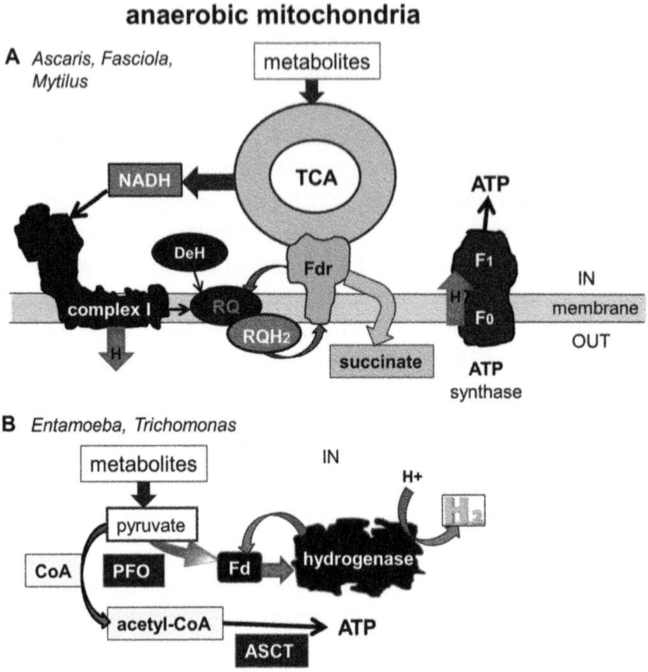

anaerobic mitochondria

Fig. 2. Functional simplification of anaerobic mitochondria and MRO. (A) Mitochondria of adult nematodes and bivalves. These mitochondria do not express anymore the bc1 complex and other cytochromes of the normal respiratory chain, substituting complex II with the paralogue enzyme Fumarate Reductase (Fdr) which re-oxidizes the rhodoquinol (RQH2) produced by complex I and other dehydrogenases (DeH) reducing RQ. This simple redox circuit produces enough protonmotive force to drive ATP synthesis via ATP synthase. Mitochondrial protonmotive reactions lead to an accumulation of succinate, the product of Fdr, which is then excreted. (B) The matrix of hydrogenosomes and some other MRO contains the key elements normally associated with rTCA (see Chapter three) to produce substrate-level ATP via the reaction of pyruvate-ferredoxin oxidoreductase (PFO), which forms the high-energy compound acetyl-CoA that is subsequently hydrolyzed to produce ATP. The process is linked to the reduction of ferredoxins (Fd), which are then re-oxidized by [FeFe]-hydrogenases that are homologs to those present in strictly anaerobes such as Clostridia and *Azospirillum* taxa found in human guts (Degli Esposti et al. 2016).

some Rhodospirillaceae. Hence, the transition from aerobic to anaerobic mitochondria of nematodes can be considered a modular adaptation that was inborn in the alphaproteobacterial ancestors of these mitochondria.

The same alphaproteobacterial ancestors, most likely, also contained the metabolic traits now found in the mitosomes and hydrogenosomes of anaerobically adapted eukaryotes, which essentially include pyruvate-ferredoxin oxidoreductase (PFO), ferredoxins (Fd), [FeFe]-hydrogenases and acetate:succinyl-CoA transferase (ASCT), the latter enzyme catalyzing the phosphorylation of ADP at the substrate level by hydrolyzing the acetyl-CoA produced by PFO (Fig. 2B, cf. Müller et al. 2012, Stairs et al. 2015). These enzymes are present in members of the Rhodospirillaceae family that are either strictly or facultatively anaerobes (Degli

Esposti et al. 2016), as well as a few Rhizobiales such as *Pleomorphomonas* (Im et al. 2006) that also contain other anaerobic traits that are not present in eukaryotes (see Fig. 2 in Chapter seven). Hence, members of the Rhodospirillaceae family with the above traits could be considered the closest relatives of the ancestral lineage of alphaproteobacteria from which protomitochondria originated (Degli Esposti et al. 2016, Martin 2017a).

This evidence, however, has been dismissed in two recent reviews (Roger et al. 2017, Eme et al. 2017), the authors of which continue to prefer the alternative scenario of lateral gene transfer (LGT) of the above traits among eukaryotes, despite its improbability (Martin 2017a,b). This LGT scenario will be subsequently ignored here, together with other conjectures put forward by Roger et al. (2017) which are highly improbable, for instance the idea that 'photosynthetic physiology might be ancestral to all alphaproteobacteria' (see Chapter one, two and seven for solid evidence to the contrary). This decision has the additional advantage of simplifying the possible pattern for the evolution of proto-mitochondria into the mitochondrial organelles and their derivatives that are present in contemporary eukaryotes. Essentially, such evolution followed diverse paths of differential loss of genes and functional traits, besides the transfer of many genes from the proto-mitochondrial genome to that of the nascent eukaryotic cell (Ku et al. 2015). In some organisms adapted to micro-oxic or anaerobic conditions, terminal oxidases of low affinity for O_2 have been immediately lost while retaining the anaerobic systems shown in Fig. 2B. In organisms that adapted instead to permanently aerobic environments, such as the photic zone of the oceans, the anaerobic traits shown in Fig. 2B were soon lost, as in strictly aerobes of the alphaproteobacteria class (Chapter seven), while additional traits were developed to reduce the damage by oxygen radicals inside and outside the mitochondrial organelles.

The number of proteins coded by the mtDNA of these fully aerobic organisms also decreased by their progressive transfer to the nuclear genome or, as in the case of Apicomplexan organisms, almost entirely lost because not necessary anymore in their specialized lifestyles. Yet, about a dozen proteins are still coded by the mtDNA of most animals, including us, thereby suggesting that there may be a natural limit to the transfer of the original alphaproteobacterial complement of genes to the nucleus. Probably the best explanation for this is provided by the hypothesis of Colocation for redox regulation, CoRR (Allen 2015). The fulfilled predictions and other aspects of the CoRR hypothesis have been nicely discussed by Martin et al. (2015).

Phylogeny—Phylogenetic Features of Mitochondria from the Host Angle

Before discussing the approach of investigating the origin of eukaryotic cells from the perspective of the host in the symbiogenic process that nearly everybody believes produces such cells, a brief survey on the nature of this host is necessary. Nowadays, there is a strong consensus that the host was a member of a new clade of archaea that has been recently named Asgard (Zaremba-Niedzwiedzka et al. 2017, Eme et al. 2017), which includes a previously reported lineage called Lokiarchaeota (Spang et al. 2015, Eme et al. 2017). These prokaryotes harbor genes that have clearly eukaryotic character, not only in the systems devoted to the replication of RNA and DNA as found earlier (see below), but also in membrane traffic pathways that are present only in eukaryotes (Spang et al. 2015, Zaremba-Niedzwiedzka et al. 2017). While the original endosymbiotic theory of Lynn Margulis did not specify the nature of the host partner of the proto-mitochondrion (Sagan, 1967, Margulis 1970), for a while the host has been considered to be a form of ancestral eukaryote without mitochondria (Margulis et al. 2000), as reviewed by Martin (2017a). No evidence for such organisms has been found, while an impressive amount of data has progressively indicated that *all* eukaryotes have mitochondria, in one form or the other (Embley and Williams 2015, Martin et al. 2015). Then phylogenetic analysis began to show that many eukaryotic proteins interacting with RNA and DNA cluster with their archaeal homologs rather than with their bacterial homologs (Martin et al. 2015, Martin 2017a), generating the now consensual concept that the exquisite functions of the eukaryotic nucleus originated among archaea (McInerney et al. 2014, Ettema 2016, Martin 2017a, Eme et al. 2017). Ettema and coworkers have recently found the closest living relatives to such archaea in deep ocean sediments and other extreme environments, collectively classified among the Asgard clade (Zaremba-Niedzwiedzka et al. 2017, Eme et al. 2017). Despite ongoing controversies regarding the physiological profile of these yet unseen archaea cells (Sousa et al. 2016, Burns et al. 2018), today it is well accepted that some lineage within the Asgard clade was the progenitor of the host in the eukaryogenesis process (Ettema 2016, Martin 2017a, Eme et al. 2017).

Of note is that the contribution of the archaeal host to the genome of LECA and subsequent eukaryotic lineages is not large, say about one-fourth of the total complement of protein-coding genes. The rest of LECA genome was predominantly made up by genes that originally were part of the genome of the bacterial endosymbiont and had migrated to the nucleus (Esser et al. 2004, McInerney et al. 2014, Martin 2017a,b). Consequently, it is possible to infer the nature of the bacterial endosymbiont by examining the phylogenesis of the nuclear encoded proteins that are not related to archaea, starting from those that form the mitochondrial proteome in diverse eukaryotes. This approach was pioneered by Karlberg et al. (2000), then expanded by various other studies (Esser et al. 2004, Atteia et al. 2009, Abhishek et al. 2011, Thiergart et al. 2012, Rochette et al. 2014, Gray 2015, Wang and Wu 2015) and taken to the ultimate level by Pittis and Gabaldon (2016). Pittis and Gabaldon (2016) reached the provocative conclusion that eukaryotic proteins affiliated to alphaproteobacteria residing in mitochondria were acquired after those affiliated to other bacterial lineages, which are associated with diverse membrane compartments. These findings support the concept that mitochondria might have been acquired late along the process of eukaryogenesis (Ettema 2016, Eme et al. 2017), which is diametrically opposite to the idea that mitochondria were acquired early and defined the whole process of eukaryogenesis (Martin and Müller 1998, Lane and Martin 2010).

The debate of mitochondria late vs. mitochondrial early reflects a long history of opposing views on how eukaryotic cells came about; several reviews summarize these views and argue for one or the other, or maintain a neutral position (Martin et al. 2015, Ettema 2016, Degli Esposti 2016, Martin 2017a, Lane 2017, Roger et al. 2017, Eme et al. 2017). Lately, such a debate is at risk of turning stale by running into circular arguments and strongly held positions. However, the application of what can be called 'biological common sense', a combination of factual evidence and accumulated hand-on experience on the biochemistry and biology of cellular organelles, clearly resolves the debate in favor of the mitochondria early idea. One reason for this resolution is that many results reported by Pittis and Gabaldon (2016) actually defy biological common sense. According to these authors, the proportion of eukaryotic proteins found to be affiliated to alphaproteobacteria is, for example, lower than that of deltaproteobacteria, which are distant from the proteobacterial lineage evolved into proto-mitochondria, as discussed in Chapters four and seven. Detailed re-examination of these results indicated that the bioinformatics approach used by Pittis and Gabaldon (2016) largely underestimated the proportion of eukaryotic proteins that are effectively clustering with alphaproteobacteria (Degli Esposti 2006). It is not necessary to dwell here in technical details of phylogenetic analysis that sustain this conclusion. It is sufficient to say that the basic approach used by Pittis and Gabaldon (2016) had already been found to be inaccurate by the work of Rochette et al. (2014), which has essentially inspired their study. Quantitative aspects of these conflicting results pertain to a central problem in the analysis of nuclear-encoded proteins associated with mitochondria, namely the mosaic nature of the mitochondrial proteome (Gray 2012, 2015).

The assembly of the ca. 1000 proteins that are present in mitochondria of eukaryotic cells clearly has a mosaic nature, since only about 20% of all such proteins can be traced back to alphaproteobacterial homologs (Gray 2012, 2015). This proportion has remained more or less constant since the initial studies on the yeast genome (Karlberg et al. 2000, Esser et al. 2004), as analyzed in recent works (Rochette et al. 2014, Gray 2015, Martin 2017a, Eme et al. 2017). The problem arises from the expectation that a much greater proportion of eukaryotic proteins should be affiliated to their alphaproteobacterial homologs, if an alphaproteobacterium were the sole originator of the bacterial component of the nuclear genome (Gray 2012, 2015, Eme et al. 2017). The value around 20% of alphaproteobacterial ancestry for eukaryotic may well be underestimated, due to technical problems similar to those just discussed with regard to the work by Pittis and Gabaldon (2016). However, even if this value were to be corrected upwards by a few percentage points, it would not necessarily mean that other bacteria may have contributed to the original genome of LECA, as it is often concluded (Gray 2015, Pittis and Gabaldon 2016, Ettema 2016, Eme et al. 2017). The reason is that it could reflect instead the ancestral nature of the bacterial progenitors of mitochondria within the alphaproteobacterial lineage. We have previously seen that chimeric genomes are found even in type genera of bacterial *phyla* which are closely related to other bacterial lineages, for example *Nitrospina* (Chapter three) and *Acidithiobacillus* (Chapter five). Indeed, the deepest branching genus of alphaproteobacteria, *Magnetococcus*, has a strongly chimeric genome in which the alphaproteobacterial component accounts for no more than one third of the total number of proteins (Esser et al. 2007; see also the initial part of Chapter seven and below). By considering that proto-mitochondria were not too distant, in phylogenetic terms, from *Magnetococcus* and other ancestral alphaproteobacteria, as discussed in Chapter seven, then one would expect a relatively small proportion of truly alphaproteobacterial genes in those that subsequently migrated into the nuclear genome of primordial eukaryotes and can still be retrieved today.

The above considerations render the mosaic nature of the mitochondrial proteome no more a problem as it has been perceived so far; rather, it can be viewed as a biological consequence, in the main, of the ancestral phylogenetic position of proto-mitochondria within the whole group of proteobacteria producing ubiquinone (Fig. 3B, cf. Fig. 6 in Chapter four and Degli Esposti 2016). Or, in the language of Martin and colleagues, it would reflect the inherited chimaerism of the original genome of the bacterial endosymbiont (Ku et al. 2015). Moreover, Asgard genomes have many bacterial-like genes (Eme et al. 2017). Following these points and some wise application of Occam's razor, the major findings of the analysis of mitochondrial proteins from the host angle can be approximated to a single major source of genes from the ancestral alphaproteobacterial symbiotic partner that ended up into the nuclear genome of LECA. So, problem solved; consequently, the debate on mitochondria late vs. mitochondria early (Ettema 2016, Eme et al. 2017) evaporates to irrelevance. However, the next question becomes: what is the relation of that ancestral alphaproteobacterial symbiotic partner with extant alphaproteobacteria? The answer is discussed below.

The Living Relatives of the Ancestral Symbionts which evolved into Mitochondria

The answer to the question of which extant alphaproteobacteria are likely to be related to the ancestral alphaproteobacterial symbiotic partner has two parts. The first part is negative: such alphaproteobacteria cannot be found among Rickettsiales. There are multiple reasons for this exclusion, which are dispersed along various sections of this chapter and previous Chapter seven. First, the estimated age of *Wolbachia*, a major genus of the Rickettsiales order, is only around 500 million years (section on *Wolbachia* in Chapter seven). Because this genus is basal to one of the families of Rickettsiales, which are closely clustered together in phylogenetic trees (Wu et al. 2004, Williams et al. 2007, Sassera et al. 2011, Wang and Wu 2015), it follows that the whole order may be far too young to be the originator of proto-mitochondria. Second, Rickettsiales generally have none of the anaerobic traits that are present in contemporary eukaryotes (Fig. 2 in Chapter seven, cf. Atteia et al. 2009, Degli Esposti et al. 2014). Third, the genome of Rickettsiales does not code

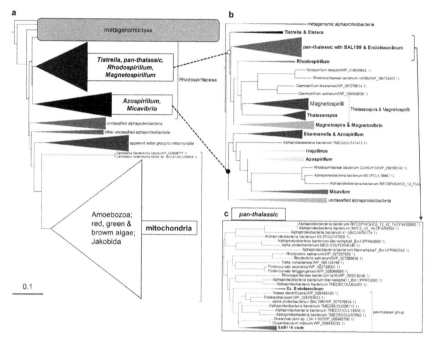

Fig. 3. Taxonomy of Rhodospirillaceae. (a) NJ tree obtained from a search of the cytochrome *b* protein extended to all the Rhodospirillaceae taxa that are known today, classified and unclassified, together with unclassified alphaproteobacteria and the mitochondrial genomes of the indicated eukaryotes. Note how Rhodospirillaceae segregates into two major clades, which are expanded in b as indicated by the dashed lines. Note also that the branches lying just before or in apparent sister position vs. the large mitochondrial clade contain unclassified or metagenomic alphaproteobacteria, including, for instance, *Thermopetrobacter*. (b) Enlarged subtree with the major taxa of the two clades of Rhodospirillaceae. Note that the species having the most ancestral set of anaerobic traits (Fig. 2 in Chapter seven) cluster in the branch containing *Rhodospirillum rubrum* and Magnetospirilli. Conversely, *Micavibrio* clusters with *Azospirillum* in a separate branch at the bottom. (c) The sub-tree is an expansion of the pan-thalassic group (Degli Esposti et al. 2014) shown at the top of the subtree in b. This group includes also the endosymbiont *Ca.* Endolissolinum (Kwan et al. 2012) and various members of the SAR116 cluster of oceanic taxa, which are distinct from the SAR11 clade of *Pelagibacter* and affiliated instead to the Rhodospirillaceae family (Oh et al. 2010). The tree topology is fundamentally consistent with 16S rRNA trees that have been published previously with more limited set of taxa (Wisniewski-Dyé et al. 2011, Kwan et al. 2012, Bazylinski et al. 2013, Cai et al. 2018).

for MICOS proteins regulating mitochondrial cristae and equivalent membrane invaginations in bacteria (this chapter, section on cytology). Fourth, Rickettsiales do not have the zeta subunit of the ATP synthase of other alphaproteobacteria, which is functionally equivalent to the inhibitory subunit of mitochondrial ATP synthase (last section of Chapter seven, cf. F. Mendoza-Hoffmann and J.J. Garcia-Trejo unpublished data; see also Mendoza-Hoffmann et al. 2018). Additionally, Rickettsiales have been excluded from the origin of mitochondria on the basis of a complex network approach (Carvalho et al. 2015). Hence, other taxa of alphaproteobacteria must be considered.

Non photosynthetic members of the Rhodospirillaceae are at present the most likely candidates for the positive part of the answer, primarily because of their complement of anaerobic traits that are shared with eukaryotes (Fig. 2 in Chapter seven). The following reasons additionally favor Rhodospirillaceae taxa. They evolved the mitochondrial pathway for the biosynthesis of ubiquinone (Degli Esposti 2017) and also produce rhodoquinone (Table 1 in Chapter seven). They have all the traits mentioned above that Rickettsiales do not have. They are among the most ancient representatives of the alphaproteobacterial lineage (initial part of Chapter seven) and evolved the basic operon for the mitochondrial type of cytochrome *c* oxidase that has low affinity for oxygen (Chapters two and seven). Their type species, *R. rubrum*, has emerged as the most frequent alphaproteobacterial relative of mtDNA- and nuclear-encoded proteins in separate phylogenetic studies (Esser et al. 2004, Atteia et al. 2009, Abhishek et al. 2011, Thiergart et al. 2012). They include ectoparasites such as *Micavibrio* (Davidov et al. 2006) and endosymbionts such as *Ca.* Endolissoclinum (Kwan et al. 2012, Kwan and Schmidt 2013). Ultimately, no other family or group of alphaproteobacteria has such a consilient set of evidence pointing to a relation with proto-mitochondria. However, the current phylogenetic profile of the Rhodospirillaceae family is not known in detail. To fill this gap in the phylogeny of the potential ancestors of proto-mitochondria, the latest taxonomy profile of Rhodospirillaceae has been delineated by the detailed analysis presented below.

The Latest Taxonomy of Rhodospirillaceae

Several new genera and isolates have been recently reported to be affiliated to the Rhodospirillaceae (see Cai et al. 2018 for the latest report), but a taxonomy study of the family has not been reported since several years ago (Wisniewski-Dyé et al. 2011, Kwan et al. 2012). The latest taxonomic profile of the family has been undertaken first by using the cytochrome *b* protein (Fig. 3), for consistency with the phylogenetic profile of the whole alphaproteobacteria class reported earlier with the same protein marker (Fig. 1 in Chapter seven). This protein has a strong phylogenetic signal, it is present in most Rhodospirillaceae taxa and is often used as a taxonomic indicator for eukaryotic organisms too (Degli Esposti et al. 1993, Baharum and Nurdalila 2012).

Figure 3A presents the phylogenetic tree encompassing all unclassified alphaproteobacteria, Rhodospirillaceae and many mitochondria from various supergroups of eukaryotes. Because the tree derives from a broad DeltaBLAST search extended to all taxa currently available in the protein database of NCBI (https://www.ncbi.nlm.nih.gov/, accessed on 14 March 2018), there is neither a bias nor arbitrary choice in the taxonomic sampling of the cytochrome *b* proteins, as discussed earlier (Degli Esposti 2016, 2017). The upper part of the tree is dominated by unclassified taxa reported in recent metagenomic studies, similar to the trees that encompass the whole class of alphaproteobacteria (Fig. 1 in Chapter seven and M.D.E, unpublished data). This means that the deep branching lineages of alphaproteobacteria include only metagenomic unclassified taxa, followed by Rhodospirillaceae and a few other taxa (Fig. 3A cf. Fig. 1B in Chapter seven).

All taxa already classified as members of the Rhodospirillaceae family cluster in two sister clades (Fig. 3A,B), which also includes several unclassified and metagenomic taxa, as shown in Fig. 3C for one large group of marine bacteria that is tentatively called 'pan-thalassic' for it contains species such as alphaproteobacterium BAL199 and *Ca.* Endolissoclinum (Kwan and Schmidt 2013) which have been previously labelled with this term (Degli Esposti et al. 2014). The pan-thalassic group also includes the SAR116 clade (Oh et al. 2000) and may reside in the same wide group that contains *R. rubrum, Phaeospirillum, Magnetospirillum* and *Magnetovibrio* (Fig. 3B), namely the species showing the largest set of anaerobic traits among the whole order of Rhodospirillales (Fig. 2 in Chapter seven). *Tistrella*, which is non photosynthetic and has a complex set of respiratory enzymes (Degli Esposti et al. 2014), occupies the deepest branch of the clade, in agreement with Wisniewski-Dyé et al. (2011).

The other Rhodospirillaceae clade is dominated by terrestrial diazotrophic taxa of the genus *Azospirillum* and also includes marine genera such as *Skermanella* and, remarkably, *Micavibrio* (Fig. 3B). So far, ectoparasites of the genus *Micavibrio* have been considered as unclassified alphaproteobacteria (Davidov et al. 2006), hence the novelty of their clustering within the *Azospirillum* clade (Fig. 3), which was anticipated in Chapter seven and has been confirmed also with other protein markers (M.D.E., unpublished data). Indeed, the separation of the Rhodospirillaceae into two major clades, one dominated (in terms of strains) by *Magnetospirillum* and the other by *Azospirillum* (Fig. 3), is consistent with previous studies that used 16S rRNA as phylogenetic marker (Wisniewski-Dyé et al. 2011, Kwan et al. 2012).

Interesting observations emerge by integrating the latest taxonomic profile shown in Fig. 3 with phenotypic properties reported in the literature and metabolic information deduced by genomic analysis. Anoxygenic photosynthesis is dispersed along the whole family of Rhodospirillaceae, including the pan-thalassic group. Phylogenetic trees of signature proteins such as the L and M subunit of the photosynthetic reaction center do not show the same topology as that illustrated in Fig. 3 or in previously reported trees (Kwan et al. 2012); their pattern, moreover, is not correlated with the profile of anaerobic traits (cf. Fig. 2 in Chapter seven). For example, photosynthesis is present in *Phaeospirillum*, which contains most of such traits, as well as in *Oceanibaculum*, which has only one of the same traits (Fig. 2 in Chapter seven and M.D.E., unpublished data).

Conversely, the character of parasitism and endophytic association with plants is diffused particularly in the *Azospirillum* clade, considering also that it includes the genus *Inquilinus*, which has been originally found as an opportunistic pathogen in cystic fibrosis patients (McHugh et al. 2016). However, the propensity for intimate association with eukaryotes is present in the pan-thalassic group of the other Rhodospirillaceae clade also, since it includes *Ca.* Endolissoclinum, a mutualistic endosymbiont of tunicates which shows genomic features that differ from those typically found in proteobacterial endosymbionts of insects (Kwan and Schmidt 2013—see also Chapters five and six). Hence, a symbiotic or parasitic character may be scattered along the whole family of Rhodospirillaceae.

Unfortunately, the taxonomic profile of Rhodospirillaceae is incomplete when obtained from phylogenetic trees of common bioenergetic proteins such as cytochrome *b* (Fig. 3), because anaerobic taxa that may be affiliated with the family (Degli Esposti et al. 2016, Degli Esposti 2017) do not have these proteins in their genome (Fig. 2 in Chapter seven). Two complementary approaches have thus been followed to complete the taxonomic profile of the family. The first approach is based upon the concatenated alignment of the largest possible number of shared ribosomal proteins (Probst et al. 2017), a method that has been routinely used for the taxonomic assignment of metagenomic assembled genomes in recent studies (see Chapter one, four and seven for a critical discussion of equivalent methods). The application of this method to representatives of the Rhodospirillaceae family and various classes of proteobacteria, with a member of the Nitrospirae as outgroup, produces the results exemplified in Fig. 4A. The tree confirms the split of the family into the *Magnetospirillum* and the *Azospirillum* clade (Fig. 4A cf. Fig. 3). Interestingly, strictly anaerobes derived from metagenomic studies cluster with one or the other of these clades, irrespective of their original taxonomic assignment.

For instance, *Acetobacter* sp. CAG:977 found in human guts (cf. Degli Esposti et al. 2016) is closely related to a taxon discovered in a subterranean aquifer metagenome, together forming a deep branching group within the *Magnetospirillum* clade (Fig. 4A). This positioning is consistent with a complete blast analysis of all the proteins coded in the genome of *Acetobacter* sp. CAG:977, which shows a few proteins affiliated with the Acetobacteraceae family (3.5% of the total examined) in comparison to those showing strong affiliation with the Rhodospirillaceae family (55.8%, M.D.E. unpublished results). On the other hand, alphaproteobacterium GWF2_58_2, which has been found in another subsurface aquifer metagenome (Probst et al. 2017), clusters with the *Azospirillum* clade (Fig. 4, cf. Degli Esposti 2017). The same may apply to the *Azospirillum* strains showing strictly anaerobic character that are listed at the top of Fig. 2 in Chapter seven (M.D.E., unpublished results).

An observation of general interest emerges by inspecting the tree shown in Fig. 4A. Although the phylogenetic sequence of the alphaproteobacteria is fully congruent with that obtained using protein markers such as cytochrome *b* (Fig. 3, see also Fig. 1B in Chapter seven), the relative order of the proteobacterial classes is incorrect. The phylogenetic sequence of these classes has been established by integrated approaches and is shown in Fig. 6 of Chapter four. For the proteobacterial classes contained in the tree of Fig. 4A, alphaproteobacteria should come first after the *Nitrospira* outgroup, at least with the deep-branching *Magnetococcus*, which is instead positioned in a branch that follows the other classes (Fig. 4A). Clearly, phylogenetic trees obtained from the concatenation of multiple ribosomal proteins are unable to resolve deep phylogenetic relationships among bacteria, as noted

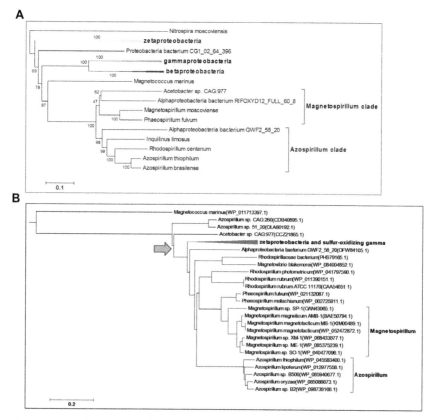

Fig. 4. Phylogenetic trees to include the strictly anaerobes in the taxonomy profile of Rhodospirillaceae. (A) NJ tree obtained with the alignment of 28 concatenated ribosomal proteins, as described by Banfield and coworkers (Probst et al. 2017). The percentage value of the bootstraps (500) is shown by each node. Equivalent results were obtained with ML trees and different programs (RaxML). Zetaproteobacteria include *Mariprofundus ferrooxydans* and zetaproteobacerium TAG-1 (*Ghiorsea bivora*, chapter four), gammaproteobacteria include endosymbiont of *Ridgeia piscesae* and *E. coli* K-12, while betaproteobacteria include betaproteobacterium RGB_16_56_24 and Methylotenera RIFCSPLOWO2_02_FULL_45_14. (B) NJ tree obtained with a selection of sequences of pyruvate-ferredoxin(flavodoxin)-oxidoreductase, PFO (Charon et al. 1999), which are distributed among strictly and facultative anaerobes of the Rhodospirillaceae family (see Fig. 2 in chapter seven). Note the monophyletic pattern of this complex multidomain protein that has been transmitted from *Magnetococcus* to alphaproteobacterial anaerobes, then to zetaproteobacteria and gammaproteobacteria (here represented by clade 2 taxa of sulfur-oxidizing symbionts, see chapter five) and then to the deep branching taxa such as *R. rubrum* as well as the two clades of Rhodospirillaceae defined in other phylogenetic trees (Fig. 3). Eukaryotes proteins stem as a sister clade from between *Acetobacter* sp. CAG:977 and zetaproteobacteria as indicated by the grey arrow, somehow as expected. The presence of multiple forms of PFO in several eukaryotes confuses the overall tree including all such proteins (cf. Stairs et al. 2015).

before (see Chapter one, cf. Thiergart et al. 2014). This is the basic problem that has led to approximate, and in some cases erroneous, taxonomic assignment of bacteria such as the above mentioned *Acetobacter* strain.

However, a solution for this taxonomic problem is available in the case of Rhodospirillaceae. Taxa from this family have deep phylogenetic relationships with other proteobacterial classes and at the same time may include the progenitors of proto-mitochondria, as discussed previously in this chapter. Such relationships can be visualized by trees constructed with PFO, a large protein that is crucial for the anaerobic metabolism of all these taxa, from zetaproteobacteria (chapter four) to alphaproteobacteria (Fig. 2 in chapter two) and MRO organelles of anaerobic eukaryotes (Fig. 2B). PFO is a modular ensemble of different conserved domains forming a protein of over 1100 residues (Charon et al. 1999) and thus has a good phylogenetic signal, despite the presence of multiple structural and genomic variants in various eukaryotes (Stairs et al. 2015). A phylogenetic tree constructed with a selection of taxa matching that in Fig. 4A shows a monophyletic pattern of inheritance that conforms to the established phylogenetic sequence of alphaproteobacteria and other proteobacterial classes (Fig. 4B). *Magnetococcus*, which has a shorter version of the normal PFO protein, lies in the deepest branch while the two clades of the Rhodospirillaceae deduced from other markers (Figs. 3 and 4A) represent the later branching groups (Fig. 4B). This pattern also matches the distribution of anaerobic traits in Rhodospirillaceae and other alphaproteobacterial taxa discussed in Chapter seven (Fig. 2).

When the phylogenetic analysis of PFO proteins is extended to the eukaryotic homologs present in diverse supergroups, the proteins from the majority of Rhodospirillaceae taxa form a sister clade vs. the clade containing various forms of eukaryotic PFO; notably, both clades have the PFO of *Acetobacter* sp. CAG:977 as common ancestor (M.D.E., unpublished results). The branching position of the eukaryotic clade is indicated by the arrow in Fig. 4B, after deep branching proteins from strictly anaerobes such as *Acetobacter* sp. CAG:977, which are thus older than the rest of the Rhodospirillaceae family, as discussed earlier in Chapter seven.

In sum, the combination of diverse phylogenetic markers (Figs. 3 and 4) delineates the contours of the current taxonomy profile of the Rhodospirillaceae family, which includes anaerobes that existed before the separation of the lineage leading to proto-mitochondria, just before the branching of the class of zetaproteobacteria and then that of gammaproteobacteria. This taxonomy profile can tentatively define the phylogenetic space from which the ancestral lineage of proto-mitochondria emerged, as indicated in Fig. 4B. When all the genomes already sequenced but not yet deposited in NCBI repositories will be available for analysis, this space will probably contain new taxa that will further narrow the search for possible extant relatives of proto-mitochondria.

Remaining Challenges to Understand the Process of Eukaryogenesis

Today, some major challenges remain to understand how the eukaryogenesis process occurred, independently of the nature of either the bacterial endosymbiont or the host. The first challenge regards the coming together of the two separate prokaryotic cells. Were they fused by an as yet unknown physical process analogous to that leading to the gammaproteobacteria engulfed within betaproteobacterial endosymbionts of insects (Chapter six)? This idea comes from the field of insect endosymbionts (Husník and McCutcheon 2016), but has not been followed much in studies regarding eukaryogenesis (Gould et al. 2016, Martin 2017a, Martin et al. 2017, Eme et al. 2017). Or had the host eaten proto-mitochondria exploiting its membrane traffic capacity (Ettema 2016)? This is a very old idea that has been proposed several times, as reviewed by Martin et al. (2015, 2017); its last version follows the finding of diverse component of the eukaryotic machinery of membrane traffic in the proposed host from the Asgard clade (Ettema 2016, Zaremba-Niedzwiedzka et al. 2017, Eme et al. 2017). An alternative possibility, also popular, is that the bacterial endosymbiont was initially a parasite of the host and was subsequently enslaved to become a useful organelle (Wang and Wu 2015). The cell biology of such a possibility would also involve the membrane traffic traits of the Asgard host and might have the known system of *Legionella* infection of human cells as a paradigm (Degli Esposti 2014 - see picture in the front cover). However, the possibility of a parasitic nature of the bacterial partner in eukaryogenesis is strongly weakened by the evidence that endocellular parasites such as Rickettsiales cannot be the source of such a bacterial partner, as discussed earlier in the chapter (see also Gould 2016).

In any case, both the eating (phagocytic) idea and the parasite idea stumble onto a major problem that regards the different chemical structure of the membrane lipids of the archaeal host and the bacterial endosymbiont, the so-called lipid divide (Caforio and Driessen 2017, Eme et al. 2017). Eukaryotes only have bacterial lipids in their membranes and therefore the problem becomes two-fold: (i) where did the archaeal lipids of the host go, and (ii) how did the bacterial lipids become the components of eukaryotic membranes. There was no solution to the problem until very recently, when the enzyme responsible for the different chemistry of archaeal lipids has not been found in the genome of Lokiarchaeota (Villanueva et al. 2017). Therefore, the lineage of the archaeal host might have already possessed the capacity of producing bacterial-like lipids, thus representing a missing link between archaea and eukaryotes that would bridge the lipid divide. However, Asgard genomes do have the enzyme responsible for the archaeal lipids, thereby confusing the current picture (Eme et al. 2017). An alternative possibility to overcome the lipid divide has been envisaged by Gould et al. (2016), who have proposed that lipid vesicles excreted by the bacterial endosymbiont might have fueled the proliferation of internal membrane that followed the insertion into the host, ultimately overtaking the entire production of membrane lipids by the same host. Although compelling, this scenario lacks biological precedents that could sustain it further and has been criticized lately (Eme et al. 2017).

In sum, the major challenges that remain for rationalizing the entire process of eukaryogenesis regard the mechanism of fusion of the two prokaryotic cells into a proto-eukaryotic cell, and the lipid divide between the bacterial endosymbiont and the archaeal host. The ultimate challenge, however, is to isolate and characterize some cells of the Asgard lineage, so to experimentally verify their capacity of membrane traffic and the lipid composition of their membranes. Rumor has it that such cells have already been seen but do not meet the expectations for the host. However, so much progress has been obtained in recent years in crucial aspects of eukaryogenesis, including the identification of the likely relatives to the prokaryotes which made such a process, that hope is growing that the remaining challenges will be dealt with in the future.

Concluding Remarks

At this point, some remarks need to be made to conclude not only this chapter, but the entire book. The reader has been taken on a scientific journey into the evolution of bacterial phylogeny and function of mitochondria, starting with current ideas on how and where life started and then diversified its ancestral physiology (Chapter one). Following a crash-course in bacterial physiology to appreciate how ancestral bacteria exploited environmental sources of bioenergy (Chapter two), the journey changes pace after oxygen became abundant in primordial earth and the elements of aerobic metabolism were put together in diverse bacterial lineages (Chapter two, three and four). From one of such lineages, the alphaproteobacteria, some bacteria emerged (Chapter seven) that then became endosymbionts of an Asgard archaeal host, together forming the first eukaryotic cell (this chapter). The extant relatives of this bacterial endosymbiont are identified here to reside within the Rhodospirillaceae family, a strong proposition that will probably excite the imagination of the reader, and stimulate colleagues who have different ideas to sharpen their point, or change their idea. It is hoped that this will constitute the major take-home message at the end of the journey undertaken in reading the book.

Acknowledgements

I thank Diego Golzalez-Halphen, Francisco Mendoza-Hoffmann and José Garcia-Trejo (UNAM, Mexico City) for enthusiastic discussion and exchange of information. I warmly thank Luis Lozano (CCG, Cuernavaca) for technical help in genomic analysis and Patricia Lievens plus Marta d'Amora (IIT, Genoa) for their assistance in microscopic images.

References

Abhishek, A., A. Bavishi, A. Bavishi and M. Choudhary. 2011. Bacterial genome chimaerism and the origin of mitochondria. Can. J. Microbiol. 57: 49–61. doi:10.1139/w10-099.

Allen, J.F. 2015. Why chloroplasts and mitochondria retain their own genomes and genetic systems: Colocation for redox regulation of gene expression. Proc. Natl. Acad. Sci. USA 112: 10231–10238. doi:10.1073/pnas.1500012112.

Atteia, A., A. Adrait, S. Brugière, M. Tardif, R. van Lis, O. Deusch et al. 2009. A proteomic survey of *Chlamydomonas reinhardtii* mitochondria sheds new light on the metabolic plasticity of the organelle and on the nature of the alpha-proteobacterial mitochondrial ancestor. Mol. Biol. Evol. 26: 1533–1548. doi:10.1093/molbev/msp068.

Atteia, A., R. van Lis, J.J. van Hellemond, A.G. Tielens, W. Martin and K. Henze. 2004. Identification of prokaryotic homologues indicates an endosymbiotic origin for the alternative oxidases of mitochondria (AOX) and chloroplasts (PTOX). Gene 330: 143–148.

Baharum, S.N. and A.A. Nurdalila. 2012. Application of 16s rDNA and cytochrome b ribosomal markers in studies of lineage and fish populations structure of aquatic species. Mol. Biol. Rep. 39: 5225–5232. doi:10.1007/s11033-011-1320-2.

Bazylinski, D.A., T.J. Williams, C.T. Lefèvre, D. Trubitsyn, J. Fang, T.J. Beveridge et al. 2013. *Magnetovibrio blakemorei gen.* nov., sp. nov., a magnetotactic bacterium (Alphaproteobacteria: Rhodospirillaceae) isolated from a salt marsh. Int. J. Syst. Evol. Microbiol. 63: 1824–1833. doi:10.1099/ijs.0.044453-0.

Burki, F. 2014. The eukaryotic tree of life from a global phylogenomic perspective. Cold Spring Harb. Perspect. Biol. 6: a016147.

Burns, J.A., A.A. Pittis and E. Kim. 2018. Gene-based predictive models of trophic modes suggest Asgard archaea are not phagocytotic. Nat. Ecol. Evol. doi:10.1038/s41559-018-0477-7. [Epub ahead of print].

Caforio, A. and A.J.M. Driessen. 2017. Archaeal phospholipids: Structural properties and biosynthesis. Biochim. Biophys. Acta. 1862: 1325–1339. doi:10.1016/j.bbalip.2016.12.006.

Cai, H., H. Cui, Y. Zeng. Y. Wang and H. Jiang. 2018. *Niveispirillum lacus* sp. nov., isolated from cyanobacterial aggregates in a eutrophic lake. Int. J. Syst. Evol. Microbiol. 68: 507–512. doi:10.1099/ijsem.0.002526.

Carvalho, D.S., R.F. Andrade, S.T. Pinho, A., Góes-Neto, T.C. Lobão, G.C. Bomfim et al. 2015. What are the evolutionary origins of mitochondria? A complex network approach. PLoS One 10: e0134988. doi:10.1371/journal.pone.0134988.

Charon, M.H., A. Volbeda, E. Chabriere, L. Pieulle and J.C. Fontecilla-Camps. 1999. Structure and electron transfer mechanism of pyruvate: Ferredoxin oxidoreductase. Curr. Opin. Struct. Biol. 9: 663–669.

Cristea, I.M. and M. Degli Esposti. 2004. Membrane lipids and cell death: An overview. Chem. Phys. Lipids 129: 133–160.

Davidov, Y., D. Huchon, S.F. Koval and E. Jurkevitch. 2006. A new alpha-proteobacterial clade of Bdellovibrio-like predators: Implications for the mitochondrial endosymbiotic theory. Environ. Microbiol. 8: 2179–2188.

Davies, K.M., M. Strauss, B. Daum, J.H. Kief, H.D. Osiewacz, A. Rycovska et al. 2011. Macromolecular organization of ATP synthase and complex I in whole mitochondria. Proc. Natl. Acad. Sci. USA 108: 14121–14126. doi:10.1073/pnas.1103621108.

Degli Esposti, M., S. De Vries, M. Crimi, A. Ghelli, T. Patarnello and A. Meyer. 1993. Mitochondrial cytochrome b: evolution and structure of the protein. Biochim. Biophys. Acta. 1143: 243–271.

Degli Esposti, M. 2002. Lipids, cardiolipin and apoptosis: a greasy licence to kill. Cell Death Differ. 9: 234–236.

Degli Esposti, M. 2014. Bioenergetic evolution in proteobacteria and mitochondria. Genome Biol. Evol 6: 3238–3251. doi:10.1093/gbe/evu257

Degli Esposti, M., B. Chouaia, F. Comandatore, E. Crotti, D. Sassera, P.M. Lievens et al. 2014. Evolution of mitochondria reconstructed from the energy metabolism of living bacteria. PLoS One 9: e96566. doi:10.1371/journal.pone.0096566.

Degli Esposti, M. 2016. Late mitochondrial acquisition, really? Genome Biol. Evol. 8: 2031–2035. doi:10.1093/gbe/evw130.

Degli Esposti, M., D. Cortez, L. Lozano, S. Rasmussen, H.B. Nielsen and E. Martinez Romero. 2016. Alpha proteobacterial ancestry of the [Fe-Fe]-hydrogenases in anaerobic eukaryotes. Biol. Direct. 11: 34. doi:10.1186/s13062-016-0136-3.

Degli Esposti, M. 2017. A journey across genomes uncovers the origin of ubiquinone in cyanobacteria. Genome Biol. Evol. 9: 3039-3053. doi: 10.1093/gbe/evx22.

Dudkina, N.V., G.T. Oostergetel, D. Lewejohann, H.P. Braun and E.J. Boekema. 2010. Row-like organization of ATP synthase in intact mitochondria determined by cryo-electron tomography. Biochim. Biophys. Acta. 1797: 272–277. doi:10.1016/j.bbabio.2009.11.004.

Embley, T.M. and T.A. Williams. 2015. Evolution: Steps on the road to eukaryotes. Nature 521: 169–170. doi:10.1038/nature14522.

Eme, L., A. Spang, J. Lombard, C.W. Stairs and T.J.G. Ettema. 2018. Archaea and the origin of eukaryotes. Nat. Rev. Microbiol. 15: v711–723. doi:10.1038/nrmicro.2017.133.

Esser, C., N. Ahmadinejad, C. Wiegand, C. Rotte, F. Sebastiani, G. Gelius-Dietrich et al. 2004. A genome phylogeny for mitochondria among alpha-proteobacteria and a predominantly eubacterial ancestry of yeast nuclear genes. Mol. Biol. Evol. 21: 1643–1660.

Esser, C., W. Martin and T. Dagan. 2007. The origin of mitochondria in light of a fluid prokaryotic chromosome model. Biol. Lett. 3: 180–184.

Ettema, T.J. 2016. Evolution: Mitochondria in the second act. Nature 531: 39–40.

Ferla, M.P., J.C. Thrash, S.J. Giovannoni and W.M. Patrick. 2013. New rRNA gene-based phylogenies of the Alphaproteobacteria provide perspective on major groups, mitochondrial ancestry and phylogenetic instability. PLoS One 8: e83383. doi:10.1371/journal.pone.0083383.

Gould, S.B., S.G. Garg and W.F. Martin. 2016. Bacterial vesicle secretion and the evolutionary origin of the eukaryotic endomembrane system. Trends Microbiol. 24: 525–534. doi:10.1016/j.tim.2016.03.005.

Gould, S.B. 2016. Infection and the first eukaryotes. Science 352. doi:10.1126/science.aaf6478.

Gray, M.W. 2012. Mitochondrial evolution. Cold Spring Harb. Perspect. Biol. 4: a011403. doi:10.1101/cshperspect.a011403.

Gray, M.W. 2015. Mosaic nature of the mitochondrial proteome: Implications for the origin and evolution of mitochondria. Proc. Natl. Acad. Sci. USA 112: 10133–10138.

Husník, F. and J.P. McCutcheon. 2016. Repeated replacement of an intrabacterial symbiont in the tripartite nested mealybug symbiosis. Proc. Natl. Acad. Sci. USA 113: E5416–E5424. doi:10.1073/pnas.1603910113.

Im, W.T., S.H. Kim, M.K. Kim, L.N. Ten and S.T. Lee. 2006. *Pleomorphomonas koreensis* sp. nov., a nitrogen-fixing species in the order Rhizobiales. Int. J. Syst. Evol. Microbiol. 56: 1663–1666.

Karlberg, O., B. Canbäck, C.G. Kurland and S.G. Andersson. 2000. The dual origin of the yeast mitochondrial proteome. Yeast. 17: 170–187.

Ku, C., S. Nelson-Sathi, M. Roettger, F.L. Sousa, P.J. Lockhart, D. Bryant et al. 2015. Endosymbiotic origin and differential loss of eukaryotic genes. Nature 524: 427–432. doi:10.1038/nature14963.

Kwan, J.C. and E.W. Schmidt. 2013. Bacterial endosymbiosis in a chordate host: long-term co-evolution and conservation of secondary metabolism. PLoS One 8: e80822. doi:10.1371/journal.pone.0080822.

Kwan, J.C., M.S. Donia, A.W. Han, E. Hirose, M.G. Haygood and E.W. Schmidt. 2012. Genome streamlining and chemical defense in a coral reef symbiosis. Proc. Natl. Acad. Sci. USA 109: 20655–20660. doi:10.1073/pnas.1213820109.

Lane, N. 2006. Power, sex, suicide: Mitochondria and the meaning of life. Oxford University Press, Oxford.

Lane, N. and W. Martin. 2010. The energetics of genome complexity. Nature 467: 929–934. doi:10.1038/nature09486.

Lane, N. 2017. Serial endosymbiosis or singular event at the origin of eukaryotes? J. Theor. Biol. 434: 58–67. doi:10.1016/j.jtbi.2017.04.031.

Lazcano, A. and J. Peretó. 2016. On the origin of mitosing cells: A historical appraisal of Lynn Margulis endosymbiotic theory. J. Theor. Biol. pii: S0022-5193(17)30322-3. doi:10.1016/j.jtbi.2017.06.036.

Mannella, C.A., M. Marko, P. Penczek, D. Barnard and J. Frank. 1994. The internal compartmentation of rat-liver mitochondria: Tomographic study using the high-voltage transmission electron microscope. Microsc. Res. Tech. 27: 278–283.

Mannella, C.A. 2006. Structure and dynamics of the mitochondrial inner membrane cristae. Biochim. Biophys. Acta. 1763: 542–548.

Margulis, L. 1970. Origin of Eukaryotic Cells. Yale University Press, New Haven, CT.

Margulis. L., M.F. Dolan and R. Guerrero. 2000. The chimeric eukaryote: Origin of the nucleus from the karyomastigont in amitochondriate protists. Proc. Natl. Acad. Sci. USA 97: 6954–6959.

Martin, W.F. and M. Müller. 1998. The hydrogen hypothesis for the first eukaryote. Nature 392: 37–41.

Martin, W.F., S. Garg and V. Zimorski. 2015. Endosymbiotic theories for eukaryote origin. Philos. Trans. R. Soc. Lond. B Biol. Sci. 370: 20140330. doi:10.1098/rstb.2014.0330.

Martin, W.F. 2017a. Physiology, anaerobes, and the origin of mitosing cells 50 years on. J. Theor. Biol. pii: S0022-5193(17)30004-8. doi:10.1016/j.jtbi.2017.01.004.

Martin, W.F. 2017b. Too much eukaryote LGT. Bioessays 39. doi:10.1002/bies.201700115.

Martin, W.F., A.G.M. Tielens, M. Mentel, S.G. Garg and S.B. Gould. 2017. The physiology of phagocytosis in the context of mitochondrial origin. Microbiol. Mol. Biol. Rev. 81(3). pii: e00008-17. doi:10.1128/MMBR.00008-17.

Martin, W.F., D.A. Bryant and J.T. Beatty. 2018. A Physiological Perspective on the Origin and Evolution of Photosynthesis 42: 205–231. doi:10.1093/femsre/fux056.

McInerney, J.O., M.J. O'Connell and D. Pisani. 2014. The hybrid nature of the Eukaryota and a consilient view of life on Earth. Nat. Rev. Microbiol. 12: 449–455. doi:10.1038/nrmicro3271.

McHugh, K.E., D.D. Rhoads, D.A. Wilson, K.B. Highland, S.S. Richter and G.W. Procop. 2016. Inquilinus limosus in pulmonary disease: case report and review of the literature. Diagn. Microbiol. Infect. Dis. 86: 446–449. doi:10.1016/j.diagmicrobio.2016.09.006.

Mendoza-Hoffmann, F., Á. Pérez-Oseguera, M.Á. Cevallosn, M. Zarco-Zavala, R. Ortega, C. Peña-Segura et al. 2018. The Biological Role of the ζ Subunit as Unidirectional Inhibitor of the F(1)F(O)-ATPase of Paracoccus denitrificans. Cell Rep. 22: 1067–1078. doi:10.1016/j.celrep.2017.12.106.

Minauro-Sanmiguel, F., S. Wilkens and J.J. García. 2005. Structure of dimeric mitochondrial ATP synthase: novel F0 bridging features and the structural basis of mitochondrial cristae biogenesis. Proc. Natl. Acad. Sci. USA 102: 12356–12358.

Mühleip, A.W., F. Joos, C. Wigge, A.S. Frangakis, W. Kühlbrandt and K.M. Davies. 2016. Helical arrays of U-shaped ATP synthase dimers form tubular cristae in ciliate mitochondria. Proc. Natl. Acad. Sci. USA 113: 8442–8447. doi:10.1073/pnas.1525430113.

Müller, M., M. Mentel, J.J. van Hellemond, K. Henze, C. Woehle, S.B. Gould et al. 2012. Biochemistry and evolution of anaerobic energy metabolism in eukaryotes. Microbiol. Mol. Biol. Rev. 76: 444–495.

Muñoz-Gómez, S.A., C.H. Slamovits, J.B. Dacks, K.A. Baier, K.D. Spencer and J.G. Wideman. 2015. Ancient homology of the mitochondrial contact site and cristae organizing system points to an endosymbiotic origin of mitochondrial cristae. Curr. Biol. 25: 1489–1495. doi:10.1016/j.cub.2015.04.006.

Muñoz-Gómez, S.A., J.G. Wideman, A.J. Roger and C.H. Slamovits. 2017. The Origin of Mitochondrial Cristae from Alphaproteobacteria. Mol. Biol. Evol. 34: 943–956. doi:10.1093/molbev/msw298.

Oh, H.M., K.K. Kwon, I. Kang, S.G. Kang, J.H. Lee, S.J. Kim et al. 2010. Complete genome sequence of "Candidatus Puniceispirillum marinum" IMCC1322, a representative of the SAR116 clade in the Alphaproteobacteria. J. Bacteriol. 192: 3240–3241. doi:10.1128/JB.00347-10.

Pittis, A.A. and T. Gabaldón. 2016. Late acquisition of mitochondria by a host with chimaeric prokaryotic ancestry. Nature 531: 101–104. doi:10.1038/nature16941.

Probst, A.J., C.J. Castelle, A. Singh, C.T. Brown, K. Anantharaman, I. Sharon et al. 2017. Genomic resolution of a cold subsurface aquifer community provides metabolic insights for novel microbes adapted to high CO(2) concentrations. Environ. Microbiol. 19: 459–474. doi:10.1111/1462-2920.13362.

Rochette, N.C., C. Brochier-Armanet and M. Gouy. 2014. Phylogenomic test of the hypotheses for the evolutionary origin of eukaryotes. Mol. Biol. Evol. 31: 832–845. doi: 10.1093/molbev/mst272.

Roger, A.J., S.A. Muñoz-Gómez and R. Kamikawa. 2017. The Origin and Diversification of Mitochondria. Curr. Biol. 27: R1177–R1192. doi:10.1016/j.cub.2017.09.015.

Sagan, L. 1967. On the origin of mitosing cells. J. Theor. Biol. 14: 255–274.

Sassera, D., N. Lo, S. Epis, G. D'Auria, M. Montagna, F. Comandatore, D. Horner et al. 2011. Phylogenomic evidence for the presence of a flagellum and cbb3 oxidase in the free-living mitochondrial ancestor. Mol. Biol. Evol. 28: 3285–3296.

Sousa, F.L., S. Neukirchen, J.F. Allen, N. Lane and W.F. Martin. 2016. Lokiarchaeon is hydrogen dependent. Nat. Microbiol. 1: 16034. doi:10.1038/nmicrobiol.2016.34.

Spang, A., J.H. Saw, S.L. Jørgensen, K. Zaremba-Niedzwiedzka, J. Martijn, A.E. Lind et al. 2015. Complex archaea that bridge the gap between prokaryotes and eukaryotes. Nature 521: 173–179. doi:10.1038/nature14447.

Stairs, C.W., M.M. Leger and A.J. Roger. 2015. Diversity and origins of anaerobic metabolism in mitochondria and related organelles. Philos. Trans. R. Soc. Lond. B Biol. Sci. 370: 20140326. doi:10.1098/rstb.2014.0326.

Thiergart, T., G. Landan, M. Schenk, T. Dagan and W.F. Martin. 2012. An evolutionary network of genes present in the eukaryote common ancestor polls genomes on eukaryotic and mitochondrial origin. Genome Biol. Evol. 4: 466–485.

Thiergart, T., G. Landan and W.F. Martin. 2014. Concatenated alignments and the case of the \ disappearing tree. BMC Evol. Biol. 14: 266. doi:10.1186/s12862-014-0266-0.

Viklund, J., T.J. Ettema and S.G. Andersson. 2012. Independent genome reduction and phylogenetic reclassification of the oceanic SAR11 clade. Mol. Biol. Evol. 29: 599–615.

Villanueva, L., S. Schouten and J.S. Damsté. 2017. Phylogenomic analysis of lipid biosynthetic genes of Archaea shed light on the 'lipid divide'. Environ. Microbiol. 19: 54–69. doi:10.1111/1462-2920.13361.

von der Malsburg, K., J.M. Müller, M. Bohnert, S. Oeljeklaus, P. Kwiatkowska, T. Becker et al. 2011. Dual role of mitofilin in mitochondrial membrane organization and protein biogenesis. Dev. Cell. 21: 694–707. doi:10.1016/j.devcel.2011.08.026.

Wang, Z. and M. Wu. 2015. An integrated phylogenomic approach toward pinpointing the origin of mitochondria. Sci. Rep. 5: 7949. doi:10.1038/srep07949.

Williams, K.P., B.W. Sobral and A.W. Dickerman. 2007. A robust species tree for the *Alphaproteobacteria*. J. Bacteriol. 189: 4578–4586.

Wisniewski-Dyé, F., K. Borziak, G. Khalsa-Moyers, G. Alexandre, L.O. Sukharnikov, K. Wuichet et al. 2011. Azospirillum genomes reveal transition of bacteria from aquatic to terrestrial environments. PLoS Genet. 7: e1002430. doi:10.1371/journal.pgen.1002430.

Wollweber, F., K. von der Malsburg and M. van der Laan. 2017. Mitochondrial contact site and cristae organizing system: A central player in membrane shaping and crosstalk. Biochim. Biophys. Acta. 1864: 1481–1489. doi:10.1016/j.bbamcr.2017.05.004.

Wu, M., L.V. Sun, J. Vamathevan, M. Riegler, R. Deboy, J.C. Brownlie et al. 2004. Phylogenomics of the reproductive parasite *Wolbachia pipientis* wMel: a streamlined genome overrun by mobile genetic elements. PLoS Biol. 2: E69.

Youle, R.J. and A.M. van der Bliek. 2012. Mitochondrial fission, fusion, and stress. Science 337: 1062–1065. doi:10.1126/science.1219855.

Zaremba-Niedzwiedzka, K., E.F. Caceres, J.H. Saw, D. Bäckström, L. Juzokaite, E. Vancaester et al. 2017. Asgard archaea illuminate the origin of eukaryotic cellular complexity. Nature 541: 353–358. doi:10.1038/nature21031.

Index

Printed and bound by CPI Group (UK) Ltd, Croydon, CR0 4YY

24/10/2024

01778304-0001